Partial Differential Equations

Avner Friedman
The Ohio State University

Dover Publications, Inc.
Mineola, New York

Bibliographical Note

This Dover edition, first published in 2008, is an unabridged republication of
the work originally published by Holt, Rinehart & Winston, New York, in 1969.

Library of Congress Cataloging-in-Publication Data

Friedman, Avner.
    Partial differential equations / Avner Friedman. — Dover ed.
        p. cm.
    Originally published: New York : Holt, Rinehart and Winston, [1969]
    ISBN-13: 978-0-486-46919-5
    ISBN-10: 0-486-46919-0
    1. Differential equations, Partial. I. Title.

QA374.F89 2008
515'.353—dc22

                                                        200802844

Manufactured in the United States of America
Dover Publications, Inc., 31 East 2nd Street, Mineola, N.Y. 11501

To Alissa, Joel, Naomi, and Tamara

PREFACE

This book describes in a unified manner some of the main directions now taken in the field of Partial Differential Equations. Part I develops the theory of elliptic equations, centering especially on the solution of the Dirichlet problem. The theory of weak derivatives is developed, various inequalities and imbedding theorems are proven, and smoothness theorems also are derived. As a by-product, we obtain some inequalities for operators A associated with elliptic differential operators and with the Dirichlet zero boundary conditions. These inequalities show, in fact, that A generates an analytic semigroup.

In Part II we develop the theory of semigroups. We then solve the Cauchy problem for evolution equations $du/dt + A(t)u = f(t)$ in a Banach space. The results are applicable to the case where $A(t)$, for each t, generates an analytic semigroup. Thus, as a by-product, we solve the initial-boundary value problem for parabolic equations. We also derive, in Part II, various other results for evolution equations, such as smoothness and uniqueness theorems, and asymptotic bounds at infinity.

Part III includes several independent topics directly related to the methods and results of the previous material.

As for prerequisites, the reader is expected to be familiar with the theory of Lebesgue integration and with the very basic concepts of Banach spaces. Otherwise, the book is self-contained. In particular, no preliminary knowledge of Partial Differential Equations is required. However, the reader who is familiar with some of the classical methods for solving the Laplace equation and the heat equation under the usual boundary conditions should, naturally, be better motivated.

In most sections there are problems, usually related directly to the material of the text. However, in separate sections at the end of Parts I and II we have collected various problems, most of which are not related to the material of the book, with the purpose of introducing the reader to some other topics.

Evanston, Illinois
February 1969

Avner Friedman

CONTENTS

Part 1 | ELLIPTIC EQUATIONS   1

Part 2 | EVOLUTION EQUATIONS   91

PARTIAL DIFFERENTIAL EQUATIONS

Part 1 | ELLIPTIC EQUATIONS

1 | DEFINITIONS

The following standard notation will be used throughout this book. $x = (x_1, \ldots, x_n)$ is a variable point in the real n-dimensional Euclidean space R^n, $D_j = \partial/\partial x_j$, $D^\alpha = D_1^{\alpha_1} \cdots D_n^{\alpha_n}$, where $\alpha = (\alpha_1, \ldots, \alpha_n)$, $\alpha! = \alpha_1! \cdots \alpha_n!$, $|\alpha| = \alpha_1 + \cdots + \alpha_n$, $x^\alpha = x_1^{\alpha_1} \cdots x_n^{\alpha_n}$. Given an open set Ω in R^n, we denote by $C^m(\Omega)$ $(C^m(\overline{\Omega}))$ the set of all complex-valued functions that are continuous (uniformly continuous) in Ω together with all their first m-derivatives. We denote by $C_0^m(\Omega)$ the subset of $C^m(\Omega)$ consisting of those functions which have a compact support in Ω.

A *linear partial differential operator* is an expression of the form

$$\sum_{|\alpha| \leq m} a_\alpha(x)D^\alpha \equiv \sum_{k=0}^{m} \sum_{\alpha_1 + \cdots + \alpha_n = k} a_{\alpha_1 \cdots \alpha_n}(x)D_1^{\alpha_1} \cdots D_n^{\alpha_n}, \qquad (1.1)$$

where the (complex) coefficients $a_\alpha(x)$ are defined in some open set Ω. It may be considered as an operator from one space of functions into another. If the operator

$$\sum_{|\alpha| = m} a_\alpha(x)D^\alpha \qquad (1.2)$$

is not identically zero (that is, if not all its coefficients vanish identically), then it is called the *principal part* (or the *leading part*) of the operator defined in

(1.1), and m is called the *order* of the operator (1.1). The coefficients a_α with $|\alpha| = m$ are then called the *principal* (or *leading*) *coefficients*.

The operator (1.1) is said to be *elliptic* (or of *elliptic type*) at a point $x^0 \in \Omega$ if $\sum_{|\alpha|=m} a_\alpha(x^0)\xi^\alpha \neq 0$ for any real $\xi \neq 0$. If the a_α are real, then m is necessarily an even integer. Suppose now that $m = 2p$, p an integer. The operator (1.1) is said to be *strongly elliptic* at x^0 if

$$(-1)^p \, \text{Re} \left\{ \sum_{|\alpha|=2p} a_\alpha(x^0)\xi^\alpha \right\} > 0 \qquad \text{for any real } \xi \neq 0.$$

If the operator is elliptic (strongly elliptic) at each point of Ω, then it is said to be elliptic (strongly elliptic) in Ω. If

$$(-1)^p \, \text{Re} \left\{ \sum_{|\alpha|=2p} a_\alpha(x)\xi^\alpha \right\} \geq c_0 |\xi|^{2p}$$

for all real ξ and $x \in \Omega$, where $c_0 > 0$ is independent of $x \in \Omega$ and if the principal coefficients are bounded in Ω, then we say that the operator (1.1) is *uniformly strongly elliptic* in Ω. We call c_0 a *module of strong ellipticity*.

The Laplace operator

$$\Delta = \sum_{i=1}^{n} \frac{\partial^2}{\partial x_i^2}$$

is clearly an elliptic operator. The operator $-\Delta$ is strongly elliptic. More generally, the polyharmonic operator Δ^k is elliptic, and $(-1)^k \Delta^k$ is strongly elliptic.

PROBLEM.   (1) Substitute in (1.1) $y = f(x)$ where f is in C^m with nonvanishing Jacobian $(\partial f/\partial x)$. Prove that the new operator $\sum b_\beta D^\beta$ $(D_i = \partial/\partial y_i)$ is elliptic (strongly elliptic) if and only if the same is true of the operator (1.1).

2 | GREEN'S IDENTITY

If u, v are two smooth functions in an open set Ω, and $D = D_i$, then

$$vD^k u = D(vD^{k-1}u) - Dv \cdot D^{k-1}u$$

$$= D(vD^{k-1}u) - D(Dv \cdot D^{k-2}u) + D^2 v \cdot D^{k-2}u$$

$$= \cdots$$

$$= D\{vD^{k-1}u - Dv \cdot D^{k-2} + \cdots + (-1)^{k-1}D^{k-1}v \cdot Du\} + (-1)^k D^k v \cdot u.$$

Hence

$$vD^k u - (-1)^k u D^k v = DB(u, v), \qquad (2.1)$$

where $B(u, v)$ is a bilinear in u, v and in their derivatives up to order $\leq k - 1$. Similarly one proves, for any kth order partial derivative D^α ($|\alpha| = k$),

$$vD^\alpha u - (-1)^{|\alpha|} u D^\alpha v = \sum_{j=1}^{n} \frac{\partial}{\partial x_j} B_j(u, v). \qquad (2.2)$$

Let

$$L = \sum_{|\alpha| \leq m} a_\alpha(x) D^\alpha$$

be a differential operator with coefficients $a_\alpha(x)$ in $C^{|\alpha|}(\Omega)$. Using (2.2), one finds

$$\bar{v} L u - u \overline{L^* v} = \sum_{j=1}^{n} \frac{\partial}{\partial x_j} B_j[u, \bar{v}], \qquad (2.3)$$

where $B_j[u, v]$ are bilinear expressions in u, v and their derivatives up to order $\leq m - 1$, and

$$L^* v = \sum_{|\alpha| \leq m} (-1)^{|\alpha|} D^\alpha \big(\overline{a_\alpha(x)} v\big). \qquad (2.4)$$

The relation (2.3) is called *Green's identity* and the operator L^* is called the *formal adjoint* of L. If $L^* = L$, then we say that L is *formally self-adjoint*.

Integrating (2.3) over a bounded domain R with smooth boundary ∂R ($\bar{R} \subset \Omega$), so that integration by parts is valid, we get

$$\int_R \big(\bar{v} L u - u \overline{L^* v}\big) \, dx = \int_{\partial R} B[u, \bar{v}] \, dS_x, \qquad (2.5)$$

where dS_x is the surface element on ∂R and $B[u, \bar{v}] = \sum v_j B_j[u, \bar{v}]$, v_j being the jth cosine direction of the outward normal to ∂R at x ($x \in \partial R$).

PROBLEMS. (1) Prove that if $Lu = 0$ for any $u \in C^m(\Omega)$, then $L \equiv 0$—that is, all the coefficients of L vanish identically.

(2) Prove that the assertion of the previous problem is still true if it is only assumed that $Lu = 0$ for all $u \in C_0^m(\Omega)$.

(3) Prove that $L^{**} = L$.

(4) Let M be a linear differential operator of order h, and let $k = \max(m, h)$. Suppose that $(Lu, v) = (u, Mv)$ holds for all u, v in $C_0^k(\Omega)$, where $(\ , \ )$ is the scalar product in $L^2(\Omega)$. Prove that $M = L^*$.

3 | FUNDAMENTAL SOLUTIONS

A *fundamental solution* of the differential operator L in Ω is a function $K(x, z)$ defined for $x \in \Omega$, $z \in \Omega$, $x \neq z$ and satisfying the following property: For any bounded domain R with smooth boundary ∂R such that $\bar{R} \subset \Omega$, and for any $z \in R$,

$$v(z) = \int_R K(x, z)\overline{L^*v(x)} \, dx \tag{3.1}$$

for every $v \in C_0^m(R)$. We say that $K(x, z)$ has a *pole* at $x = z$. It is assumed that $K(x, z)$ is in $C^m(\Omega - \{z\})$ and that it is integrable in R.

In view of (2.5), (3.1) is equivalent to

$$v(z) = \lim_{\varepsilon \to 0} \left\{ \int_{R_\varepsilon} v(x)LK(x, z) \, dx - \int_{|x-z|=\varepsilon} B[K(x, z), v(x)] \, dS_x^\varepsilon \right\}, \tag{3.2}$$

where R_ε is the complement in R of the ball with center z and radius ε, and dS_x^ε is the surface element on the boundary of this ball. Taking in (3.2) functions v which vanish in a neighborhood of z, we easily deduce that

$$LK(x, z) = 0 \qquad \text{for all } x \in \Omega, x \neq z. \tag{3.3}$$

Consequently (3.2) reduces to

$$v(z) = -\lim_{\varepsilon \to 0} \int_{|x-z|=\varepsilon} B[K(x, z), v(x)] \, dS_x^\varepsilon. \tag{3.4}$$

This condition implies a certain behavior of $K(x, z)$ near the pole z.

Since the condition (3.4) depends on the function v only in a neighborhood of z, it must hold for all $v \in C^m(\Omega)$ (and not just for $v \in C_0^m(R)$). Using (2.5), we then conclude that the relation

$$v(z) = \int_R K(x, z)\overline{L^*v(x)} \, dx + \int_{\partial R} B[K(x, z), v(x)] \, dS \tag{3.5}$$

holds for any $v \in C^m(\Omega)$.

If the coefficients of L are constants, then we consider only fundamental solutions with $K(x, z) = K(x - z)$.

It is not difficult to verify that

$$K(x) = \begin{cases} c_n|x|^{2-n} & \text{if } n \geq 3, \\ c_2 \log |x| & \text{if } n = 2 \end{cases} \tag{3.6}$$

is a fundamental solution of the Laplace operator, where

$$c_n = -\frac{\Gamma(n/2)}{2(n-2)\pi^{n/2}} = -\frac{1}{(n-2)\Omega_n}, \qquad c_2 = \frac{1}{2\pi},$$

and Ω_n is the surface area of the unit sphere in R^n.

PROBLEMS. (1) Prove that $K(x)$ defined by (3.6) is a fundamental solution of Δ.

(2) Prove that $K(x) = |x|^{2p-n}(A_{pn} \log |x| + B_{pn})$ is a fundamental solution of Δ^p. Here A_{pn}, B_{pn} are constants, and $A_{pn} = 0$ if $2p < n$ or if n is odd, whereas $B_{pn} = 0$ if $2p \geq n$ and n is even.

(3) Prove that for the elliptic operator with real constant coefficients $\sum a_{ij} D_i D_j$ ((a_{ij}) positive matrix), a fundamental solution is given by (3.6) with $|x|$ replaced by $(\sum A_{ij} x_i x_j)^{1/2}$, where A_{ij} is the cofactor of a_{ij} in the matrix (a_{ij}).

4 | CONSTRUCTION OF FUNDAMENTAL SOLUTIONS

THEOREM 4.1. Let L *be an elliptic operator of order* $m \geq 2$, *with constant coefficients and with only highest-order terms. Then there exists a fundamental solution* $K(x)$, *defined for all* $x \neq 0$, *and it has the following properties:*

(i) $K(x)$ *is real analytic for* $x \neq 0$.

(ii) *If* n *is odd or if* n *is even and* $n > m$, *then*

$$K(x) = |x|^{m-n} K_0(x) \tag{4.1}$$

where $K_0(x)$ *satisfies:* $K_0(tx) = K_0(x)$ *if* $x \neq 0$, $t > 0$. *If* n *is even and* $n \leq m$, *then*

$$K(x) = |x|^{m-n} K_0(x) + K_1(x) \log |x|, \tag{4.2}$$

where $K_1(x)$ *is a homogeneous polynomial of degree* $m - n$.

We shall only give the proof in case the coefficients of L are real, n is odd, and $n > m$. The proof in the general case is quite involved, and the reader is referred to John [3].

Writing $L = \sum_{|\alpha|=m} a_\alpha D^\alpha$, we introduce the *characteristic form Q* of L, given by

$$Q(\xi) = \sum_{|\alpha|=m} a_\alpha \xi^\alpha.$$

Consider the function

$$J(x) = \int_{|\xi|=1} \frac{|x \cdot \xi|^{m+1}}{Q(\xi)} \, dS_\xi, \tag{4.3}$$

where $x \cdot \xi = x_1 \xi_1 + \cdots + x_n \xi_n$. We claim that $J(x)$ is real analytic for $x \neq 0$. Indeed, write

$$J(x) = \int_{S(x)} \frac{(x \cdot \xi)^{m+1}}{Q(\xi)} \, dS_\xi + \int_{T(x)} \frac{(-1)^{m+1}(x \cdot \xi)^{m+1}}{Q(\xi)} \, dS_\xi \equiv J_1(x) + J_2(x),$$

where $x \cdot \xi > 0$ on $S(x)$ and $x \cdot \xi < 0$ on $T(x)$.

Introduce polar coordinates for x—that is, $x = (r, \theta_1, \ldots, \theta_{n-1}) = (r, \theta)$, where

$$x_1 = r \sin \theta_{n-1} \sin \theta_{n-2} \cdots \sin \theta_2 \sin \theta_1,$$
$$x_2 = r \sin \theta_{n-1} \sin \theta_{n-2} \cdots \sin \theta_2 \cos \theta_2,$$
$$x_3 = r \sin \theta_{n-1} \sin \theta_{n-2} \cdots \cos \theta_2,$$
$$\vdots$$
$$x_n = r \cos \theta_{n-1}.$$

Similarly we introduce polar coordinates for ξ, with $\xi = (1, \psi_1, \ldots, \psi_{n-1}) = (1, \psi)$. If ξ varies on the unit sphere, then $\psi_2, \ldots, \psi_{n-2}$ vary each in the interval $(0, \pi)$, whereas ψ_1 varies in the interval $(0, 2\pi)$. Also,

$$dS_\xi = \sin^{n-2} \psi_{n-1} \sin^{n-3} \psi_{n-2} \cdots \sin \psi_2 \, d\psi_1 \, d\psi_2 \cdots d\psi_{n-1}.$$

It seems clear that the set $S(x)$ can be described in the form

$$a_1 < \psi_1 < b_1, \quad a_2 < \psi_2 < b_2, \ldots, a_{n-1} < \psi_{n-1} < b_{n-1}, \tag{4.4}$$

where a_i, b_i are smooth and, in fact, real analytic in the variables $\theta_1, \ldots, \theta_{n-1}$, $\psi_1, \ldots, \psi_{i-1}$.

Writing

$$J_1(x) = r^{m+1} \int_{a_{n-1}}^{b_{n-1}} \cdots \int_{a_1}^{b_1} F(\theta, \psi) \, d\psi_1 \cdots d\psi_{n-1},$$

where $F(\theta, \psi)$ is real analytic in θ, ψ, we can now conclude, by iteration, that $J(x)$ is real analytic with respect to θ. Thus $J_1(x)$ is analytic in x, if $x \neq 0$. The proof for $J_2(x)$ is similar.

Observe next that the integrand of $J(x)$ is in C^m, as a function of (x, ξ). Hence

$$LJ(x) = \int_{|\xi| = 1} \frac{1}{Q(\xi)} L(|x \cdot \xi|^{m+1}) \, dS_\xi.$$

Computing the integrand for $\xi \in S(x)$ and for $\xi \in T(x)$ separately we find that it is equal to $(m + 1)! |x \cdot \xi|$. Introducing polar coordinates, we also find that

$$\int_{|\xi| = 1} |x \cdot \xi| \, dS_\xi = \text{const. } |x|$$

Hence

$$LJ(x) = c_0 |x| \qquad (c_0 \text{ constant} \neq 0). \tag{4.5}$$

We now define

$$K(x) = c\Delta^{(n+1)/2} J(x), \tag{4.6}$$

where c is a suitable constant. This is a real analytic function for $x \neq 0$.

Next, in polar coordinates,

$$J(x) = r^{m+1} J_0(\theta) \qquad (J_0(\theta) \text{ analytic in } \theta)$$

$$\Delta = \frac{\partial^2}{\partial r^2} + \frac{n-1}{r} \frac{\partial}{\partial r} + \frac{1}{r^2} A(\theta, D_\theta), \tag{4.7}$$

where $A(\theta, D_\theta)$ is a second-order linear differential operator with the independent variables $\theta_1, \ldots, \theta_{n-1}$. From this we easily get the condition (ii) of the theorem.

It remains to prove (3.1)—that is,

$$v(z) = (-1)^m \int_R K(z - x) Lv(x) \, dx. \tag{4.8}$$

Using (4.6) and integration by parts, we find that the right-hand side of (4.8) is equal to

$$(-1)^m c \int_R \Delta^{(n-1)/2} LJ(z - x) \cdot \Delta v(x) \, dx.$$

Using (4.5) and noting that

$$\Delta^{(n-1)/2} |x| = \text{const. } |x|^{2-n}, \tag{4.9}$$

we conclude that (4.8) is equivalent to

$$v(z) = \text{const.} \int_R |z - x|^{2-n} \, \Delta v(x) \, dS_x.$$

But this relation is true by Problem 1, Section 3.

PROBLEMS.   (1) Compute the a_i, b_i in (4.4) for $n = 3$.
   (2) Verify the relation (4.7).
   (3) Prove (4.9).
   We shall describe a method to construct a fundamental solution for general elliptic operators with variable coefficients. This method can always be carried out if $m = 2$ or (when $m > 2$) if the diameter of Ω is sufficiently small. It is called the *parametrix method*. It supposes that $K(x, z)$ has the form

$$K(x, z) = K_z(x - z) + \int_\Omega K_z(x - y)\Phi(y, z) \, dy + \sum_{i=1}^{s} \phi_i(x)\psi_i(z), \quad (4.10)$$

where $K_z(x)$ is the fundamental solution of the operator

$$L_z \equiv \sum_{|\alpha|=m} a_\alpha(z)D^\alpha$$

and Φ, ϕ_i, ψ_i are to be determined. Setting

$$\alpha(x, z) = LK_z(x - z) = (L - L_z)K_z(x - z),$$

the condition (3.1) leads to the integral equation (for Φ)

$$\alpha(x, z) + \Phi(x, z) + \int_\Omega \alpha(x, y)\Phi(y, z) \, dy + \sum_{i=1}^{s} \psi_i(z)\phi_i(x) = 0. \quad (4.11)$$

   Note that

$$\alpha(x, z) = 0(|x - z|^{1-n}).$$

   The task is now (i) to choose ψ_i, ϕ_i such that (4.11) has a solution Φ; (ii) to study the smoothness of Φ and its behavior near $x = z$, and, finally, (iii) to prove that (4.11) implies (3.1) with K defined by (4.10). For details we refer the reader to John [1].

PROBLEM.   (4) Assume that a fundamental solution $K(x, z)$ has the form (4.10) with $\Phi(x, y) = 0(|x - y|^{1-n})$ and with $\psi_i \in C(\Omega)$, $\phi_i \in C^m(\Omega)$. Prove that Φ satisfies (4.11).

5 | PARTITION OF UNITY

THEOREM 5.1.   Let $\{\Omega_i\}$ be a countable covering of an open set Ω in R^n by open sets Ω_i. Then there exist functions α_i having the following properties:
(i) $\alpha_i \in C_0^\infty(\Omega_j)$, for some $j = j(i)$;
(ii) $\alpha_i \geq 0$, $\sum \alpha_i = 1$ in Ω;
(iii) *every compact subset of Ω intersects only a finite number of the supports of the α_i.*
    If each $\overline{\Omega}_i$ is a compact subset of Ω, or if $\overline{\Omega}$ is bounded and contained in $\cup \, \Omega_i$, then one can take $j(i) = i$ in (i).
    One says that the $\{\alpha_i\}$ constitutes a *partition of unity subordinate* to the covering $\{\Omega_i\}$.
    For a proof of this well-known theorem, the reader is referred to the literature (for instance, Friedman [3]). We shall prove here, however, a useful lemma which is used in the proof of Theorem 5.1.

LEMMA 5.1.   Let A be a compact set in R^n, and let B be an open set in R^n which contains A. Then there exists a C^∞ function $h(x)$ in R^n having the following properties: (i) $h(x) = 0$ *outside* B, (ii) $h(x) = 1$ *in* A, *and* (iii) $0 \leq h(x) \leq 1$ *in* $B - A$.

Proof.   Introduce the function

$$\rho(x) = \begin{cases} 0 & \text{if } |x| \geq 1, \\ c \exp\left\{\dfrac{1}{|x|^2 - 1}\right\} & \text{if } |x| < 1, \end{cases} \qquad (5.1)$$

where c is a constant such that $\int_{R^n} \rho(x) \, dx = 1$. Let G be a bounded open set satisfying: $A \subset G \subset \overline{G} \subset B$. The function

$$h(x) = \int_G \rho\left(\frac{x - y}{\varepsilon}\right) dy \qquad (\varepsilon > 0)$$

satisfies the assertions of the lemma if ε is sufficiently small.
    We shall give an application of Theorem 5.1.
    Let Ω be a bounded domain with boundary $\partial\Omega$. We say that $\partial\Omega$ is of class C^m if for each point $x^0 \in \partial\Omega$ there exists a ball B_0 with center x^0 such that $\partial\Omega \cap B_0$ can be represented in the form

$$x_i = h(x_1, \ldots, x_{i-1}, x_{i+1}, \ldots, x_m)$$

for some i, with h in C^m. We call $x_1, \ldots, x_{i-1}, x_{i+1}, \ldots, x_n$ *local parameters* of $\partial\Omega$.

LEMMA 5.2. *Let Ω be a bounded domain with $\partial\Omega$ in C^m and let Ω_0 be an open set containing $\bar{\Omega}$. Given any function $u \in C^m(\bar{\Omega})$ there exists a function U in $C_0^m(\Omega_0)$, such that $U = u$ in Ω and*

$$\sum_{|\alpha| \leq m} |D^\alpha U|_{L^p(\Omega_0)} \leq C \sum_{|\alpha| \leq m} |D^\alpha u|_{L^p(\Omega)} ;$$

the constant C being independent of u.

Proof. Let $x^0 \in \partial\Omega$. We may suppose that for some neighborhood B_0 of x^0, $B_0 \cap \partial\Omega$ is given by $x_n = h(x_1, \ldots, x_{n-1})$ with $h \in C^m$ and that $x_n > h(x_1, \ldots, x_{n-1})$ in $B_0 \cap \Omega$. The transformation

$$y_j = x_j \quad \text{if } 1 \leq j \leq n-1, \quad y_n = x_n - h(x_1, \ldots, x_{n-1})$$

is a C^m one-to-one map of $B_0 \cap \bar{\Omega}$ onto the closure $\bar{E}_0$ of some open set E_0 in $y_n > 0$, having a portion $\partial_0 E_0$ of its boundary on $y_n = 0$. Let us suppose that E_0 is a half-ball. If v is in $C^m(\bar{E}_0)$, then it can be continued into the reflection of $\bar{E}_0$ with respect to $y_n = 0$, by the formula

$$v(y_1, \ldots, y_{n-1}, y_n) = \sum_{j=1}^{m+1} c_j v\left(y_1, \ldots, y_{n-1}, \frac{-y_n}{j}\right) \quad (y_n < 0), \quad (5.2)$$

where the c_j satisfy

$$\sum_{j=1}^{m+1} c_j \left(-\frac{1}{j}\right)^k = 1 \quad (0 \leq k \leq m). \quad (5.3)$$

This yields a C^m continuation of functions in $B_0 \cap \bar{\Omega}$ into a neighborhood $\tilde{B}_0$ of x^0, where $\tilde{B}_0$ is some ball with center x^0 and radius sufficiently small.

Now cover $\partial\Omega$ by a finite number of balls of the form $\tilde{B}_0$, and add another subset S of Ω ($\bar{S} \subset \Omega$) so as to obtain an open covering of Ω. Denote the sets of the covering by Ω_i, with $\Omega_1 = S$. Let $\{\alpha_i\}$ be a partition of unity subordinate to this covering.

Denote by u_i the C^m continuation of u into $\Omega_i (i > 1)$ and set

$$U = u\alpha_1 + \sum_{i>1} u_i \alpha_i,$$

where $u_i \alpha_i = 0$ if $\alpha_i = 0$. It is easy to verify that U satisfies all the assertions.

PROBLEMS. (1) Let Ω be a bounded domain with $\partial\Omega$ in C^m and let ϕ be a function defined on $\partial\Omega$ and having m continuous derivatives with respect to the local parameters of $\partial\Omega$. Prove that there exists a function Φ in $C^m(R^n)$ such that $\Phi = \phi$ on $\partial\Omega$.

(2) Prove the existence of a partition of unity on R^n of the form $\{\phi(x + l)\}$, where l runs over a suitable subset of R^n.

6 | WEAK AND STRONG DERIVATIVES

Let Ω be a bounded domain, and let u, v be two locally integrable functions defined in Ω. We say that $D^\alpha u = v$ in the *weak sense* (and call v the αth *weak derivative* of u) if, for every $\phi \in C_0^\infty(\Omega)$,

$$\int_\Omega u D^\alpha\phi \, dx = (-1)^{|\alpha|} \int_\Omega v\phi \, dx. \qquad (6.1)$$

We say that $D^\alpha u = v$ in the *strong L^p sense* ($p \geq 1$) (and call v the αth L^p *strong derivative* of u) if for any compact subset K of Ω there exists a sequence $\{\phi_j\}$ of functions in $C^{|\alpha|}(\Omega)$ such that

$$\int_K |\phi_j - u|^p \, dx \to 0, \quad \int_K |D^\alpha\phi_j - v|^p \, dx \to 0 \qquad \text{as} \quad j \to \infty. \qquad (6.2)$$

It is naturally assumed that u and v belong to L^p locally.

PROBLEMS. (1) Prove: if $D^\alpha u = v$ in the strong L^p sense, then $D^\alpha u = v$ in the weak sense.

THEOREM 6.1. If u *has a weak derivative* $D^\alpha u$, *and if* $D^\alpha u$ *has a weak derivative* $D^\beta(D^\alpha u)$, *then* u *has a weak derivative* $D^{\beta+\alpha}u$ *and* $D^{\beta+\alpha}u = D^\beta(D^\alpha u)$.

Proof. Let $\phi \in C_0^\infty(\Omega)$. Then

$$\int_\Omega u \cdot D^\alpha\phi \, dx = (-1)^{|\alpha|} \int_\Omega \phi \cdot D^\alpha u \, dx.$$

Since $D^\alpha u$ has a weak derivative $D^\beta(D^\alpha u)$,

$$\int_\Omega D^\alpha u \cdot D^\beta\psi \, dx = (-1)^{|\beta|} \int_\Omega D^\beta(D^\alpha u) \cdot \psi \, dx$$

for any $\psi \in C_0^\infty(\Omega)$. Taking $\phi = D^\beta \psi$ and combining the previous two equations, we get

$$\int_\Omega u \cdot D^{\beta+\alpha} \psi \, dx = (-1)^{|\beta|+|\alpha|} \int_\Omega D^\beta(D^\alpha u) \cdot \psi \, dx,$$

and the assertion follows.

We shall use the following notation: $\alpha \leq \beta$ if $\alpha_i \leq \beta_i$ for all i, and

$$\binom{\beta}{\alpha} = \binom{\beta_1}{\alpha_1} \cdots \binom{\beta_n}{\alpha_n}.$$

PROBLEM. (2) Let u have weak derivatives in Ω of any order $\leq m$ and let $v \in C^m(\Omega)$. Prove that uv has weak derivatives in Ω of any order $\leq m$ and

$$D^\beta(uv) = \sum_{\alpha \leq \beta} \binom{\beta}{\alpha} D^\alpha u \cdot D^{\beta-\alpha} v \qquad (0 \leq |\beta| \leq m).$$

We shall now prove the converse of Problem 1:

THEOREM 6.2. If u, v *belong to* L^p *locally in* Ω $(p \geq 1)$ *and if* (6.1) *holds for any* $\phi \in C_0^\infty(\Omega)$, *then for any compact set* $K \subset \Omega$ *there exist* $C^\infty(\Omega)$ *functions* ϕ_j *such that* (6.2) *holds.*

Thus, the concepts of weak and strong derivatives are equivalent.

Proof. Let $\rho(x)$ be a C^∞ function in R^n with support in $|x| \leq 1$, such that $\rho(x) \geq 0$, $\int_{R^n} \rho(x) \, dx = 1$; for example, the function defined in (5.1).

Introduce, for $\varepsilon > 0$, ε smaller than the distance from K to $\partial\Omega$, the function

$$J_\varepsilon u(x) = \varepsilon^{-n} \int_\Omega \rho\left(\frac{x-y}{\varepsilon}\right) u(y) \, dy. \tag{6.3}$$

This is a C^∞ function and it is called the *mollifier* of u. Obviously,

$$J_\varepsilon u(x) = \varepsilon^{-n} \int_{|y-x| \leq \varepsilon} \rho\left(\frac{x-y}{\varepsilon}\right) u(y) \, dy = \int_{|z| \leq 1} \rho(z) u(x - \varepsilon z) \, dz. \tag{6.4}$$

Hence,

$$J_\varepsilon u(x) - u(x) = \int_{|z| < 1} \rho(z)[u(x - \varepsilon z) - u(x)] \, dz. \tag{6.5}$$

Using Hölder's inequality, we get from (6.4)

$$|J_\varepsilon u(x)|^p \leq \left(\int_{|z| < 1} \rho(z) \, dz \right)^{p-1} \int_{|z| < 1} \rho(z) |u(x - \varepsilon z)|^p \, dz.$$

Hence,

$$\int_K |J_\varepsilon u(x)|^p \, dx \leq \int_{|z| < 1} \left[\int_K |u(x - \varepsilon z)|^p \, dx \right] \rho(z) \, dz.$$

It follows that

$$|J_\varepsilon u|_{L^p(K)} \leq |u|_{L^p(K_0)}, \tag{6.6}$$

where K_0 is any compact subset of Ω whose interior contains K, provided $\varepsilon < \text{dist.} \, (K, \Omega - K_0)$.

Given any $\delta > 0$, let w be any function in $C^{|\alpha|}(\Omega)$ such that

$$\int_{K_0} |u - w|^p \, dx < \delta.$$

From (6.6) it follows that

$$\int_K |J_\varepsilon u - J_\varepsilon w|^p \, dx < \delta.$$

From (6.5) it is also clear that $J_\varepsilon w(x) \to w(x)$ uniformly on K, if $\varepsilon \to 0$. Hence

$$\int_K |J_\varepsilon w - w|^p \, dx < \delta$$

if ε is sufficiently small. Combining the last three inequalities, we get

$$\int_K |J_\varepsilon u - u|^p \, dx \to 0 \qquad \text{if} \quad \varepsilon \to 0. \tag{6.7}$$

Observe next that, for fixed $x \in K$,

$$\rho\left(\frac{x - y}{\varepsilon} \right)$$

is in $C_0^\infty(\Omega)$. From (6.3) and the definition of a weak derivative we then obtain

$$D^\alpha J_\varepsilon u(x) = \varepsilon^{-n} \int_\Omega D_x^\alpha \rho\left(\frac{x - y}{\varepsilon} \right) u(y) \, dy$$

$$= \varepsilon^{-n} \int_\Omega \rho\left(\frac{x - y}{\varepsilon} \right) v(y) \, dy = J_\varepsilon v(x).$$

Applying (6.7) with u replaced by v, it follows that

$$\int_K |D^\alpha J_\varepsilon u - v|^p \, dx \to 0 \qquad \text{if} \quad \varepsilon \to 0. \tag{6.8}$$

The relations (6.7), (6.8) yield the assertion (6.2) of the theorem.

We introduce the following norm, which will be used throughout this book:

$$|u|_{j,p}^\Omega = \left\{ \sum_{|\alpha| \leq j} \int_\Omega |D^\alpha u(x)|^p \, dx \right\}^{1/p} \tag{6.9}$$

for any nonnegative integer j and a real number p, $1 \leq p < \infty$; Ω will usually denote a bounded open domain. We denote by $\hat{C}^{j,p}(\Omega)$ the linear normed space consisting of all functions in $C^j(\Omega)$ that have a finite norm (6.9). We denote by $H^{j,p}(\Omega)$ the completion of $\hat{C}^{j,p}(\Omega)$ with respect to the norm (6.9). When there is no confusion, we may abbreviate $\hat{C}^{j,p}(\Omega)$, $H^{j,p}(\Omega)$, $|u|_{j,p}^\Omega$ to $\hat{C}^{j,p}$, $H^{j,p}$, $|u|_{j,p}$, respectively.

$\{u_m\}$ is a Cauchy sequence in $\hat{C}^{j,p}$ if and only if

$$\int_\Omega |D^\alpha u_m - D^\alpha u_k|^p \, dx \to 0 \qquad \text{as} \quad m, k \to \infty \qquad (0 \leq |\alpha| \leq j). \tag{6.10}$$

Since $L^p(\Omega)$ is a complete space, there exist functions $u^\alpha \in L^p(\Omega)$ such that

$$\int_\Omega |D^\alpha u_m - u^\alpha|^p \, dx \to 0 \qquad \text{as} \quad m \to \infty. \tag{6.11}$$

Since any two Cauchy sequences determine the same element u^α, we can identify the elements of $H^{j,p}$ with the vectors $\{u^\alpha; 0 \leq |\alpha| \leq j\}$.

Clearly $u^\alpha = D^\alpha u^0$ in the strong sense. Hence, if $u^0 = 0$, then for any $\phi \in C_0^\infty(\Omega)$,

$$\int_\Omega u^\alpha \phi \, dx = (-1)^{|\alpha|} \int_\Omega u^0 D^\alpha \phi \, dx = 0$$

—that is, $u^\alpha = 0$. It follows that the u^α are uniquely determined by u^0. We can therefore identify the elements of $H^{j,p}$ with the functions u^0 in L^p. We sum up:

LEMMA 6.1.   *A function* $u^0 \in L^p(\Omega)$ *belongs to* $H^{j,p}(\Omega)$ *if and only if there exists a sequence* $\{u_m\}$ *of functions in* $\hat{C}^{j,p}(\Omega)$ *such that* $|u_m - u^0|_{0,p} \to 0$ *and such that (6.10) holds.*

We introduce another linear space $W^{j,p}(\Omega)$, or simply $W^{j,p}$. A function u is said to belong to $W^{j,p}$ if u belong to $L^p(\Omega)$ and if all its weak derivatives of order $\leq j$ exist and belong to $L^p(\Omega)$.

THEOREM 6.3. $W^{j,p} = H^{j,p}$.

Proof.   The inclusion $H^{j,p} \subset W^{j,p}$ is obvious. Now let $u \in W^{j,p}$. If we show that for any $\varepsilon > 0$ there exists a function $w \in C^j(\Omega)$ such that, for some constant c independent of ε,

$$|D^\alpha w - D^\alpha u|_{0,p} < c\varepsilon \qquad (0 \leq |\alpha| \leq j), \tag{6.12}$$

where $D^\alpha u$ is the αth weak derivative of u, then it follows that $u \in H^{j,p}$. Let

$$\Omega = \bigcup_{m=1}^{\infty} \Omega_m, \qquad \overline{\Omega}_m \subset \Omega_{m+1},$$

where $\{\Omega_{m+2} - \overline{\Omega}_m\}$ form an open covering of Ω, and let $\{\psi_i\}$ be a partition of unity subordinate to this covering. Each function $u\psi_i$ has j weak derivatives and its support lies in $\Omega_{i+2} - \overline{\Omega}_i$. If ε_i is a sufficiently small positive number, then $w_i = J_{\varepsilon_i}(u\psi_i)$ has its support in $\Omega_{i+3} - \Omega_{i-1}$ and

$$|w_i - u\psi_i|_{j,p} < \frac{\varepsilon}{2^j}. \tag{6.13}$$

Take $w = \sum_{i=1}^{\infty} w_i$. This is clearly a C^∞ function in Ω, and

$$D^\alpha w = \sum_i D^\alpha(w_i - u\psi_i) + \sum_i D^\alpha(u\psi_i) = \sum_i D^\alpha(w_i - u\psi_i) + D^\alpha u.$$

Since for each $x \in \Omega$ there are at most four terms in the series $\sum D^\alpha(w_i - u\psi_i)$ that do not vanish at x, we get

$$|D^\alpha w - D^\alpha u|^p \leq C \sum_i |D^\alpha(w_i - u\psi_i)|^p \qquad (C = C(p)).$$

Integrating this inequality with respect to $x \in \Omega$, and using (6.13), we obtain (6.12).

The proof that $W^{j,p} \subset H^{j,p}$ gives more, namely, $W^{j,p}$ is contained in the space $\tilde{H}^{j,p}$ obtained by completing the space $\hat{C}^{\infty,j,p}(\Omega)$ (of all $C^\infty(\Omega)$ functions with finite norm (6.9)) with respect to the norm (6.9). Since $\hat{C}^{\infty,j,p}(\Omega) \subset \hat{C}^{j,p}(\Omega)$, we also have $\tilde{H}^{j,p} \subset H^{j,p} = W^{j,p}$. Hence:

COROLLARY. $\tilde{H}^{j,p} = H^{j,p}$.

PROBLEMS.   (3) Let $g(t)$ and its weak derivative $g'(t)$ belong to $L^1(a, b)$, and assume that $g(t) \geq 0$, $g'(t) \leq 0$ almost everywhere, and that $g(t) \equiv 0$ in some interval (a, c) with $a < c < b$. Prove that $g(t) = 0$ almost everywhere in (a, b).

(4) Is the mollifier of an entire analytic function also an entire analytic function?

We conclude this section with a useful lemma.

LEMMA 6.2.   Let u *be a function in* $L^p(\Omega)$ *and let* $\{u_k\}$ *be a sequence in* $W^{j,p}(\Omega)$ *that is bounded in* $W^{j,p}(\Omega)$ *and converges weakly to* u *in* $L^p(\Omega)$. *Then* $u \in W^{j,p}(\Omega)$ *and, for every* α, $0 \leq |\alpha| \leq j$, $\{D^\alpha u_k\}$ *converges weakly to* $D^\alpha u$, *in* $L^p(\Omega)$.

We recall that a sequence $\{g_k\}$ (in $L^p(\Omega)$) is said to be *weakly convergent* to g ($g \in L^p(\Omega)$) in $L^p(\Omega)$ if for any $f \in L^q(\Omega)$ $(1/p + 1/q = 1)$

$$\int_\Omega f g_k \, dx \to \int_\Omega f g \, dx \qquad \text{as } k \to \infty.$$

We also recall the well-known fact that if $\{g_k\}$ is a bounded sequence in $L^p(\Omega)$, then there is a subsequence $\{g_{k_i}\}$ that is weakly convergent to some g in $L^p(\Omega)$.

Proof.   There is clearly a subsequence $\{u_{k_i}\}$ such that, for each α with $0 \leq |\alpha| \leq j$, $\{D^\alpha u_{k_i}\}$ is weakly convergent in $L^p(\Omega)$. Denote the weak limit by u^α. Then, for any $\phi \in C_0^\infty(\Omega)$,

$$\int_\Omega u^\alpha \phi \, dx = \lim_{k_i \to \infty} \int_\Omega D^\alpha u_{k_i} \phi \, dx = \lim_{k_i \to \infty} (-1)^{|\alpha|} \int_\Omega u_{k_i} D^\alpha \phi \, dx = (-1)^{|\alpha|} \int_\Omega u^\alpha D^\alpha \phi \, dx.$$

Thus u^0 has weak derivatives $D^\alpha u^0 = u^\alpha$ for all $|\alpha| \leq j$. It follows that $u^0 \in W^{j,p}$. Since $\{u_{k_i}\}$ is weakly convergent to u, we have $u = u^0$. We have thus proved that $u \in W^{j,p}$ and that $\{u_k\}$ has a subsequence $\{u_{k_i}\}$ such that $\{D^\alpha u_{k_i}\}$ converges weakly in $L^p(\Omega)$ to $D^\alpha u$ $(0 \leq |\alpha| \leq j)$. Since the same is true if we start with any subsequence of $\{u_k\}$, it follows that $\{D^\alpha u_k\}$ converges weakly in $L^p(\Omega)$ to $D^\alpha u$. This completes the proof.

7 | STRONG DERIVATIVE AS A LOCAL PROPERTY

LEMMA 7.1.   Let $\Omega_1', \ldots, \Omega_s'$ *be domains with* $\overline{\Omega} \subset \bigcup_{i=1}^s \Omega_i'$ *and set* $\Omega_i = \Omega_i' \cap \Omega$. *If a function* u *belongs to* $H^{j,p}(\Omega_i)$ *for every* i, *then* $u \in H^{j,p}(\Omega)$.

Proof. Let $\{\phi_i\}$ be a partition of unity subordinate to the covering $\{\Omega_i'\}$. Since $u \in H^{j,p}(\Omega_i)$, there exists a sequence $\{u_{i,m}\}$ in $C^j(\Omega_i)$ such that

$$|u_{i,m} - u|_{j,p}^{\Omega_i} \to 0 \qquad \text{as } m \to \infty.$$

Set $u_m = \sum_{i=1}^s \phi_i u_{i,m}$. Then

$$D^\alpha(u_m - u_k) = \sum_{i=1}^s D^\alpha[\phi_i(u_{i,m} - u_{i,k})] \qquad (0 \le |\alpha| \le j).$$

It easily follows that $|u_m - u|_{0,p}^\Omega \to 0$ and

$$|u_m - u_k|_{j,p}^\Omega \le \text{const.} \sum_{i=1}^s \sum_{|\alpha| \le j} |D^\alpha u_{i,m} - D^\alpha u_{i,k}|_{0,p}^{\Omega_i} \to 0 \qquad \text{as } m, k \to \infty.$$

Hence $u \in H^{j,p}(\Omega)$.

COROLLARY. *If for each* $x \in \Omega$ *there exists a neighborhood of* x *in which* u *has* j *weak (or strong) derivatives, then* u *has weak (or strong) derivatives in* Ω.

LEMMA 7.2. *Let* Ω *be a domain whose boundary consists of two disjoint sets,* Γ_1 *and* Γ_2, *where* Γ_2 *is an open domain on a hyperplane* $x_n = 0$. *If* u *has weak derivatives in* Ω *of all orders* $\le j$ *and if they all belong to* $L^p(\Omega)$, *then for any subdomain* A *of* Ω *whose closure lies in* $\Omega + \Gamma_2$ *there exists a sequence* $\{u_m\}$ *of functions in* $C^j(\bar{A})$ *such that* $|u_m - u|_{j,p}^A \to 0$ *as* $m \to \infty$.

By Theorem 6.3 $u \in H^{j,p}(A)$ so that there exists a sequence of functions v_m in $C^j(A)$ with $|v_m - u|_{j,p}^A \to 0$ as $m \to \infty$. Thus the interest of the lemma is in the assertion that $u_m \in C^j(\bar{A})$.

Proof. Let B be a subdomain of Ω whose boundary lies in $\Omega + \Gamma_2$, such that $A \subset B$, and let the boundary of B consist of two disjoint sets, ∂B_1 and ∂B_2, where ∂B_2 is an open set on $x_n = 0$. We may assume that the boundary of A decomposes similarly into two sets ∂A_1 and ∂A_2, ∂A_2 being an open set on $x_n = 0$. We take B such that $\overline{\partial A_2} \subset \partial B_2$, $\overline{\partial A_1} \cap \overline{\partial B_1} = \emptyset$. Finally, we may assume that B lies in the half-space $x_n > 0$.

We now define mollifiers

$$(J_\varepsilon' u)(x) = \frac{1}{\varepsilon^n} \int_\Omega \rho\left(\frac{x_\varepsilon - y}{\varepsilon}\right) u(y) \, dy \tag{7.1}$$

where $x_\varepsilon = (x_1, \ldots, x_{n-1}, x_n + 2\varepsilon)$ if $x = (x_1, \ldots, x_{n-1}, x_n)$ and $0 < \varepsilon < \varepsilon_0/3$, $\varepsilon_0 = \text{dist.}\ (\partial A_1, \partial B_1)$.

Similarly to the proof of Theorem 6.2, one shows that

$$|J'_\varepsilon u|^A_{0, p} \le |u|^B_{0, p} \tag{7.2}$$

$$|J'_\varepsilon u - u|^A_{0, p} \to 0 \qquad \text{as } \varepsilon \to 0, \tag{7.3}$$

and, finally, that

$$|J'_\varepsilon u - u|^A_{j, p} \to 0 \qquad \text{as } \varepsilon \to 0. \tag{7.4}$$

Now take $u_m = J'_{1/m} u$.

Denote by $\hat{H}^{j,p}(\Omega)$, or $\hat{H}^{j,p}$, the completion of the space $C^j(\bar{\Omega})$ with respect to the norm (6.9). Clearly $\hat{H}^{j,p} \subset H^{j,p}$. We shall now prove:

THEOREM 7.1. *If $\partial\Omega$ is in C^j, then $\hat{H}^{j,p} = H^{j,p}$.*

Proof. We have to show that if $u \in H^{j,p}$, then $u \in \hat{H}^{j,p}$. From the proof of Lemma 7.1 it follows that it suffices to show that for every $x^0 \in \partial\Omega$ there is a neighborhood N_0 such that u belongs to $\hat{H}^{j,p}(N_0 \cap \Omega)$. We may assume that for some neighborhood N of x^0, $N \cap \partial\Omega$ can be represented in the form $x_n = h(x_1, \ldots, x_{n-1})$ with $h \in C^j$ and $N \cap \Omega$ lying on the side $x_n > h(x_1, \ldots, x_{n-1})$. Perform the transformation

$$y_i = x_i \quad (1 \le i \le n-1), \qquad y_n = x_n - h(x_1, \ldots, x_{n-1}). \tag{7.5}$$

If $V(y) = v(x)$, where $v \in C^1$, then

$$\frac{\partial V}{\partial y_i} = \sum_{k=1}^n \frac{\partial v}{\partial x_k} \frac{\partial x_k}{\partial y_i}, \qquad \frac{\partial v}{\partial x_i} = \sum_{k=1}^n \frac{\partial V}{\partial y_k} \frac{\partial y_k}{\partial x_i}. \tag{7.6}$$

It follows that the L^p norm of $\sum |\partial v/\partial x_i|$ is bounded from above and below by positive constants times the L^p norm of $\sum |\partial V/\partial y_i|$. Consequently $\{v_m(x)\}$ is a Cauchy sequence in $\hat{H}^{1,p}$ of C^1 functions if and only if $\{V_m\}$ is a Cauchy sequence in $\hat{H}^{1,p}$ of C^1 functions. Thus the map $v \to V$ is an isomorphism between the corresponding $\hat{H}^{1,p}$ spaces, and (7.6) holds with strong derivatives $\partial v/\partial x_k$, $\partial V/\partial y_k$.

The considerations above extend to spaces $\hat{H}^{j,p}$ for any $j \ge 1$. Hence, if we show that $U(y)\ (=u(x))$ belongs to $\hat{H}^{j,p}(M)$, where M is the image of $N_0 \cap \Omega$ in the y-space, then it would follow that $u \in \hat{H}^{j,p}(N_0 \cap \Omega)$. But the assertion about U follows from Lemma 7.2.

PROBLEMS. (1) Prove Lemma 7.1, using only the concept of weak deriva-tives (and replacing $H^{j,p}$ by $W^{j,p}$).

(2) Denote by B the unit ball in R^n and let $1 < p < \infty$. Find the largest integer k such that the function $|x|$ belongs to $H^{k,p}(B)$.

8 | CALCULUS INEQUALITIES

THEOREM 8.1.   Let Ω be a bounded domain with $\partial\Omega$ in C^2 and let j be a positive integer and p *a real number* ≥ 1. *Then there exists a constant* $\varepsilon_0 = \varepsilon_0(\Omega, p, j) > 0$ *and, for any* ε *with* $0 < \varepsilon < \varepsilon_0$, *a constant* $C = C(\Omega, p, j, \varepsilon)$ *such that the inequality*

$$|u|^\Omega_{j-1, p} \leq \varepsilon |u|^\Omega_{j, p} + C |u|^\Omega_{0, p} \tag{8.1}$$

holds for all $u \in C^j(\overline{\Omega})$.

Proof. (8.1) is a consequence of the inequality

$$\sum_{|\alpha|=i} \int_\Omega |D^\alpha u|^p \, dx \leq \varepsilon \sum_{|\beta|=j} \int_\Omega |D^\beta u|^p \, dx + \frac{C}{\varepsilon^{i/(j-i)}} \int_\Omega |u|^p \, dx \qquad (i < j), \tag{8.2}$$

where $0 < \varepsilon \leq \varepsilon_0$, and ε_0, C depend only on p, j, Ω.

We first prove (8.2) for $i = 1, j = 2, n = 1$, and $0 < \varepsilon < 2^p |\Omega|^p$, where $|\Omega|$ is the length of the interval Ω. Divide Ω into subintervals of length $\leq \varepsilon^{1/p}$ and $\geq \varepsilon^{1/p}/2$. For each such subinterval $a < x < b$, set $\alpha = (b - a)/4$ and let x_1, x_2 be variable points in the intervals $a < x < a + \alpha$, $a + 3\alpha < x < b$, respectively. By the mean value theorem,

$$\frac{u(x_2) - u(x_1)}{x_2 - x_1} = Du(x_{12}) \qquad (x_1 < x_{12} < x_2). \tag{8.3}$$

Write, for any $x \in (a, b)$

$$Du(x) = Du(x_{12}) + \int_{x_{12}}^x D^2 u(\xi) \, d\xi.$$

Using (8.3), we then get

$$|Du(x)| \leq \frac{|u(x_1)| + |u(x_2)|}{2\alpha} + \int_a^b |D^2 u(\xi)| \, d\xi.$$

Integrating with respect to x_1, x_2 in their respective intervals $(a, a + \alpha)$, $(a + 3\alpha, b)$, we find

$$\alpha^2 |Du(x)| \leq \tfrac{1}{2} \int_a^b |u(\xi)| \, d\xi + \alpha^2 \int_a^b |D^2u(\xi)| \, d\xi.$$

Taking the pth power and using Hölder's inequality, we get

$$|Du(x)|^p \leq \frac{c}{\alpha^{p+1}} \int_a^b |u(\xi)|^p \, d\xi + c\alpha^{p-1} \int_a^b |D^2u(\xi)|^p \, d\xi,$$

where c is a constant depending only on p.

Integrating the last inequality with respect to x in the interval (a, b) ,we obtain

$$\int_a^b |Du|^p \, dx \leq \frac{c}{\alpha^p} \int_a^b |u|^p \, dx + c\alpha^p \int_a^b |D^2u|^p \, dx$$

$$\leq \frac{c'}{\varepsilon} \int_a^b |u|^p \, dx + c'\varepsilon \int_a^b |D^2u|^p \, dx,$$

with suitable constants c, c' depending only on p.

Summing over all the subintervals of Ω, we get (after replacing $c'\varepsilon$ by ε)

$$\int_\Omega |Du|^p \, dx \leq \varepsilon \int_\Omega |D^2u|^p \, dx + \frac{c}{\varepsilon} \int_\Omega |u|^p \, dx, \tag{8.4}$$

with a suitable constant c.

We next prove (8.2) for $n > 1$, $i = 1, j = 2$. Suppose first that Ω is a cube with edges parallel to the coordinate axes, and set $\partial u / \partial x_1 = Du$. Considering u as a function of x_1 with $x_2, \ldots, x_n$ as parameters, we can apply (8.4). After integrating the inequality with respect to the parameters $x_2, \ldots, x_n$, we find that $\int_\Omega |\partial u / \partial x_1|^p \, dx$ is bounded by the right-hand side of (8.4). Since a similar bound holds for $\partial u / \partial x_k$ $(2 \leq k \leq n)$, (8.2) follows for $i = 1, j = 2$.

If Ω is not a cube, then we cover it by a finite number of domains Γ_λ, Δ_μ, where Γ_λ are cubes with edges parallel to the coordinate axes, and each Δ_μ can be mapped by a one-to-one transformation $y = y(x)$ onto a cube with edges parallel to the coordinate axes in the y-space, the mapping $y = y(x)$ and its inverse having two continuous derivatives.

(8.4) for $i = 1, j = 2$ holds for each Γ_λ (with C depending only on p) and for each Δ_μ (with C depending only on p, Ω). Summing over λ, μ, we get (8.2) for $i = 1, j = 2$.

The proof for Ω not a cube can also be given by using Lemma 5.2 with Ω_0 a cube and applying (8.4) with $i = 1, j = 2$ in the cube Ω_0.

We shall now prove (8.2) for all i, j by induction on j. For $j = 2$, (8.2) was already proved for $i = 1$ and is trivial if $i = 0$. Assuming (8.2) to hold for all $j \leq k$, we shall prove it for $j = k + 1$. The notation $|D^m u|^p = \sum_{|\alpha| = m} |D^\alpha u|^p$ will be used.

If $i = k$, then applying (8.2) with $i = 1, j = 2$ to the $(k - 1)$th derivatives of u, we get, for any sufficiently small ε,

$$\int_\Omega |D^k u|^p \, dx \leq \frac{\varepsilon}{2} \int_\Omega |D^{k+1} u|^p \, dx + \frac{C_1}{\varepsilon} \int_\Omega |D^{k-1} u|^p \, dx,$$

where C_m are used to denote positive constants depending only on k, p, Ω. Using the inductive assumption with $j = k, i = k - 1$, we have

$$\int_\Omega |D^{k-1} u|^p \, dx \leq \delta \int_\Omega |D^k u|^p \, dx + \frac{C_2}{\delta^{k-1}} \int_\Omega |u|^p \, dx.$$

Substituting this into the previous inequality and taking $\delta = \varepsilon/2C_1$, we get (8.2) with $i = k, j = k + 1$.

If $i < k$, then, by the inductive assumption,

$$\int_\Omega |D^i u|^p \, dx \leq \delta \int_\Omega |D^k u|^p \, dx + \frac{C_3}{\delta^{i/(k-i)}} \int_\Omega |u|^p \, dx.$$

By (8.2) with $i = k, j = k + 1$,

$$\int_\Omega |D^k u|^p \, dx \leq \mu \int_\Omega |D^{k+1} u|^p \, dx + \frac{C_4}{\mu^k} \int_\Omega |u|^p \, dx.$$

Substituting the last inequality into the previous one and taking $\delta = \varepsilon^{(k-1)/(k+1-i)}$, $\mu = \varepsilon^{1/(k+1-i)}$, we obtain (8.2) for $j = k + 1$.

PROBLEMS. (1) Prove that the assertion of Theorem 8.1 holds for any $u \in W^{j,p}(\Omega)$.

(2) Prove: if $0 < p_1 < p < p_2$,

$$(|u|_{0,p})^p \leq (|u|_{0,p_2})^{(p-p_1)p_2/(p_2-p_1)} (|u|_{0,p_1})^{(p_2-p)p_1/(p_2-p_1)}.$$

(3) Prove: if $0 < r < p < s < \infty$, then, for any $\varepsilon > 0$,

$$|u|_{0,p} < \varepsilon |u|_{0,s} + \varepsilon^{-s(p-r)/r(s-p)} |u|_{0,r}.$$

9 | EXTENDED SOBOLEV INEQUALITIES IN R^n

Definition. A bounded domain Ω is said to have the *cone property* (or to satisfy the *cone condition*) if there exist positive constants α, h such that for any $x \in \Omega$ one can construct a right spherical cone V_x with vertex x, opening α, and height h such that it lies in Ω.

PROBLEMS. (1) Prove that if $\partial\Omega$ is of class C^1, then Ω satisfies the cone condition.

(2) Prove that a convex domain has the cone property.

The following theorem is due to Sobolev:

THEOREM 9.1. *Let Ω be a bounded domain satisfying the cone condition with constants α, h and let u be a function in $C^m(\Omega) \cap W^{m,p}(\Omega)$ for some $p > 1$. If m $> n/p$, then*

$$\sup_{\Omega} |u(x)| \leq C |u|^{\Omega}_{m,p} \tag{9.1}$$

where C is a constant depending only on α, h, n, p.

Proof. Let $g(t)$ be a C^∞ function for $-\infty < t < \infty$, such that $g(t) = 1$ if $t < \frac{1}{2}$ and $g(t) = 0$ if $t \geq 1$. For fixed $x \in \Omega$, introduce polar coordinates (r, θ) for the points y in the right spherical cone V_x with height h and opening α, which occurs in the cone condition. Then

$$u(x) = -\int_0^h \frac{\partial}{\partial r} \left[g\left(\frac{r}{h}\right) u(r, \theta) \right] dr.$$

Integrate this with respect to dS_θ and then perform integration by parts $m - 1$ times to obtain

$$u(x) = c \int_\alpha \int_0^h r^{m-1} \frac{\partial^m}{\partial r^m} \left[g\left(\frac{r}{h}\right) u \right] dr \, dS_\theta \qquad (c \text{ constant}).$$

Writing $r^{m-1} = r^{m-n} r^{n-1}$, $r^{n-1} \, dr \, dS_\theta = dV$, and using Hölder's inequality, we get

$$|u(x)| \leq c_0 \left\{ \int \left| \frac{\partial^m}{\partial r^m} \left[g\left(\frac{r}{h}\right) u \right] \right|^p dV \right\}^{1/p} \qquad (c_0 \text{ constant}),$$

from which the assertion follows.

Definition. A function $v(x)$ defined on a set S is said to be *uniformly Hölder continuous (exponent α)* on S if

$$\underset{S}{\text{l.u.b.}} \; [v]_\alpha \equiv \underset{x,\,y \in S}{\text{l.u.b.}} \; \frac{|v(x) - v(y)|}{|x - y|^\alpha} \tag{9.2}$$

is finite. If a function v is uniformly Hölder continuous (exponent α) on each compact subset of a domain Ω, then it is said to be *Hölder continuous* (exponent α) in Ω.

Sobolev has also derived bounds on $|u|_{0,r}$ in terms of $|u|_{m,p}$. On the other hand, Morrey has estimated the *Hölder coefficient* l.u.b. $[u]_\alpha$ in terms of $|u|_{1,p}$ for $p > n$:

THEOREM 9.2. *Let* u *be a function in* $C_0^1(R^n)$ *and let* r *be a real number* $> n$. *Then for any* x, y *in* R^n,

$$\frac{|u(x) - u(y)|}{|x - y|^{1 - n/r}} \le C \sum_{i=1}^{n} |D_i u|_{0,r}, \tag{9.3}$$

where C *is a constant that depends only on* n, r.

Proof. Let $d = |x - y|$ and let S_x, S_y be the balls with radius d and centers x and y, respectively. Set $S = S_x \cap S_y$. Then

$$|u(x) - u(y)| \, \text{vol. } S \le \int_S |u(x) - u(z)| \, dz + \int_S |u(z) - u(y)| \, dz. \tag{9.4}$$

Introducing polar coordinates (ρ, ψ) about x, we can estimate the first integral on the right by

$$\int_{S_x} \left\{ \int_0^\rho \left| \frac{\partial u}{\partial \rho} \right| d\rho \right\} \rho^{n-1} \, dS_\psi \, d\rho \le \text{const. } d^n \int_{S_x} \left| \frac{\partial u}{\partial \rho} \right| \frac{dz}{\rho^{n-1}}$$

$$\le \text{const. } d^n \left(\int_{S_x} \left| \frac{\partial u}{\partial \rho} \right|^r dz \right)^{1/r}$$

$$\times \left(\int_{S_x} \rho^{(1-n)r/(r-1)} \, dz \right)^{(r-1)/r},$$

where Hölder's inequality has been used. The last integral is equal to const. $d^{(r-n)/(r-1)}$. Since the second integral on the right-hand side of (9.4) can be estimated in the same way, we get the bound

$$\text{const. } d^{n+1-n/r} \left\{ \int_{S_x \cup S_y} \left| \frac{\partial u}{\partial \rho} \right|^r dz \right\}^{1/r} \le \text{const. } d^{n+1-n/r} \left\{ \int_{R^n} \sum_{i=1}^{n} |D_i u|^r dz \right\}^{1/r}$$

for the right-hand side of (9.4). Since the left-hand side of (9.4) is $\geq \lambda \, d^n |u(x) - u(y)|$ (λ positive constant), the assertion of the lemma follows.

PROBLEM. (3) Prove the assertion of Theorem 9.2 with R^n replaced by Ω, assuming that $u \in C^1(\overline{\Omega})$ and that $\partial\Omega$ is in C^1; C now depends also on Ω.

We shall derive in the following section very general inequalities of the type of Sobolev's inequalities and of Theorem 9.2. In the present section we derive such inequalities in R^n. First we introduce some notation.

For $p > 0$, $|u|_{p,\Omega} = \{\int_\Omega |u|^p \, dx\}^{1/p}$. For $p < 0$, set $h = [-n/p]$, $-\alpha = h + n/p$, and define

$$|u|_{p,\,\Omega} = \text{l.u.b.} \ |D^h u| \equiv \sum_{|\beta|=h} \text{l.u.b.} \ |D^\beta u| \qquad \text{if } \alpha = 0, \qquad (9.5)$$

$$|u|_{p,\,\Omega} = [D^h u]_{\alpha,\,\Omega} \equiv \sum_{|\beta|=h} \text{l.u.b.} \ [D^\beta u]_\alpha \qquad \text{if } \alpha > 0, \qquad (9.6)$$

where the notation (9.2) has been used.

If $\Omega = R^n$, then we write $|u|_p$ instead of $|u|_{p,\Omega}$.

THEOREM 9.3.   *Let* q, r *be any numbers satisfying* $1 \leq q, r \leq \infty$, *and let* j, m *be any integers satisfying* $0 \leq j < m$. *If* u *is any function in* $C_0^m(R^n)$, *then*

$$|D^j u|_p \leq C \, |D^m u|_r^a \, |u|_q^{1-a} \qquad (9.7)$$

where

$$\frac{1}{p} = \frac{j}{n} + a\left(\frac{1}{r} - \frac{m}{n}\right) + (1 - a)\frac{1}{q},$$

for all a *in the interval*

$$\frac{j}{m} \leq a \leq 1, \qquad (9.8)$$

where C *is a constant depending only on* n, m, j, q, r, a, *with the following exception: If* $m - j - n/r$ *is a nonnegative integer, then* (9.7) *is asserted only for* a = j/m.

Actually, if $m - j - n/r$ is a nonnegative integer, then (9.7) holds for any $j/m \leq a < 1$, but we shall not prove this here.

Some steps in the proof of Theorem 9.3 are left to the reader in the following problems.

PROBLEMS. (4) Assume that (9.7) with $a = j/m$ is true for $j = 1$, $m = 2$. Use induction on m to show that (9.7) with $a = j/m$ is then true for any $0 \leq j < m < \infty$.

(5) Assume that (9.7) with $a = 1$ is true for $j = 0$, $m = 1$, provided $n \neq r$. Use induction on m to show that (9.7) with $a = 1$ is then true for any $0 \leq j < m < \infty$, provided $m - j - n/r$ is not a nonnegative integer.

(6) Prove the interpolation inequality if: $-\infty < \lambda \leq \mu \leq \nu < \infty$, then

$$|u|_{1/\mu} \leq c |u|_{1/\lambda}^{(\nu-\mu)/(\nu-\lambda)} \cdot |u|_{1/\nu}^{(\mu-\lambda)/(\nu-\lambda)}, \tag{9.9}$$

where c is a constant independent of u.

(7) Using (9.9), show that if (9.7) holds for $a = 1$ and for $a = j/m$, then it holds for all a in the interval (9.8).

Proof of Theorem 9.3.   In view of the assertions of Problems 4, 5, and 7, it remains to prove (9.7) in the following cases: $a = 1$, $j = 0$, $m = 1$ (provided $n \neq r$) and $j = 1$, $m = 2$, $a = \frac{1}{2}$. Consider first the case $a = 1$, $j = 0$, $m = 1$. If $r > n$, then the assertion follows from Theorem 9.2. The case $r = n$ is the exceptional case which we have excluded. Thus it remains to consider the case $r < n$. We shall prove that

$$|u|_{nr/(n-r)} \leq \frac{r}{2} \frac{n-1}{n-r} \prod_{i=1}^{n} \left| \frac{\partial u}{\partial x_i} \right|_r^{1/n} ; \tag{9.10}$$

this would yield the assertion of the theorem.

Suppose first that $r = 1$. (9.10) becomes

$$|u|_{n/(n-1)} \leq \frac{1}{2} \prod_{i=1}^{n} \left| \frac{\partial u}{\partial x_i} \right|_1^{1/n} . \tag{9.11}$$

Consider the case $n = 3$. Denote by $\int_i$ the integral taken along the whole line through $x = (x_1, x_2, x_3)$ parallel to the x_i-axis. Since u has a compact support,

$$|u(x)| \leq \frac{1}{2} \int_i \left| \frac{\partial u}{\partial x_i} \right| dx_i .$$

Hence,

$$|2u(x)|^{3/2} \leq \left(\int_1 \left| \frac{\partial u}{\partial x_1} \right| dx_1 \right)^{1/2} \left(\int_2 \left| \frac{\partial u}{\partial x_2} \right| dx_2 \right)^{1/2} \left(\int_3 \left| \frac{\partial u}{\partial x_3} \right| dx_3 \right)^{1/2} .$$

Integrating with respect to x_1 and using Schwarz's inequality on the right, then doing the same thing with respect to x_2 and x_3, we get

$$\iiint |2u|^{3/2} \, dx \leq \left(\iiint \left| \frac{\partial u}{\partial x_1} \right| dx \right)^{1/2} \left(\iiint \left| \frac{\partial u}{\partial x_2} \right| dx \right)^{1/2} \left(\iiint \left| \frac{\partial u}{\partial x_3} \right| dx \right)^{1/2}$$

—that is, (9.11). For $n > 3$ the proof is similar, but we have to use the inequality

$$\int_j |\alpha_1 \alpha_2 \cdots \alpha_k| \leq \left(\int_j |\alpha_1|^k\right)^{1/k} \left(\int_j |\alpha_2|^k\right)^{1/k} \cdots \left(\int_j |\alpha_k|^k\right)^{1/k} \qquad (9.12)$$

with $k = n - 1$, which follows by successive applications of Hölder's inequality.

To prove (9.10) we note that the function $v = |u|^{(n-1)r/(n-r)}$ is continuously differentiable if $r > 1$. We can therefore apply (9.11) with u replaced by v. This immediately yields (9.10) for $r > 1$.

Consider now the case $j = 1$, $m = 2$, $a = \frac{1}{2}$. We have to prove:

$$|Du|_p \leq c |D^2 u|_r^{1/2} |u|_q^{1/2}, \qquad \text{where } \frac{2}{p} = \frac{1}{r} + \frac{1}{q}, 1 \leq q, r \leq \infty. \qquad (9.13)$$

Suppose first that $1 \leq q < \infty$, $1 < r < \infty$, $n = 1$. Then (9.13) becomes

$$\int |Du|^p \, dx \leq c^p \left(\int |D^2 u|^r \, dx\right)^{p/2r} \left(\int |u|^q \, dx\right)^{p/2q}. \qquad (9.14)$$

An argument used in the proof of Theorem 8.1 shows that for any interval λ of length $|\lambda|$,

$$\int_\lambda |Du|^p \, dx \leq C^p |\lambda|^{1 + p - p/r} \left(\int_\lambda |D^2 u|^r \, dx\right)^{p/r} + C^p |\lambda|^{-(1 + p - p/r)} \left(\int_\lambda |u|^q \, dx\right)^{p/q},$$

$$(9.15)$$

where C is a constant independent of q, r, λ.

If we prove, for any $L > 0$,

$$\int_0^L |Du|^p \, dx \leq 2C^p \left(\int_0^\infty |D^2 u|^r \, dx\right)^{p/2r} \left(\int_0^\infty |u|^q \, dx\right)^{p/2q}, \qquad (9.16)$$

then (9.14) follows. Take any integer $k \geq 1$ and consider the interval $\lambda : 0 \leq x \leq L/k$. If the first term on the right of (9.15) is larger than the second term, we take $\lambda_1 = \lambda$. If not, we increase λ (keeping the left end-point fixed) until the two terms are equal, and call the corresponding interval λ_1. (Note that λ_1 exists since we may assume that $D^2 u \not\equiv 0$ on $(0, \infty)$.) Clearly,

$$\int_{\lambda_1} |Du|^p \, dx \leq \begin{cases} 2C^p \left(\dfrac{L}{k}\right)^{1 + p - p/r} \left(\displaystyle\int_0^L |D^2 u|^r \, dx\right)^{p/r} & \text{if } \lambda_1 = \lambda, \\[2ex] 2C^p \left(\displaystyle\int_{\lambda_1} |D^2 u|^r \, dx\right)^{p/2r} \left(\displaystyle\int_{\lambda_1} |u|^q \, dx\right)^{p/2q} & \text{if } \lambda_1 \neq \lambda. \end{cases} \qquad (9.17)$$

If $|\lambda_1| \geq L$ then (9.16) already follows. If $|\lambda_1| < L$, then we proceed to λ_2, λ_3, and so on. The last λ_i is such that $|\lambda_1| + \cdots + |\lambda_i| \geq L > |\lambda_1| + \cdots + |\lambda_{i-1}|$. Note that the analog of the first inequality in (9.17) occurs at most k times, whereas the sum of the right-hand sides in the analogs of the second inequality in (9.17) is bounded (by Hölder's inequality) by the right-hand side of (9.16). Hence, if we sum up the inequalities and take $k \to \infty$, then (9.16) follows. Note that since the constant C is independent of q, r, also the constant c in (9.14) is independent of q, r.

If $n > 1$, then we apply (9.14) to each derivative $D_i u$, treating all the x_j with $j \neq i$ as parameters. We then integrate the inequality with respect to the variables x_j $(j \neq i)$ and use Hölder's inequality.

Finally, the cases $q = \infty$ and $r = 1, \infty$ are obtained by letting $q \to \infty$ and $r \to 1, \infty$ in the inequality (9.14).

10 | EXTENDED SOBOLEV INEQUALITIES IN BOUNDED DOMAINS

We shall extend the definitions (9.5), (9.6) to apply to integrable functions u having weak derivatives $D^\beta u$. This we do simply by replacing the l.u.b. by the essential supremum. We can then state the analog of Theorem 9.3 for bounded domains.

THEOREM 10.1. *Let* Ω *be a bounded domain with* $\partial\Omega$ *in* C^m, *and let* u *be any function in* $W^{m,r}(\Omega) \cap L^q(\Omega)$, $1 \leq r, q \leq \infty$. *For any integer* j, $0 \leq j < m$, *and for any number* a *in the interval* $j/m \leq a \leq 1$, *set*

$$\frac{1}{p} = \frac{j}{n} + a\left(\frac{1}{r} - \frac{m}{n}\right) + (1-a)\frac{1}{q}.$$

If $m - j - n/r$ *is not a nonnegative integer, then*

$$|D^j u|^{\Omega}_{0,p} \leq C(|u|^{\Omega}_{m,r})^a(|u|^{\Omega}_{0,q})^{1-a}. \tag{10.1}$$

If $m - j - n/r$ *is a nonnegative integer, then* (10.1) *holds for* a $= j/m$. *The constant* C *depends only on* Ω, r, q, m, j, a.

Note that the derivatives that occur in (10.1) are weak derivatives.

Proof. Suppose first that $u \in C^m(\overline{\Omega})$. By Lemma 5.2 there exists a function U in $C_0^m(R^n)$ such that $U = u$ in Ω and

$$|U|^{R^n}_{i,s} \leq \text{const.} \ |u|^{\Omega}_{i,s} \qquad (0 \leq i \leq m, 1 \leq s \leq \infty).$$

Applying Theorem 9.3 to U, the assertion (10.1) immediately follows. Suppose now that $u \in W^{m,r}(\Omega) \cap L^q(\Omega)$. From the proof of Theorem 7.1 it follows that there exists a sequence $\{u_i\}$ of functions in $C^m(\overline{\Omega})$ such that

$$|u_i - u|_{m,r} \to 0, \quad |u_i - u|_{0,q} \to 0 \qquad \text{as } i \to \infty.$$

Suppose $p > 0$. Applying (10.1) to $u_i - u_h$ $(h > i)$, we find that

$$\int_\Omega |D^\alpha u_i(x) - D^\alpha u_h(x)|^p \, dx \le \varepsilon_i \to 0 \qquad \text{if } i \to \infty \qquad (|\alpha| = j).$$

Since $D^\alpha u_h(x) \to D^\alpha u(x)$ almost everywhere as $h \to 0$, Fatou's lemma gives

$$\int_\Omega |D^\alpha u_i(x) - D^\alpha u(x)|^p \, dx \le \varepsilon_i.$$

Hence $\int |D^\alpha u_i|^p \, dx \to \int |D^\alpha u|^p \, dx$ as $i \to \infty$. Writing (10.1) for u_i, and taking $i \to \infty$, we get (10.1) for u.

For $p < 0$ the proof is similar (now Fatou's lemma is not needed).

THEOREM 10.2.   *Let Ω be a bounded domain with $\partial\Omega$ in C^1, and let* u *be any function in* $W^{m,r}(\Omega)$, $1 \le r \le \infty$. *Then, for any integer* j, $0 \le j < m$,

$$|u|_{j,p}^\Omega \le C |u|_{m,r}^\Omega \tag{10.2}$$

where $1/p = j/n + 1/r - m/n$, *provided* p > 0. *The constant* C *depends only on* Ω, m, j, r.

Proof.   The case $j = 0$, $m = 1$ follows from Theorem 10.1. Now we use induction on m. Assuming that (10.2) holds for m, with $0 \le j < m$, we shall prove it for $m + 1$, with $0 \le j < m + 1$. First we consider the case $j = m$. Applying the case $j = 0$, $m = 1$ to $D^i u$ (more precisely to $D^\beta u$, $|\beta| = i$), we get

$$|D^i u|_{0,p} \le C |D^{i+1} u|_{0,r} \le C|u|_{m+1,r} \qquad \text{if } 0 \le i \le m, \frac{1}{p} = \frac{1}{r} - \frac{1}{n}.$$

Hence

$$|u|_{m,p} \le C |u|_{m+1,r}, \tag{10.3}$$

which is the assertion for $j = m$.

Now let $0 \leq j \leq m$. By the inductive assumption,

$$|u|_{j,p} \leq C |u|_{m,s} \qquad \text{if } \frac{1}{p} = \frac{j}{n} + \frac{1}{s} - \frac{m}{n}. \tag{10.4}$$

Setting in (10.3) $p = s$ and substituting the resulting inequality in the right-hand side of (10.4), we get

$$|u|_{j,p} \leq C |u|_{m+1,r}, \qquad \text{where } \frac{1}{p} = \frac{j}{n} + \left(\frac{1}{r} - \frac{1}{n}\right) - \frac{m}{n} = \frac{j}{n} + \frac{1}{r} - \frac{m+1}{n};$$

this completes the proof.

Remark. Theorems 10.2 and 9.1 include all of Sobolev's inequalities. For $p < 0$, set

$$|u|_{p,\infty} = \sum_{|\beta| \leq h} \text{l.u.b.} |D^\beta u| + [D^h u]_{\alpha,\infty}.$$

where $-n/p = h + \alpha,\ 0 \leq \alpha < 1$.

PROBLEMS. (1) Assume that $\partial\Omega$ is in C^1. Using Theorems 9.1 and 9.2, show that for any $u \in W^{m,r}(\Omega) \cap C^m(\Omega)$ $(1 \leq r \leq \infty,\ mr > n,\ \alpha > 0)$

$$|u|_{p,\infty} \leq C |u|_{m,r}, \qquad \text{where } \frac{1}{p} = \frac{1}{r} - \frac{m}{n}.$$

(2) Extend the result of Problem 1 for any function u in $W^{m,r}(\Omega)$. This result is the analog of Theorem 10.2 for $p < 0$ (and $j = 0$).

(3) A special case of (10.1) gives, for $1 \leq p < \infty$:

$$|u|_{j,p} \leq C(|u|_{m,p})^{j/m}(|u|_{o,p})^{(m-j/m)}. \tag{10.5}$$

Give another proof of (10.5), that is based on (8.2).

11 | IMBEDDING THEOREMS

Theorem 10.2 shows that the identity map $u \to u$ from $W^{m,r}$ into $W^{j,p}$ is well defined and is furthermore a bounded operator, provided $1/p = j/n + 1/r - m/n$. This result is an example of an *imbedding theorem* and the *imbedding is bounded.* We shall give another example based upon Theorem 9.1.

THEOREM 11.1.   Let Ω be a bounded domain satisfying the cone con-dition. If a function u belongs to $W^{j,p}$ with $j > m + n/p$ for some nonnegative integer m, then $u \in C^m(\Omega)$ (that is, u is equivalent to a function in $C^m(\Omega)$).

Proof.   Let $\{u_k\}$ be a sequence of functions in $C^j(\Omega)$ such that $|u_k - u_h|_{j,p} \to 0$ as $k, h \to \infty$ and $|u_k - u|_{0,p} \to 0$ as $k \to \infty$. Applying (9.1) to $D^\alpha(u_k - u_h)$, where $0 \leq |\alpha| \leq m$, we see that

$$\sup_\Omega |D^\alpha u_k - D^\alpha u_h| \to 0 \qquad \text{if } k, h \to \infty.$$

It follows that $\tilde{u}(x) = \lim u_k(x)$ is in $C^m(\Omega)$. Since $\tilde{u}(x) = u(x)$ almost everywhere, the assertion follows.

Denote by $C^m_*(\Omega)$ the normed linear space consisting of all the functions u in $C^m(\Omega)$ with finite norm

$$\|u\|_m = \sum_{|\beta| \leq m} \text{l.u.b.}_\Omega |D^\beta u(x)|. \tag{11.1}$$

We have:

COROLLARY.   If $j > m + n/p$, then $u \to u$ defines a bounded imbedding of $W^{j,p}(\Omega)$ into $C^m_*(\Omega)$.

PROBLEM.   (1) If $\partial\Omega \in C^m$, then the corollary holds with $C^m_*(\Omega)$ replaced by $C^m(\overline{\Omega})$ (normed by (11.1)).

DEFINITION.   An imbedding $u \to u$ from one Banach space X into another Banach space Y is called *compact*, if from any bounded sequence in X one can extract a subsequence that converges in Y.

We introduce the norm

$$\|u\|_{m+\alpha} = \|u\|_m + \text{l.u.b.}_{|\beta| \leq m} \ \text{l.u.b.}_{x,y \in \Omega} \frac{|D^\beta u(x) - D^\beta u(y)|}{|x - y|^\alpha} \tag{11.2}$$

for $0 < \alpha < 1$, and denote by $C^{m+\alpha}(\Omega)$ the space of all functions u with finite norm $\|u\|_{m+\alpha}$.

PROBLEMS.   (2) Prove that $C^{m+\alpha}(\Omega)$ is a Banach space.

(3) Prove that the imbedding $u \to u$ from $C^{m+\alpha}(\Omega)$ into $C^{m+\beta}(\Omega)$, where $0 < \beta < \alpha < 1$, is a compact imbedding.

THEOREM 11.2.   Let Ω be a bounded domain with $\partial\Omega$ in C^1. Let r be a positive number, $1 \leq r < \infty$, and let j, m be integers, $0 \leq j < m$. If p is any positive number ≥ 1 satisfying

$$\frac{1}{p} > \frac{j}{n} + \frac{1}{r} - \frac{m}{n},$$

then the imbedding $u \to u$ of $W^{m,r}(\Omega)$ into $W^{j,p}(\Omega)$ is compact.

Proof.   It suffices to prove the theorem in case $j = 0$, $m = 1$. We shall need the following lemma.

LEMMA 11.1   Let Z be a set of functions u in $L^s(\Omega)$ $(1 \leq s < \infty)$ and extend each u into $R^n - \Omega$ by 0. If

$$|J_\varepsilon u - u|_{0,s}^{R^n} \to 0 \qquad as\ \varepsilon \to 0 \tag{11.3}$$

uniformly with respect to $u \in Z$, and if

$$|u|_{0,s}^{R^n} \leq C \qquad for\ all\ u \in Z, \tag{11.4}$$

where C is a constant independent of u, then Z is contained in a compact subset of $L^s(\Omega)$.

To prove the lemma it suffices to show that for any $\delta > 0$ there exists a finite δ-covering of Z—that is, a finite number of balls with radius δ which form a covering of Z. Let $\varepsilon_0 > 0$ be such that $|J_{\varepsilon_0} u - u|_{0,s} < \delta/2$ for all $u \in Z$. It suffices to show that the set $W = \{J_{\varepsilon_0} u; u \in Z\}$ has a finite $(\delta/2)$-covering. This follows if we show that W is contained in a compact subset of $L^s(\Omega)$. Since convergence in $C(\overline{\Omega})$ [with norm $\| \ \|_0$; see (11.1)] implies convergence in $L^s(\Omega)$, it suffices to show that W is contained in a compact subset of $C(\overline{\Omega})$. By the lemma of Ascoli-Arzela, one only needs to verify that the functions of W are uniformly bounded and equicontinuous. The verification of these properties is left to the reader.

We return to the proof of Theorem 11.2. It suffices to show that from any bounded sequence $Z_0 = \{u_i\}$ of functions in $W^{1,r}(\Omega)$, with $u_i \in C^1(\overline{\Omega})$, one can extract a subsequence that is a Cauchy sequence in $L^p(\Omega)$. We extend each u_i into R^n as in Lemma 5.2. Denote by U_i the extension of u_i and set $Z = \{U_i\}$. The set Z is a subset of $W^{1,1}$ and (11.4) is clearly satisfied for $s = 1$. We claim that also (11.3) holds for $s = 1$. Indeed, by (6.5),

$$J_\varepsilon u(x) - u(x) = \int_{|z| < 1} \rho(z) \left\{ \int_0^{-z} \frac{\partial}{\partial \zeta} u(x + \varepsilon\zeta)\, d\zeta \right\} dz.$$

Hence,

$$\int_{R^n} |J_\varepsilon u - u|\, dx \leq C \int_{|z|<1} \int_0^{\varepsilon|z|} \left\{ \int_{R^n} \left| \frac{\partial}{\partial \zeta} u(x+\zeta) \right| dx \right\} |d\zeta|\, dz$$

$$\leq C_0 \varepsilon |u|_{1,1} \leq C_1 \varepsilon \qquad (u \in Z),$$

where C, C_0, C_1 are constants.

Applying Lemma 11.1, we conclude that there exists a subsequence of $\{U_i\}$ (denote it again by $\{U_i\}$) that is convergent in $L^1(R^n)$.

Now, there is a unique value of a, $0 < a < 1$, such that

$$\frac{1}{p} = a \left(\frac{1}{r} - \frac{1}{n} \right) + 1 - a.$$

Applying (10.1) to $U_i - U_h$ with $q = 1$ and with this value of a, we get

$$|u_i - u_h|_{0,p}^\Omega \leq C(|u_i - u_h|_{1,r}^\Omega)^a (|u_i - u_h|_{0,1}^\Omega)^{1-a} \leq C'(|u_i - u_h|_{0,1}^\Omega)^{1-a} \to 0$$

as $i, h \to \infty$. This completes the proof of the theorem.

The special case of Theorem 11.2 where $j = 0$, $m = 1$, $p = 2$, $r = 2$ is known as *Rellich's lemma*.

PROBLEMS. (4) Prove the converse of Lemma 11.1.

(5) Complete the proof of Theorem 11.2 for any integers j, m with $0 \leq j < m$.

12 | GÅRDING'S INEQUALITY

Consider a differential operator of order $2m$, in a bounded domain Ω,

$$Lu = \sum_{|\alpha| \leq 2m} a_\alpha(x) D^\alpha u. \tag{12.1}$$

If the coefficients $a_\alpha(x)$ belong to $C^{|\alpha|-m}$ for $m < |\alpha| \leq 2m$, then one can rewrite the operator in the *divergence form*

$$Lu = \sum_{0 \leq |\rho|, |\sigma| \leq m} (-1)^{|\rho|} D^\rho(a^{\rho\sigma}(x) D^\sigma u). \tag{12.2}$$

The condition of strong ellipticity takes the form:

$$\text{Re}\left\{ \sum_{|\rho| = |\sigma| = m} \xi^\rho a^{\rho\sigma}(x)\xi^\sigma \right\} \geq c_0 |\xi|^{2m} \qquad (c_0 > 0) \tag{12.3}$$

for any real ξ. If L is strongly elliptic at each point of $\overline{\Omega}$, and if the leading coefficients (that is, $a^{\rho\sigma}$ with $|\rho| = |\sigma| = m$) are continuous in $\overline{\Omega}$, then c_0 can be taken to be a constant independent of x in $\overline{\Omega}$.

The formal adjoint of Lu is easily seen to be the operator

$$L^*v = \sum_{0 \le |\rho|, |\sigma| \le m} (-1)^{|\rho|} D^\rho(\overline{a^{\sigma\rho}}(x)D^\sigma v). \tag{12.4}$$

Furthermore, if $u, v \in C_0^\infty(\Omega)$, then

$$(v, Lu) - (L^*v, u) = B[v, u], \tag{12.5}$$

where $(f, g) = \int_\Omega f(x)\overline{g(x)}\, dx$ and

$$B[v, u] = \sum_{0 \le |\rho|, |\sigma| \le m} (D^\rho v, a^{\rho\sigma}D^\sigma u). \tag{12.6}$$

We call $B[v, u]$ the *bilinear form associated with* L. $B[v, u]$ is well defined even if the coefficients $a^{\rho\sigma}$ are merely integrable (but then, of course, L (in (12.2)) is not a differential operator but merely a formal expression).

We shall need the following conditions:

(A_1) L is strongly elliptic in Ω with a module of strong ellipticity c_0 independent of x in Ω.

(A_2) The coefficients $a^{\rho\sigma}$ of L are bounded (in Ω) by a constant c_1.

(A_3) The principal coefficients of L have a modulus of continuity $c_2(t)$—that is,

$$|a^{\rho\sigma}(x) - a^{\rho\sigma}(y)| \le c_2(|x - y|) \text{ if } |\rho| = |\sigma| = m, x \in \Omega, y \in \Omega,$$

and $c_2(t) \searrow 0$ if $t \searrow 0$.

In Section 6 we have introduced the spaces $H^{j,p}$, $W^{j,p}$, $\tilde{H}^{j,p}$ and proved that they are all equal. We now introduce the spaces $\tilde{H}_0^{j,p}(\Omega)$ and $H_0^{j,p}(\Omega)$ as the completion space of all the functions in $C_0^\infty(\Omega)$ and $C_0^j(\Omega)$, respectively, with respect to the norm (6.9). The reader will easily verify that $\tilde{H}_0^{j,p}(\Omega) = H_0^{j,p}(\Omega)$.

Since we shall deal mostly with spaces $H^{j,p}$, $H_0^{j,p}$, where $p = 2$, we introduce the special notation:

$$H^j(\Omega) = H^{j,2}(\Omega), \qquad H_0^j(\Omega) = H_0^{j,2}(\Omega), \qquad \| \ \|_j^\Omega = | \ |_{j,2}^\Omega. \tag{12.7}$$

Note that $H^j(\Omega)$ is a Hilbert space. Its scalar product is denoted by $(\ , \ )_j^\Omega$ and, for $j = 0$, by $(\ , \ )^\Omega$. When there is no confusion, we may omit the symbol Ω.

We can now state the following result, known as *Gårding's inequality*:

THEOREM 12.1.   If the assumptions (A_1)–(A_3) *hold, then there exist constants* $c > 0$ *and* k_0, *depending only on* c_0, c_1, c_2, *and* Ω, *such that*

$$\text{Re } B[\phi, \phi] \geq c \|\phi\|_m^2 - k_0 \|\phi\|_0^2 \tag{12.8}$$

for all $\phi \in H_0^m(\Omega)$.

Proof.   We introduce the seminorm

$$|\hat{\phi}|_j = \left\{ \sum_{|\alpha| = j} \int_\Omega |D^\alpha \phi(x)|^2 \, dx \right\}^{1/2}. \tag{12.9}$$

It suffices to prove (12.8) for $\phi \in C_0^\infty(\Omega)$. For then, given any $\phi \in H_0^m(\Omega)$, take a sequence $\{\phi_i\}$ in $C_0^\infty(\Omega)$ such that $\|\phi - \phi_i\|_m \to 0$ as $i \to \infty$. Note that $B[\phi_i, \phi_i] \to B[\phi, \phi]$ as $i \to \infty$. Applying (12.8) to each ϕ_i and taking $i \to \infty$, (12.8) thereby follows.

We consider first the case where L is homogeneous (that is, $a^{\rho\sigma} = 0$ if $|\rho| + |\sigma| < 2m$) and its coefficients are constants. We shall employ the tool of Fourier transforms. We therefore recall that the Fourier transform of $\phi(x)$ is defined as

$$\tilde{\phi}(\xi) = \int_{R^n} e^{-ix \cdot \xi} \phi(x) \, dx. \tag{12.10}$$

As easily verified,

$$\widetilde{D^\alpha u}(\xi) = (i\xi)^\alpha \tilde{u}(\xi). \tag{12.11}$$

The Plancherel theorem states that

$$\int_{R^n} |\tilde{\phi}(\xi)|^2 \, d\xi = (2\pi)^n \int_{R^n} |\phi(x)|^2 \, dx. \tag{12.12}$$

Using (12.11), (12.12), we have

$$B[\phi, \phi] = \sum_{|\rho| = |\sigma| = m} (D^\rho \phi, a^{\rho\sigma} D^\sigma \phi) = (2\pi)^{-n} \sum_{|\rho| = |\sigma| = m} ((i\xi)^\rho \tilde{\phi}, a^{\rho\sigma} (i\xi)^\sigma \tilde{\phi})$$

$$= (2\pi)^{-n} \int_{R^n} |\tilde{\phi}(\xi)|^2 \left[\sum_{|\rho| = |\sigma| = m} \xi^\rho a^{\rho\sigma} \xi^\sigma \right] d\xi. \tag{12.13}$$

Using (12.3), we get

$$\text{Re } B[\phi, \phi] \geq (2\pi)^{-n} c_0 \int_{R^n} |\tilde{\phi}(\xi)|^2 \left[\sum_{|\rho| = m} \xi^\rho \xi^\rho \right] d\xi.$$

Since the integral on the right-hand side coincides with the integral on right-hand side of (12.13) when $a^{\rho\sigma} = \delta^{\rho\sigma}$ ($\delta^{\rho\sigma} = 0$ if $\rho \neq \sigma$, $\delta^{\rho\rho} = 1$) we get, upon using (12.13) with $a^{\rho\sigma} = \delta^{\rho\sigma}$,

$$\operatorname{Re} B[\phi, \phi] \geq c_0 \sum_{|\rho| = m} (D^\rho \phi, D^\rho \phi),$$

—that is,

$$\operatorname{Re} B[\phi, \phi] \geq c_0 |\hat{\phi}|_m^2. \tag{12.14}$$

Consider next the case where L is still homogeneous but its coefficients are not assumed to be constants. We shall then estimate $\operatorname{Re} B[\phi, \phi]$ in case the support of ϕ lies in an open neighborhood W of a point $x^0 \in \Omega$, and the diameter $|W|$ of W is sufficiently small.

Write

$$B[\phi, \phi] = \sum_{|\rho| = |\sigma| = m} (D^\rho \phi, a^{\rho\sigma}(x^0) D^\sigma \phi) + \sum_{|\rho| = |\sigma| = m} (D^\rho \phi, [a^{\rho\sigma}(x) - a^{\rho\sigma}(x^0)] D^\sigma \phi)$$

$$\equiv I + J.$$

By (12.14),

$$\operatorname{Re} I \geq c_0 |\hat{\phi}|_m^2.$$

If $|W| \leq \delta$ (δ depending only on c_0, c_2), then

$$|J| \leq \tfrac{1}{2} c_0 |\hat{\phi}|_m^2.$$

We conclude that

$$\operatorname{Re} B[\phi, \phi] \geq \tfrac{1}{2} c_0 |\hat{\phi}|_m^2. \qquad \text{for all } \phi \in C_0^\infty(W). \tag{12.15}$$

We finally consider the general case. We cover $\bar{\Omega}$ by a finite number of open sets W_j, each having a diameter $\leq \delta$. Let $\{\alpha_j\}$ be a partition of unity subordinate to the covering $\{W_j\}$ of Ω, and set $\beta_j = \sqrt{\alpha_j}$. We may assume that $\beta_j \in C^\infty$, for otherwise we can replace the α_j by $\alpha_j^2 / (\sum \alpha_i^2)$. Write

$$B[\phi, \phi] = \sum_{|\rho| = |\sigma| = m} \int_\Omega D^\rho \phi \cdot \overline{a^{\rho\sigma} D^\sigma \phi} \, dx + \sum_{\substack{|\rho|, |\sigma| \leq m \\ |\rho| + |\sigma| < 2m}} \int_\Omega D^\rho \phi \cdot \overline{a^{\rho\sigma} D^\sigma \phi} \, dx$$

$$\equiv H + K. \tag{12.16}$$

Clearly

$$|K| \leq C \|\phi\|_m \|\phi\|_{m-1}, \tag{12.17}$$

where C is a constant depending only on c_1. Next,

$$
\begin{aligned}
H &= \sum_j \sum_{|\rho|=|\sigma|=m} \int_\Omega [\beta_j D^\rho \phi][\beta_j \overline{a^{\rho\sigma} D^\sigma \bar{\phi}}]\, dx \\
&= \sum_j \sum_{|\rho|=|\sigma|=m} \int_\Omega D^\rho(\beta_j \phi) \cdot \overline{a^{\rho\sigma} D^\sigma(\beta_j \bar{\phi})}\, dx \\
&\quad - \sum_j \sum \int_\Omega c_{\rho_1 \rho_2 \sigma_1 \sigma_2} D^{\rho_1}\beta_j \cdot D^{\rho_2}\phi \cdot \overline{a^{\rho\sigma} D^{\sigma_1}\beta_j} \cdot D^{\sigma_2}\bar{\phi}\, dx \\
&\equiv H_1 - H_2.
\end{aligned}
\tag{12.18}
$$

In H_2 the c's are bounded functions and $\rho_1 + \sigma_1 > 0$, so that $\rho_2 + \sigma_2 < 2m$. Hence

$$
|H_2| \le C_0 \|\phi\|_m \|\phi\|_{m-1},
\tag{12.19}
$$

where C_0 is a constant depending only on c_1, β_j.

H_1 is a sum of terms

$$
\sum_{|\rho|=|\sigma|=m} \int_\Omega D^\rho \psi_j \cdot \overline{a^{\rho\sigma} D^\sigma \bar{\psi}_j}\, dx \qquad \text{with } \psi_j = \phi\beta_j \in C_0^\infty(W_j).
$$

By the result of (12.15) with $W = W_j$ we get

$$
\operatorname{Re} H_1 \ge \tfrac{1}{2} c_0 \sum_j |\hat{\psi}_j|_m^2 = \tfrac{1}{2} c_0 \sum_j \sum_{|\rho|=m} \int_\Omega D^\rho(\beta_j \phi) \cdot D^\rho(\beta_j \bar{\phi})\, dx.
$$

The double sum on the right coincides with the double sum of H_1 on the right-hand side of (12.18) when $a^{\rho\sigma} = \delta^{\rho\sigma}$. Hence

$$
\operatorname{Re} H_1 \ge \tfrac{1}{2} c_0 \sum_j \sum_{|\rho|=m} \int_\Omega [\beta_j D^\rho \phi][\beta_j D^\rho \bar{\phi}]\, dx + H_2^1,
\tag{12.20}
$$

where H_2^1 coincides with H_2 when $a^{\rho\sigma} = \delta^{\rho\sigma}$. Employing (12.19) for $a^{\rho\sigma} = \delta^{\rho\sigma}$, we get

$$
|H_2^1| \le C_1 \|\phi\|_m \|\phi\|_{m-1},
\tag{12.21}
$$

where C_1 is a constant depending only on β_j.

Combining (12.16)–(12.21), we obtain

$$
\operatorname{Re} B[\phi, \phi] \ge \tfrac{1}{2} c_0 |\hat{\phi}|_m^2 - C_2 \|\phi\|_m \|\phi\|_{m-1}
\tag{12.22}
$$

where C_2 is a constant depending only on c_0, c_1, c_2, Ω.

Using the inequality

$$C_2 \|\phi\|_m \|\phi\|_{m-1} \leq \eta \|\phi\|_m^2 + \frac{C_2}{4\eta} \|\phi\|_{m-1}^2 \qquad (\eta > 0)$$

and Theorem 8.1, we get

$$C_2 \|\phi\|_m \|\phi\|_{m-1} \leq \eta \|\phi\|_m^2 + \frac{C_2}{4\eta} (\varepsilon \|\phi\|_m^2 + C_3 \|\phi\|_0^2) \qquad (\varepsilon > 0), \quad (12.23)$$

where C_3 is a constant depending only on ε, m, Ω. By Theorem 8.1 we also have

$$\tfrac{1}{2}c_0 |\hat{\phi}|_m^2 \geq \tfrac{1}{3}c_0 \|\phi\|_m^2 - C_4 \|\phi\|_0^2, \qquad (12.24)$$

where C_4 is a constant depending only on c_0, m, Ω. Taking $\eta = c_0/9$, $\varepsilon \leq (4\eta/C_2)(c_0/9)$, and substituting (12.23), (12.24) into (12.22), the inequality (12.8) follows.

PROBLEMS. (1) Denote by $B^*[u, v]$ the bilinear form associated with L^*. Prove that $B^*[u, v] = \overline{B[v, u]}$.

(2) Prove the converse of Theorem 12.1—that is, if (12.8) holds for any $\phi \in C_0^\infty(\Omega)$ and if (A_2), (A_3) hold, then L is strongly elliptic in Ω.

13 | THE DIRICHLET PROBLEM

Let Ω be a bounded domain with boundary $\partial\Omega$ in C^{m-1}, and let L be an elliptic operator of order $2m$ with coefficients defined in Ω. Given any function f in Ω and functions $g_0, \ldots, g_{m-1}$ on $\partial\Omega$, the *Dirichlet problem* consists of finding a function u satisfying

$$Lu = f \text{ in } \Omega, \qquad (13.1)$$

$$\frac{\partial^j u}{\partial v^j} = g_j \quad \text{on } \partial\Omega \quad (j = 0, 1, \ldots, m-1), \qquad (13.2)$$

where $\partial/\partial v$ is differentiation with respect to the outward normal to the boundary.

If a function u belongs to $C^{2m}(\Omega)$ and satisfies (13.1), then it is called a *classical solution* of (13.1). If a function u belongs to $C^{m-1}(\overline{\Omega})$ and satisfies (13.2), then it is said to satisfy (13.2) in the *classical sense*. If $u \in C^{2m}(\Omega) \cap C^{m-1}(\overline{\Omega})$ and satisfies (13.1), (13.2), then it is called a *classical solution of the Dirichlet problem* (13.1), (13.2).

LEMMA 13.1.   Let $\partial\Omega$ be of class $C^{k+1}(k \geq m - 1)$ and assume that the $g_j (0 \leq j \leq m - 1)$ are of class C^k in the local parameters of $\partial\Omega$. Then there exists a function Φ in $C^k(\bar{\Omega})$ satisfying $\partial^j\Phi/\partial v_j = g_j (0 \leq j \leq m - 1)$ on $\partial\Omega$.

Proof.   At each point $x^0 \in \partial\Omega$ draw an inward normal $v(x^0)$ and measure on it a length δ. Denote the end-point by $v(x^0, \delta)$. The function $x = v(x^0, \delta)$ is given by

$$x_i = x_i^0(s) + \frac{\delta p_i(s)}{p(s)} \qquad (1 \leq i \leq n), \tag{13.3}$$

where $s = (s_1, \ldots, s_{n-1})$ are local parameters on $\partial\Omega$, $x_i = x_i^0(s)$ are the equations of $\partial\Omega$, $p_i(s)$ is $(-1)^{i-1}$ times the determinant of the matrix obtained from the matrix $(\partial x_i^0/\partial s_j)$ (i indicates rows, j indicates columns) by suppressing the ith row, and

$$p(s) = \left\{ \sum_{i=1}^{n} (p_i(s))^2 \right\}^{1/2}.$$

Using the implicit function theorem, it follows that if $\delta_0 > 0$ is sufficiently small, then $x = v(x^0, \delta)$ establishes a one-to-one C^k map from the set $\{(x^0, \delta); x^0 \in \partial\Omega, 0 \leq \delta \leq \delta_0\}$ onto some $\bar{\Omega}$-neighborhood Ω_0 of $\partial\Omega$. It can also be verified that the distance $\sigma(x)$ from any $x = v(x^0, \delta)$ to $\partial\Omega$ is equal to δ. Thus, in particular, Ω_0 consists of all the points in $\bar{\Omega}$ whose distance to $\partial\Omega$ is $\leq \delta_0$. Now take

$$\Phi(x) = \zeta(x) \sum_{j=0}^{m-1} \frac{(\sigma(x))^j}{j!} g_j(x^0)$$

in Ω_0, $\Phi = 0$ in $\Omega - \Omega_0$, where ζ is a C^∞ function satisfying: $\zeta = 1$ in a neighborhood of $\partial\Omega$, $\zeta = 0$ in a neighborhood of $\Omega \cap \partial\Omega_0$ ($\partial\Omega_0 = $ boundary of Ω_0).

Suppose $\partial\Omega$ is of class C^{2m+1} and the g_j belong to $C^{2m}(\partial\Omega)$. Setting $v = u - \Phi$, the system (13.1), (13.2) reduces to $Lv = \hat{f}$, $\partial^j v/\partial v^j = 0 (0 \leq j \leq m - 1)$, where $\hat{f} = f - L\Phi$. Thus, without loss of generality one may consider the Dirichlet problem for homogeneous boundary data—that is, the problem of finding a solution u of (13.1) satisfying

$$\frac{\partial^j u}{\partial v^j} = 0 \qquad \text{on } \partial\Omega \qquad (0 \leq j \leq m - 1). \tag{13.4}$$

DEFINITION.   A function u locally integrable in Ω is called a *weak solution* of (13.1) if

$$(L^*\phi, u) = (\phi, f) \qquad \text{for any } \phi \in C_0^\infty(\Omega). \tag{13.5}$$

If $u \in H^m(\Omega_0)$ for any subdomain Ω_0 of Ω with closure in Ω, and if

$$B[\phi, u] = (\phi, f) \qquad \text{for any } \phi \in C_0^\infty(\Omega), \tag{13.6}$$

then u is called a *strong solution* of (13.1)

PROBLEMS. (1) Assume that the $a^{\rho\sigma}$ belong to $C^{|\sigma|}$ and let u be a function in $H^m(\Omega_0)$ for any subdomain Ω_0 of Ω with $\overline{\Omega_0} \subset \Omega$. Prove: u is a weak solution of (13.1) if and only if u is a strong solution of (13.1).

(2) Assume that the $a^{\rho\sigma}$ belong to $C^{|\rho|}$ and let u be a function in $C^{2m}(\Omega)$. Prove that u is a classical solution of (13.1) if and only if u is a strong solution of (13.1).

So far we have only given generalized notions of a solution of (13.1). Now we give a generalized notion of the conditions (13.4):

If $u \in H_0^m(\Omega)$, then we say that u *satisfies the conditions* (13.4) *in the generalized sense.*

To justify this definition we state the following results:

LEMMA 13.2.   *Assume that $\partial\Omega$ is of class* C^k. *If* u *belongs to* $H_0^k(\Omega)$ *and to* $C^{k-1}(\overline{\Omega})$, *then* $\partial^j u / \partial v^j = 0$ *on* $\partial\Omega$ *for* $0 \le j \le k - 1$.

LEMMA 13.3.   *Assume that $\partial\Omega$ is of class* C^k. *If a function* u *belongs to* $C^k(\overline{\Omega})$, *and if* $\partial^j u / \partial v^j = 0$ *on* $\partial\Omega$ *for* $0 \le j \le k - 1$, *then* u $\in H_0^k(\Omega)$.

Proof of Lemma 13.2.   It suffices to consider the case $k = 1$, for the general case then follows by induction. Take $x^0 \in \partial\Omega$ and let N be a neighborhood of x^0 such that $N \cap \partial\Omega$ can be represented, say, in the form $x_n = h(x_1, \ldots, x_{n-1})$ with $x_n > h(x_1, \ldots, x_{n-1})$ in $N \cap \Omega$. We use the C^1 mapping (7.5), and introduce in the image of $N \cap \Omega$ a right circular cylinder S_k^* of height k for which y^0 (the image of x^0) is the center of its lower base B^*.

Denote by S_k, B the sets whose images in the y-space are S_k^* and B^*, respectively. Consider first the case where u is in $C^1(\overline{\Omega})$ and $u = 0$ on B. Then the function $v(y) = u(x)$ satisfies

$$|v(y)|^2 \le \left(\int_0^{y_n} \left| \frac{\partial v}{\partial y_n} \right| dy_n \right)^2 \le k \int_0^{y_n} \left| \frac{\partial v}{\partial y_n} \right|^2 dy_n$$

for any $y \in S_k^*$. Integrating over S_k^*, we get

$$\int_{S_k^*} |v|^2 \, dy \leq k^2 \int_{S_k^*} \left| \frac{\partial v}{\partial y_n} \right|^2 \, dy.$$

This implies

$$\int_{S_k} |u|^2 \, dx \leq Ck^2 \int_{S_k} \sum_{i=1}^n \left| \frac{\partial u}{\partial x_i} \right|^2 \, dx, \tag{13.7}$$

where C is a constant independent of u, k.

Now let u be any function $H_0^1(\Omega)$. Then there exists a sequence $\{u_i\}$ of functions in $C_0^\infty(\Omega)$ such that $|u_i - u|_{1,2} \to 0$ as $i \to \infty$. Applying (13.7) to each u_i and taking $i \to \infty$, we obtain the inequality (13.7) for any u in $H_0^1(\Omega)$.

Using the transformation (7.5), we get from (13.7)

$$\frac{1}{k} \int_{S_k^*} |v(y)|^2 \, dy \leq Ck \int_{S_k} \sum_{i=1}^n \left| \frac{\partial u}{\partial x_i} \right|^2 \, dx$$

with another constant C. Taking $k \to 0$ and using the continuity of v, it follows that $v(y) = 0$ on B^*. Hence $u(x^0) = 0$. Since x^0 is arbitrary, the proof is complete.

PROBLEM. (3) Prove Lemma 13.3. [*Hint:* If $\Omega = \{x; |x| < R, x_n > 0\}$ and $u(x) = 0$ for $R - \delta < |x| < R$, define $u(x) = 0$ if $x_n < 0$ and $J_\varepsilon' u$ by (7.1) with $x_\varepsilon = (x_1, \ldots, x_{n-1}, x_n - 2\varepsilon)$, and prove that $J_\varepsilon' u \to u$ in $H_0^m(\Omega)$.]

DEFINITION. A function u that belongs to $H_0^m(\Omega)$ and that satisfies (13.6) is called a *generalized solution of the Dirichlet problem* (13.1), (13.4). The problem of finding such a function u is called the *generalized Dirichlet problem of* (13.1), (13.4).

We can now summarize some of the results of this section in the following theorem:

THEOREM 13.1. *Let $\partial \Omega$ be of class C^m and assume that $a^{p\sigma}$ belong to $C^{|p|} \cap C^{|\sigma|}$. If u is a classical solution of the Dirichlet problem* (13.1), (13.4) *and if $u \in C^m(\bar{\Omega})$, then u is a generalized solution of the Dirichlet problem. Conversely, if u is a generalized solution of the Dirichlet problem* (13.1), (13.4) *and if $u \in C^{2m}(\Omega) \cap C^{m-1}(\bar{\Omega})$, then u is a classical solution of the Dirichlet problem.*

Our plan for the next few sections is to prove existence theorems for the generalized Dirichlet problem and then show that the generalized solution is a smooth function in Ω and (what is more difficult) in $\bar{\Omega}$. In conjunction with Theorem 13.1 we shall then have obtained existence theorems for the classical Dirichlet problem.

14 | EXISTENCE THEORY

In this section Ω is an arbitrary bounded domain (with no conditions on $\partial\Omega$).

We shall need the following theorem, known as the Lax-Milgram lemma.

THEOREM 14.1.   Let B[x, y] *be a bilinear form (that is, linear in* x *and antilinear in* y*) in a Hilbert space* H *with norm* $\|\ \ \|$ *and scalar product* (,) *and assume that* B[x, y] *is bounded—that is,*

$$|B[x, y]| \leq \text{const. } \|x\|\, \|y\| \qquad \text{for all x, y in H.} \tag{14.1}$$

Suppose further that

$$|B[x, x]| \geq c\|x\|^2 \qquad \text{for all x} \in H \tag{14.2}$$

for some positive constant c. *Then every bounded linear functional* F(x) *in* H *can be represented in the form*

$$F(x) = B[x, v] = \overline{B[w, x]} \tag{14.3}$$

for some elements v, w *in* H *that are uniquely determined by* F.

Proof.   For fixed v, $B[x, v]$ is a bounded linear functional. Hence there exists a unique y such that

$$B[x, v] = (x, y).$$

Set $y = Av$. A is a bounded linear operator in H. Since

$$c\|v\|^2 \leq |B[v, v]| \leq |(v, y)| \leq \|v\|\, \|y\|$$

—that is, $\|v\| \leq \|y\|/c$—A has a bounded inverse. It follows that the range of A, $R(A)$, is a closed linear subspace of H. We claim that $R(A) = H$. Indeed, if $R(A) \neq H$, then there exists an element $z \neq 0$ that is orthogonal to $R(A)$—that is, $(z, Av) = 0$ for all $v \in H$. This implies that $B[z, v] = (z, Av) = 0$. Taking $v = z$, we get $B[z, z] = 0$. Hence, by (14.2), $z = 0$—a contradiction.

Consider now the functional $F(x)$. By a well-known theorem of Riesz, there is an element b in H such that $F(x) = (x, b)$ for all $x \in H$. Since $R(A) = H$, there is an element v such that $Av = b$. Hence,

$$F(x) = (x, b) = (x, Av) = B[x, v].$$

v is uniquely determined. Indeed, if $B[x, v'] = F(x)$ for some $v' \in H$ and for all $x \in H$ then $B[x, v - v'] = 0$ for all x. Taking $x = v - v'$ and using (14.2), we get $v = v'$.

The representation $F(x) = \overline{B[w, x]}$ follows by applying the previous result to the bilinear form $\overline{B[(y, x]}$.

Theorem 14.1 may be conceived of as an existence theorem—that is, for every bounded linear functional F there exist elements v, w such that (14.3) holds. We shall need also the existence theory of Fredholm-Riesz-Schauder for Hilbert space. We recall that a continuous operator from a Banach space X into a Banach space Y is called *completely continuous* (or *compact*) if it maps bounded sets in X into compact subsets of Y.

Example.   Let $K(x, y)$ be a continuous function in $\bar{D} \times \bar{D}$ (D a bounded domain) and let T be defined by

$$Tf(x) = \int_D K(x, y)f(y) \, dy.$$

Then T is a completely continuous linear operator from $L^p(D)$ $(1 \leq p \leq \infty)$ into any $L^q(D)$ $(1 \leq q \leq \infty)$ or into $C(\bar{D})$ (with the uniform topology).

Let T be a linear operator with domain D_T dense in H (H Hilbert space). Consider the set of all the elements y for which there exists an element z such that

$$(Tx, y) = (x, z) \qquad \text{for all } x \in D_T.$$

We write $T^*y = z$ and this defines a linear operator T^*. If T is bounded (completely continuous) then T^* is also bounded (completely continuous) and $\|T^*\| = \|T\|$.

We shall state the main results of the Fredholm-Riesz-Schauder theory in the following two theorems.

> **THEOREM 14.2.**   *Let* T *be a completely continuous linear operator in a Hilbert space* H *and consider the equations*

$$x - Tx = f, \tag{14.4}$$

$$y - T^*y = g. \tag{14.5}$$

Then the following alternative holds: either (i) *there exists a unique solution of* (14.4) *and of* (14.5) *for any* f *and* g *in* H, *or* (ii) *the equation*

$$x - Tx = 0 \tag{14.6}$$

has nontrivial solutions. If (ii) *holds, then the dimension of the space* N *of solutions of* (14.6) *is finite and equals the dimension of the space* N* *of solutions of*

$$y - T^*y = 0.$$

Furthermore, (14.4) *then has a solution* (*not unique, of course*) *if and only if* $(f, x) = 0$ *for every* $x \in$ N*.

A complex number λ is called an *eigenvalue* of T if there exists a nonzero element x, called an *eigenelement* (*corresponding to* λ) such that

$$Tx = \lambda x.$$

THEOREM 14.3. *The eigenvalues of a completely continuous linear operator in a Hilbert space* H *form either a finite set or a countable sequence converging to* 0.

For a proof of Theorems 14.2, 14.3 we refer the reader to Bers-John-Schechter [1].

We now turn to existence theorems for the Dirichlet problem.

THEOREM 14.4. *Assume that* L *satisfies* (A_1)–(A_3) (*of Section* 12) *and that the bilinear form* $B[\phi, u]$ *associated with* L *satisfies, for some constant* $c > 0$,

$$\text{Re } B[\phi, \phi] \geq c \|\phi\|_m^2 \qquad \text{for all } \phi \in H_0^m(\Omega). \tag{14.7}$$

Then there exists a unique solution of the generalized Dirichlet problem (13.1), (13.4) *for any* $f \in L^2(\Omega)$.

Proof. $B[\phi, u]$ and the functional $F(\psi) = (\psi, f)$ satisfy the assumptions of Theorem 14.1 with $H = H_0^m(\Omega)$. Hence there exists a unique $u \in H_0^m(\Omega)$ such that

$$(\psi, f) = B[\psi, u] \qquad \text{for all } \psi \in H_0^m(\Omega),$$

and this is precisely the assertion of the theorem.

From Theorems 12.1, 14.4 we immediately obtain:

THEOREM 14.5. *Assume that* L *satisfies* (A_1)–(A_3). *Then there exists a constant* k_0 (*depending only on* c_0, c_1, c_2, Ω) *such that for any* $k \geq k_0$ *the generalized Dirichlet problem for* L + k, *with homogeneous boundary conditions, has a unique solution.*

Our final and most general existence theorem is the following:

THEOREM 14.6.   Assume that L *satisfies* (A_1)–(A_3). *Then the Fredholm alternative holds for the generalized Dirichlet problem. More precisely, either for any* $f \in L^2(\Omega)$ *there exists a unique solution of*

$$B[\phi, u] = (\phi, f) \qquad for\ all\ \phi \in H_0^m(\Omega),\ u \in H_0^m(\Omega), \qquad (14.8)$$

or there is a finite number of linearly independent solutions v_j ($j = 1, \ldots, h$) *of*

$$B[v, \phi] = 0 \qquad for\ all\ \phi \in H_0^m(\Omega),\ v \in H_0^m(\Omega),$$

and then there exists a solution of (14.8) *if and only if* $(f, v_j) = 0$ *for* $1 \leq j \leq h$. *In the second case the solution, if existing, is not unique.*

Proof.   Take a fixed $k \geq k_0$, $k \neq 0$, where k_0 is the constant that appears in Theorems 12.1, 14.5. Let $L_k = L + k$. The bilinear form associated with L_k is $B_k[u, v] = B[u, v] + k(u, v)$. By Theorem 14.5, for any $g \in L^2(\Omega)$ there exists a unique solution w of

$$B_k[\phi, w] = (\phi, g) \qquad for\ all\ \phi \in H_0^m(\Omega),\ w \in H_0^m(\Omega). \qquad (14.9)$$

Set $w = L_k^{-1} g$. Since (14.8) is equivalent to

$$B_k[\phi, u] = (\phi, ku + f) \qquad for\ all\ \phi \in H_0^m(\Omega),\ u \in H_0^m(\Omega),$$

u satisfies (14.8) if and only if $u = L_k^{-1}(ku + f)$—that is, if and only if

$$u - Tu = f_1 \qquad (T = kL_k^{-1},\ f_1 = L_k^{-1}f). \qquad (14.10)$$

From (14.9) we get

$$c\|w\|_m^2 \leq |B_k[w, w]| \leq |(w, g)| \leq \|w\|_m \|g\|_0$$

—that is, $c\|w\|_m \leq \|g\|_0$. Hence T maps bounded sets of $L^2(\Omega)$ into bounded sets in $H_0^m(\Omega)$. Take a bounded domain Ω_0 with C^1 boundary. As easily seen, we can imbed $H_0^m(\Omega)$ into $H_0^m(\Omega_0)$ isometrically by extending each function by zero into $\Omega_0 - \Omega$. Applying Theorem 11.2 in Ω_0, we conclude that T maps bounded sets in $L^2(\Omega)$ into compact sets in $L^2(\Omega)$. Thus, the Fredholm-Riesz-Schauder theory can be applied. It follows that:

(a) either for every f_1 in $L^2(\Omega)$ there exists a unique solution of (14.10), or

(b) there are nontrivial solutions of $u - Tu = 0$ and the equation (14.10) has a solution if and only if $(f_1, v_j) = 0$ for a finite number of functions v_j (say $1 \leq j \leq h$) which form a basis of the space of solutions of $v - T^*v = 0$

If case (a) holds, there exists a solution of (14.8) for any $f \in L^2(\Omega)$. It is unique since $f = 0$ implies $f_1 = 0$ and, consequently, $u = 0$.

To consider the case (b), note first that Theorems 12.1, 14.5 are true also for L^* with a possibly different constant k_0. Taking k larger also than this constant, we shall prove that

$$T^* = k(L^* + k)^{-1}. \tag{14.11}$$

Indeed, by the definition of T^*,

$$(T^*v, g) = (v, Tg) \qquad \text{for all } v, g \text{ in } L^2(\Omega). \tag{14.12}$$

Set $Tg = h$, $T^*v = w$. Then

$$B_k[\phi, h] = k(\phi, g) \qquad \text{for all } \phi \in H_0^m(\Omega). \tag{14.13}$$

In particular,

$$B_k[w, h] = k(w, g). \tag{14.14}$$

Set $k(L^* + k)^{-1}v = w_1$. Then w_1 satisfies

$$B_k[w_1, \psi] = k(v, \psi) \qquad \text{for all } \psi \in H_0^m(\Omega), w_1 \in H_0^m(\Omega).$$

Hence

$$B_k[w_1, h] = k(v, h) = k(v, Tg) = k(T^*v, g) = k(w, g). \tag{14.15}$$

Combining (14.14) and (14.15) and using (14.13), we get

$$0 = B_k[w - w_1, h] = k(w - w_1, g).$$

Since g is arbitrary, $w - w_1 = 0$—that is, $T^*v = k(L^* + k)^{-1}v$. This completes the proof of (14.11).

From (14.11) we conclude that $T^*v - v = 0$ if and only if

$$B[v, \phi] = 0 \qquad \text{for all } \phi \in H_0^m(\Omega), v \in H_0^m(\Omega). \tag{14.16}$$

Thus, to complete the proof of the theorem it remains to show that the conditions $(f_1, v_j) = 0$ are equivalent to the conditions $(f, v_j) = 0$. This follows from the equalities

$$(f_1, v_j) = (L_k^{-1}f, v_j) = (f, (L_k^{-1})^*v_j) = \frac{1}{k}(f, T^*v_j) = \frac{1}{k}(f, v_j).$$

If for a complex number λ there is a nontrivial generalized solution u of the Dirichlet problem

$$\begin{cases} Lu - \lambda u = 0 & \text{in } \Omega, \\ \dfrac{\partial^j u}{\partial v^j} = 0 & \text{on } \partial\Omega \quad (0 \le j \le m-1), \end{cases} \tag{14.17}$$

then we call λ an *eigenvalue* of L and we call the solution u a *generalized eigenfunction (corresponding to λ)*.

PROBLEMS. (1) Prove, under the assumptions (A_1)–(A_3) of Section 12, that the set of eigenvalues has no finite point of accumulation.

(2) Prove that the operators $(-1)^k \Delta^k + 1$ and $(1 - \Delta)^k$ satisfy the assumptions of Theorem 14.4.

(3) The equation

$$u_{xx} + 2iu_{xy} + u_{yy} = 0$$

is elliptic but not strongly elliptic. Verify that $u = (1 - |z|^2)f(z)$ $(z = x + iy)$ is a solution of the equation in the unit disc $|z| < 1$ which vanishes on $|z| = 1$, where $f(z)$ is *any* complex analytic function for $|z| \le 1$.

15 | REGULARITY IN THE INTERIOR

In Section 14 we have established existence theorems for the generalized Dirichlet problem. Our task in this and in the following two sections is to prove that the generalized solutions are classical solutions if $\partial\Omega, f$, and the coefficients of L are sufficiently smooth. We shall need the following two lemmas concerned with estimating norms of difference quotients.

LEMMA 15.1. Let A be a subdomain of Ω with $\bar{A} \subset \Omega$ and let h_0 be the distance from A to $\partial\Omega$. Let $u \in H^j(\Omega)$ for some $j \ge 1$. Then the difference quotient with respect to x_i $(1 \le i \le n)$

$$u^h(x) = \frac{1}{h}(u(x + he^i) - u(x)) \qquad (e^i = (e_1^i, \ldots, e_n^i) \text{ with } e_j^i = \delta_{ij}) \tag{15.1}$$

is well defined for $x \in A$ if $|h| < h_0$, and

$$\|u^h\|_{j-1}^A \le \|u\|_j^\Omega, \tag{15.2}$$

$$\lim_{h \to 0} \left\| u^h - \frac{\partial u}{\partial x_i} \right\|_{j-1}^A = 0. \tag{15.3}$$

LEMMA 15.2. Let Ω, j be as in Lemma 15.1 and let $u \in H_0^j(\Omega)$. Extend u outside Ω by 0. Then

$$\|u^h\|_{j-1}^\Omega \le \|u\|_j^\Omega, \tag{15.4}$$

$$\lim_{h \to 0} \left\|u^h - \frac{\partial u}{\partial x_1}\right\|_{j-1}^\Omega = 0. \tag{15.5}$$

Proof of Lemma 15.1. For simplicity we take $i = 1$, $h > 0$. If $u \in \hat{C}^{j,2}(\Omega)$ and $0 \le |\alpha| < j$, then

$$\int_A |D^\alpha u^h(x)|^2 \, dx = \int_A \left| \frac{1}{h} \int_{x_1}^{x_1+h} D^\alpha D_{\xi_1} u(\xi_1, x_2, \ldots, x_n) \, d\xi_1 \right|^2 dx.$$

Using Schwarz's inequality and then substituting $\xi_1 = \xi_1' + x_1$, we get

$$\int_A |D^\alpha u^h(x)|^2 \, dx \le \int_A \frac{1}{h} \int_0^h |D^\alpha D_{\xi_1'} u(\xi_1' + x_1, x_2, \ldots, x_n)|^2 \, d\xi_1' \, dx$$

$$= \frac{1}{h} \int_0^h \left\{ \int_A |D^\alpha D_{x_1} u(\xi_1' + x_1, x_2, \ldots, x_n)|^2 \, dx \right\} d\xi_1'.$$

Substituting x_1 for $x_1 + \xi_1'$ in the inner integral and noting that A is then mapped into a subset of Ω, we conclude that the inner integral is bounded by

$$\int_\Omega |D^\alpha D_{x_1} u(x)|^2 \, dx.$$

Hence

$$\int_A |D^\alpha u^h(x)|^2 \, dx \le \int_\Omega |D^\alpha D_{x_1} u(x)|^2 \, dx.$$

Summing over α, (15.2) follows for $u \in \hat{C}^{j,2}(\Omega)$. Now assume only that $u \in H^j(\Omega)$ and take a sequence of functions $\{u_m\}$ in $\hat{C}^{j,2}(\Omega)$ with $|u_m - u|_j^\Omega \to 0$. If $|\alpha| < j$, then

$$\int_A |D^\alpha u_m(x_1 + h, x_2, \ldots, x_n) - D^\alpha u(x_1 + h, x_2, \ldots, x_n)|^2 \, dx$$

$$\le \int_\Omega |D^\alpha u_m(x) - D^\alpha u(x)|^2 \, dx,$$

as follows by substituting $x_1 + h \to x_1$. Hence

$$\|u_m^h - u^h\|_{j-1}^A \to 0 \qquad \text{as } m \to \infty.$$

If we then write (15.2) for each u_m and take $m \to \infty$, then we obtain (15.2).

To prove (15.3) we note that it certainly holds if $u \in \hat{C}^{j,2}(\Omega)$ since $u^h \to \partial u / \partial x_1$, as $h \to 0$, together with all its first $j - 1$ derivatives, uniformly on A. If $u \in H^j(\Omega)$, then, given any $\varepsilon > 0$, we choose v in $\hat{C}^{j,2}(\Omega)$ such that $\|v - u\|_j^{\Omega} \leq \varepsilon$. By (15.2) we also have $\|v^h - u^h\|_j^A \leq \varepsilon$. Hence,

$$\left\| u^h - \frac{\partial u}{\partial x_1} \right\|_{j-1}^A \leq \|u^h - v^h\|_{j-1}^A + \left\| v^h - \frac{\partial v}{\partial x_1} \right\|_{j-1}^A + \left\| \frac{\partial v}{\partial x_1} - \frac{\partial u}{\partial x_1} \right\|_{j-1}^A$$

$$\leq 3\varepsilon$$

if h is sufficiently small.

Proof of Lemma 15.2.   For any bounded domain $G \supset \overline{\Omega}$, u belongs to $H^j(G)$. Now apply Lemma 15.1 with A, Ω replaced by Ω, G respectively.

Recalling the notation (15.1), we now state the converse of Lemma 15.1 (for a fixed i):

LEMMA 15.3.   *If* $u \in H^j(\Omega)$ *and if for any subdomain* A *of* Ω, *with* $\overline{A} \subset \Omega$, $\|u^h\|_j^A \leq C$ *for all* h *sufficiently small, where* C *is a constant independent of* A, h, *then* $D_i u \in H^j(\Omega)$ *and* $\|D_i u\|_j^{\Omega} \leq C$.

Proof.   There exists a sequence $\{h_k\}$, $h_k \to 0$ as $k \to \infty$, such that $\{u^{h_k}\}$ converges weakly in $L^2(A)$ to a function u_i. For any $\phi \in C_0^\infty(A)$,

$$\int_A u_i \phi \, dx = \lim_{k \to \infty} \int_A u^{h_k} \phi \, dx = -\lim_{k \to \infty} \int_A u \phi^{-h_k} \, dx = -\int_A u D_i \phi \, dx.$$

Hence u_i is the weak derivative $D_i u$ of u. Since the last two equations hold for any sequence $\{h_k\}$, $\{u^h\}$ converges weakly in $L^2(A)$ to $D_i u$ as $h \to 0$. By Lemma 6.2, $D_i u \in H^j(A)$ and $\{D^\alpha u^h\}$ (for $0 \leq |\alpha| < j$) converges weakly to $\{D^\alpha D_i u\}$ as $h \to 0$. Hence,

$$\|D^\alpha D_i u\|_0^A \leq \liminf_{h \to 0} \|D^\alpha u^h\|_0^A.$$

Summing over α, we get $\|D_i u\|_j^A \leq C$. Hence $\|D_i u\|_j^{\Omega} \leq C$. Finally, by Theorem 6.3, $D_i u \in H^j(\Omega)$ and the proof is complete.

Remark.   From the proof of Lemmas 15.1, 15.2, 15.3 it is clear that the lemmas remain true if the spaces H^j are replaced by the more general spaces $H^{j,p}$.

We shall need the following assumption:

(A_4)   If $|\rho| = m$, then $|a^{\rho\sigma}(x) - a^{\rho\sigma}(y)| \leq c_3|x - y|$ for $|\sigma| \leq m$, and for all x, y in Ω.

LEMMA 15.4.   Let the assumptions (A_1), (A_2), (A_4) *hold and assume that* $u \in H^m(\Omega)$ *and that for all* $\phi \in C_0^\infty(\Omega)$

$$|B[\phi, u]| \leq P\|\phi\|_{m-1} \qquad (P \text{ constant}). \tag{15.6}$$

Then, for any subdomain Ω' *of* Ω *with* $\overline{\Omega'} \subset \Omega$, $D_i u$ *belongs to* $H^m(\Omega')$ *for* $i = 1, \ldots, n$ *and*

$$\|D_i u\|_m^{\Omega'} \leq C(P + \|u\|_m^\Omega), \tag{15.7}$$

where C *is a constant depending only on* c_0, c_1, c_3 *and* Ω, Ω'.

Proof.   Without loss of generality we may assume that $a^{\rho\sigma} = 0$ if $|\rho| < m$. Indeed, if we denote by $\tilde{B}[\phi, u]$ the form obtained from $B[\phi, u]$ by deleting all the terms with $a^{\rho\sigma}$ for which $|\rho| < m$, then all the assumptions of the lemma still hold for $\tilde{B}[\phi, u]$ except that P is replaced by $P + c\|u\|_m$, where c depends only on c_1.

Let Ω'' be a domain satisfying $\overline{\Omega'} \subset \Omega''$, $\overline{\Omega''} \subset \Omega$ and take $\zeta \in C_0^\infty(\Omega)$ such that $\zeta = 1$ on Ω', $\zeta = 0$ outside Ω''. Set $v = \zeta u$ and $v^h(x) = (v(x^h) - v(x))/h$, where $x^h = x + he^i$ for some fixed i, $1 \leq i \leq n$. It is easily seen that v and v^h belong to $H_0^m(\Omega)$ if $|h| < h_0$, where $h_0 = \text{dist.}(\Omega'', \partial\Omega)$.

For any $\phi \in C_0^\infty(\Omega)$

$$B[\phi, v^h] = \sum_{\substack{|\rho| = m \\ |\sigma| \leq m}} (D^\rho\phi, a^{\rho\sigma}D^\sigma(\zeta u)^h).$$

Since, by Lemma 15.2,

$$\|(D^{\sigma-\gamma}\zeta \cdot D^\gamma u)^h\|_0^\Omega \leq \|D^{\sigma-\gamma}\zeta \cdot D^\gamma u\|_1^\Omega \qquad (\gamma \leq \sigma).$$

we have

$$|B[\phi, v^h]| \leq \left| \sum_{\substack{|\rho| = m \\ |\sigma| \leq m}} (D^\rho\phi, a^{\rho\sigma}(\zeta D^\sigma u)^h) \right| + c\|\phi\|_m \|u\|_m, \tag{15.8}$$

where c is a generic constant depending only on $c_0, c_1, c_3, \Omega, \Omega'$. Using the rule

$$f(x)g^h(x) = (f(x)g(x))^h - f^h(x)g(x^h), \tag{15.9}$$

we find that the first term on the right of (15.8) is equal to

$$\left| \sum_{|\rho|=m,\,|\sigma|\leq m} (D^\rho\phi, (a^{\rho\sigma}\zeta D^\sigma u)^h - \sum_{|\rho|=m,\,|\sigma|\leq m} (D^\rho\phi, (a^{\rho\sigma})^h\zeta(x^h)D^\sigma u(x^h))\right|$$

$$\leq \left| \sum_{|\rho|=m,\,|\sigma|\leq m} (D^\rho\phi, (a^{\rho\sigma}\zeta D^\sigma u)^h)\right| + c\,\|\phi\|_m\,\|u\|_m\,;$$

here we have used the Lipschitz continuity of $a^{\rho\sigma}$ ($|\rho|=m$) and the fact that for any g with support in Ω''

$$\int_\Omega |g(x^h)|^2\,dx \leq \int_\Omega |g(x)|^2\,dx.$$

From (15.8) we then obtain

$$|B[\phi, v^h]| \leq \left| \sum_{|\rho|=m,\,|\sigma|\leq m} (D^\rho\phi, (a^{\rho\sigma}\zeta D^\sigma u)^h)\right| + c\,\|\phi\|_m\,\|u\|_m$$

$$= \left| \sum_{|\rho|=m,\,|\sigma|\leq m} (\zeta D^\rho\phi^{-h}, a^{\rho\sigma}D^\sigma u)\right| + c\,\|\phi\|_m\,\|u\|_m, \qquad (15.10)$$

where the rule

$$(f, g^h) = -(f^{-h}, g) \qquad \text{(support of } g \text{ lies in } \Omega'')$$

has been used

Using Lemma 15.2, we find that the first term on the right of (15.10) can be estimated by

$$\left| \sum_{|\rho|=m,\,|\sigma|\leq m} (D^\rho(\zeta\phi^{-h}), a^{\rho\sigma}D^\sigma u)\right| + c\,\|\phi\|_m\,\|u\|_m.$$

Hence,

$$|B[\phi, v^h]| \leq |B[\zeta\phi^{-h}, u]| + c\|\phi\|_m\|u\|_m.$$

Using the assumption (15.6), with ϕ replaced by $\zeta\phi^{-h}$, and the inequality

$$\|\zeta\phi^{-h}\|_{m-1} \leq c\|\phi\|_m,$$

we get, for every $\phi \in C_0^\infty(\Omega)$,

$$|B[\phi, v^h]| \leq c(P + \|u\|_m)\|\phi\|_m. \qquad (15.11)$$

By completion, this holds for every $\phi \in H_0^m(\Omega)$ Taking, in particular, $\phi = v^h$, and using Gårding's inequality, we obtain

$$\|v^h\|_m^2 \leq c(P + \|u\|_m)\|v^h\|_m + c\,\|v^h\|_0^2 \leq c\,\|v^h\|_m(P + \|u\|_m),$$

since $\|v^h\|_0 \leq c\|u\|_1$ (by Lemma 15.2). Hence

$$\|v^h\|_m \leq c(P + \|u\|_m). \qquad (15.12)$$

Using Lemma 15.3, the assertion (15.7) follows.

We shall need the following assumption:

(**B$_j$**) The coefficients $a^{\rho\sigma}$, for $|\rho| + j - m > 0, |\sigma| \le m$, belong to $C^{|\rho|+j-m}(\overline{\Omega})$.

If this assumption holds, then we denote by K a bound on the first $|\rho| + j - m$ derivatives of all the coefficients $a^{\rho\sigma}$ in Ω ($|\rho| + j - m > 0, |\sigma| \le m$).

LEMMA 15.5. Let the assumptions (A$_1$), (A$_2$) *and* (B$_j$) *hold, with some* j, $1 \le j \le m$. *Assume that* $u \in H^m(\Omega)$ *and that for all* $\phi \in C_0^\infty(\Omega)$

$$|B[\phi, u]| \le P\|\phi\|_{m-j} \qquad (P \text{ constant}). \tag{15.13}$$

Then, for any subdomain Ω' *of* Ω *with* $\overline{\Omega}' \subset \Omega$, $D_i u$ *belongs to* $H^{m+j-1}(\Omega')$ *for* $1 \le i \le n$ *and*

$$\|D_i u\|_{m+j-1}^{\Omega'} \le C(P + \|u\|_m^\Omega) \tag{15.14}$$

where C *is a constant depending only on* $c_0, c_1, K, \Omega, \Omega'$.

Proof. The proof is by induction on j. For $j = 1$ the assertion coincides with Lemma 15.4. We shall now assume the lemma to hold for $j - 1$ and prove it for j ($2 \le j \le m$). As in the proof of Lemma 15.4 we may suppose that $a^{\rho\sigma} = 0$ if $|\rho| + j - m \le 0$. We introduce a domain Ω'' with $\overline{\Omega}' \subset \Omega''$, $\overline{\Omega}'' \subset \Omega$. (15.13) implies

$$|B[\phi, u]| \le P\|\phi\|_{m-(j-1)}.$$

Hence, by the inductive assumption, $D_i u \in H^{m+j-2}(\Omega'')$ for $i = 1, \dots, n$, and

$$\|D_i u\|_{m+j-2}^{\Omega''} \le c(P + \|u\|_m^\Omega); \tag{15.15}$$

c is a generic constant depending only on $c_0, c_1, K, \Omega, \Omega'$.

We have: $u \in H^{m+j-1}(\Omega'')$. Since $j \ge 2$, $u \in H^{m+1}(\Omega'')$ and we can write, for any $\phi \in C_0^\infty(\Omega'')$,

$$\begin{aligned}
B[\phi, D_i u] &= \sum_{\substack{m-j<|\rho|\le m \\ |\sigma|\le m}} (D^\rho\phi, a^{\rho\sigma}D^\sigma D_i u)^{\Omega''} \\
&= \sum_{\substack{m-j<|\rho|\le m \\ |\sigma|\le m}} (D^\rho\phi, D_i(a^{\rho\sigma}D^\sigma u))^{\Omega''} \\
&\quad - \sum_{\substack{m+j<|\rho|\le m \\ |\sigma|\le m}} (D^\rho\phi, D_i a^{\rho\sigma} \cdot D^\sigma u)^{\Omega''} \\
&= -B[D_i\phi, u] + I, \tag{15.16}
\end{aligned}$$

where I is a sum of terms of the form

$$I_{\rho\sigma} = (D^\rho\phi,\ D_i\,a^{\rho\sigma}\cdot D^\sigma u)^{\Omega''}$$

To estimate $I_{\rho\sigma}$ we integrate by parts $|\rho| - m + j - 1$ times, transferring each time one derivative from ϕ to the second factor of the scalar product (this is possible since $a^{\rho\sigma} \in C^{|\rho|+j-m}(\overline{\Omega})$ and $u \in H^{m+j-1}(\Omega'')$). We then get

$$|I_{\rho\sigma}| \le c\,\|\phi\|_{m-j+1}\,\|u\|_{m+j-1}.$$

From (15.16), the estimates on the $I_{\rho\sigma}$ and the assumption (15.13), we obtain

$$|B[\phi,\ D_i u]| \le P\,\|D_i\phi\|_{m-j} + c\,\|\phi\|_{m-j+1}\,\|u\|_{m+j-1}$$
$$\le c(P + \|u\|_{m+j-1})\,\|\phi\|_{m-j+1} \qquad (15.17)$$

for any $\phi \in C_0^\infty(\Omega'')$. We now use the inductive assumption and (15.15) and conclude that $D_i u \in H^{m+j-1}(\Omega')$ and

$$\|D_i u\|_{m+j-1}^{\Omega'} \le c(P + \|u\|_{m+j-1}^{\Omega''} + \|D_i u\|_m^{\Omega''})$$
$$\le c(P + \|u\|_m^{\Omega}).$$

COROLLARY. *Let the assumptions* (A_1), (A_2) *(of Section 12) and* (B_j) *(of Section 15) hold, for some* $1 \le j \le m$. *Let* $f \in L^2(\Omega)$ *and assume that* $u \in H^m(\Omega)$ *and*

$$B[\phi, u] = (\phi, f) \qquad \text{for all } \phi \in C_0^\infty(\Omega). \qquad (15.18)$$

Then for any subdomain Ω' *of* Ω, *with* $\overline{\Omega}' \subset \Omega$, $u \in H^{m+j}(\Omega')$ *and*

$$\|u\|_{m+j}^{\Omega'} \le C(\|f\|_0 + \|u\|_m^{\Omega}); \qquad (15.19)$$

C *is a constant depending only on* c_0, c_1, c_2, K, Ω, Ω'.

Proof. Apply Lemma 15.5, noting that

$$|B[\phi, u]| = |(\phi, f)_0| \le \|f\|_0\|\phi\|_0 \le \|f\|_0\|\phi\|_{m-j}.$$

We now state the main result on regularity in the interior of strong solutions of (13.1).

THEOREM 15.1.   Let the assumptions (A_1), (A_2), *and* (B_{m+k}) *hold for some* $k \geq 0$, *and let* $f \in H^k(\Omega)$. *If* $u \in H^m(\Omega)$ *and satisfies* (15.18), *then for any subdomain* Ω' *of* Ω *with* $\bar{\Omega}' \subset \Omega$, $u \in H^{2m+k}(\Omega')$ *and*

$$\|u\|_{2m+k}^{\Omega'} \leq C(\|f\|_k^{\Omega} + \|u\|_m^{\Omega}); \tag{15.20}$$

C *is a constant depending only on* c_0, c_1, k, K, Ω, Ω'.

Proof.   The proof is by induction on k. The case $k = 0$ follows from the corollary to Lemma 15.5. Assuming now the theorem to hold for $k - 1$, we shall prove it for k $(k \geq 1)$.

By the inductive assumption, $u \in H^{2m+k-1}(\Omega'')$ $(\bar{\Omega}' \subset \Omega''$, $\bar{\Omega}'' \subset \Omega)$, and

$$\|u\|_{2m+k-1}^{\Omega''} \leq c(\|f\|_{k-1}^{\Omega} + \|u\|_m^{\Omega}). \tag{15.21}$$

Thus, for any $\phi \in C_0^\infty(\Omega'')$,

$$B[D_i\phi, u] = - \sum_{\substack{|\rho| \leq m \\ |\sigma| \leq m}} (D^\rho\phi, D_i(a^{\rho\sigma}D^\sigma u))$$

$$= -B[\phi, D_i u] - \sum_{\substack{|\rho| \leq m \\ |\sigma| \leq m}} (D^\rho\phi, D_i a^{\rho\sigma} \cdot D^\sigma u).$$

In each term of the last sum we transfer all the derivatives of ϕ to the other factor of the scalar product. We then have

$$B[D_i\phi, u] = -B[\phi, D_i u] - (\phi, A_i u) \tag{15.22}$$

where

$$A_i u = \sum_{\substack{|\rho| \leq m \\ |\sigma| \leq m}} (-1)^{|\rho|} D^\rho(D_i a^{\rho\sigma} \cdot D^\sigma u).$$

From the assumption (B_{m+k}), $k \geq 1$, it follows that $A_i u$ is a differential operator of order $\leq 2m$ with coefficients that are continuous in $\bar{\Omega}$ together with their first $k - 1$ derivatives.

Substituting $D_i\phi$ instead of ϕ in (15.18) and combining the result thus obtained with (15.22), we find that

$$B[\phi, D_i u] = (\phi, f_i), \qquad \text{where } f_i = D_i f - A_i u. \tag{15.23}$$

By (15.21), $f_i \in H^{k-1}(\Omega'')$ and

$$\|f_i\|_{k-1}^{\Omega''} \leq \|D_i f\|_{k-1}^{\Omega''} + \|A_i u\|_{k-1}^{\Omega''} \leq c(\|f\|_k^{\Omega''} + \|u\|_{2m+k-1}^{\Omega''}) \tag{15.24}$$

$$\leq c(\|f\|_k^{\Omega} + \|u\|_m^{\Omega}).$$

We now apply the inductive assumption to $D_i u$ and conclude that $D_i u \in H^{2m+k-1}(\Omega')$ and

$$\|D_i u\|_{2m+k-1}^{\Omega'} \leq c(\|f_i\|_{k-1}^{\Omega''} + \|D_i u\|_m^{\Omega''}). \tag{15.25}$$

Using (15.24) and (15.21) (with $k = 1$), we find that the right-hand side of (15.25) is bounded by

$$c(\|f\|_k^{\Omega} + \|u\|_m^{\Omega}).$$

This completes the proof of the theorem.

16 | REGULARITY IN THE INTERIOR (Continued)

In all the results of Section 15 about the regularity of u it was assumed that $u \in H^m$. In this section we shall prove regularity theorems for $u \in H^0(\Omega)$. The first result is the following one:

THEOREM 16.1.   Let the assumption (A_1) *hold and let* $a^{\rho,\sigma} \in C^m(\bar{\Omega})$. *Assume that* $u \in H^0(\Omega)$ *and that for some integer* j, $0 \leq j \leq 2m$, *and for any* $\phi \in C_0^\infty(\Omega)$,

$$|(L^*\phi, u)| \leq P\|\phi\|_{2m-j}. \tag{16.1}$$

Then for any subdomain Ω' *of* Ω *with* $\bar{\Omega}' \subset \Omega$, $u \in H^j(\Omega')$ *and*

$$\|u\|_j^{\Omega'} \leq C(P + \|u\|_0^{\Omega}). \tag{16.2}$$

C is a constant depending only on c_0, K', Ω, Ω', *where* K' *is an upper bound on the* $D^\alpha a^{\rho\sigma}(x)$, *for* $x \in \Omega$, $|\alpha| \leq m$, $|\rho| \leq m$, $|\sigma| \leq m$.

Proof.   The proof is by induction on j. For $j = 0$ the assertion is trivial. We now assume that the assertion is true for $j - 1$ and prove it for j $(1 \leq j \leq 2m)$. Since (16.1) implies

$$|(L^*\phi, u)| \leq P\|\phi\|_{2m-j+1} \qquad \text{for all } \phi \in C_0^\infty(\Omega),$$

it follows by the inductive assumption that $u \in H^{j-1}(\Omega''')$ for any domain Ω''' with $\overline{\Omega}' \subset \Omega'''$, $\overline{\Omega}''' \subset \Omega$. Furthermore,

$$\|u\|_{j-1}^{\Omega'''} \leq c(P + \|u\|_0^{\Omega}), \tag{16.3}$$

where c is a generic constant depending only on c_0, K', Ω, Ω'.

If $j - 1 \geq m$, then $(L^*\phi, u) = B[\phi, u]$. Hence we can apply Lemma 15.5 with j replaced by $j' = j - m$ and the desired assertion follows. It thus remains to consider the case where $j - 1 < m$.

We introduce the strongly elliptic operator

$$L_q = (-1)^q \Delta^q + 1, \tag{16.4}$$

where Δ is the Laplacian and $q = m - j + 1$. From the inequality

$$\sum_{|\alpha| \leq 2q} |\xi^\alpha|^2 \leq \text{const.} \left(\sum_{|\beta| = 2q} |\xi^\beta|^2 + 1 \right)$$

and the first part of the proof of Gårding's inequality, one obtains

$$(\phi, L_q \phi) \geq c' \|\phi\|_q^2 \qquad \text{for all } \phi \in C_0^\infty(\Omega), \tag{16.5}$$

where c' is a positive constant. Theorem 14.4 can therefore be applied. It follows that there exists a generalized solution h of

$$L_q h = u \qquad \text{in } \Omega''', \qquad h \in H_0^q(\Omega''').$$

Since $u \in H^{j-1}(\Omega''')$, Theorem 15.1 implies that $h \in H^{2q+j-1}(\Omega'') = H^{m+q}(\Omega'')$ for any domain Ω'' with $\overline{\Omega}' \subset \Omega''$, $\overline{\Omega}'' \subset \Omega'''$.

Substituting $u = L_q h$ into (16.1), we get

$$|(L^*\phi, L_q h)| \leq P \|\phi\|_{2m-j} \qquad \text{for any } \phi \in C_0^\infty(\Omega''). \tag{16.6}$$

Now, LL_q is an elliptic operator of order $2m + 2q$. Writing it in the form

$$\sum_{\substack{|\rho| \leq m \\ |\sigma| \leq m}} \sum_{|\tau| = q} (-1)^{|\rho|+q} D^\rho [a^{\rho\sigma} D^\sigma D^\tau D^\tau v] + Lv$$

we find, by integration by parts, that

$$(L^*\phi, L_q \psi) = (\phi, LL_q) = \sum_{\substack{|\alpha| \leq m+q \\ |\beta| \leq m+q}} (D^\alpha \phi, b^{\alpha\beta} D^\beta \psi) + B[\phi, \psi] = \hat{B}[\phi, \psi]$$

for any ϕ, ψ in $C_0^\infty(\Omega'')$, where $\hat{B}[\phi, \psi]$ is a bilinear form associated with LL_q, and $b^{\alpha\beta}$ are sum of terms $\pm D^k a^{\rho\sigma}$, where $|\alpha| = |\rho| + q - |k|$, $|\beta| = |\sigma| + q$, and $0 \leq |k| \leq q$.

By completion, the relation

$$(L^*\phi, L_q\psi) = \hat{B}[\phi, \psi]$$

holds for any $\phi \in C_0^\infty(\Omega'')$, $\psi \in C^{m+q}(\Omega'')$. Taking a sequence $\{\psi_i\}$ in $C^{m+q}(\Omega)$ that converges to h in $H^{m+q}(\Omega'')$, we get

$$(L^*\phi, L_q h) = \hat{B}[\phi, h].$$

Recalling (16.6), we find that

$$|\hat{B}[\phi, h]| \le P\|\phi\|_{2m-j} = P\|\phi\|_{m+q-1}.$$

We are now in position to apply Lemma 15.4 (with m replaced by $m + q$ and B replaced by $\hat{B}$). We conclude that $h \in H^{m+q+1}(\Omega')$ and

$$\|h\|_{m+q+1}^{\Omega'} \le c(P + \|h\|_{m+q}^{\Omega''}).$$

Noting that $u = L_q h$ and that $(m + q + 1) - 2q = m + 1 - q = j$, we get

$$\|u\|_j^{\Omega'} \le c(P + \|u\|_{j-1}^{\Omega''}).$$

Using (16.3), the inequality (16.2) follows.
Combining Theorems 15.1, 16.1, and 11.1, we get:

THEOREM 16.2. Let

$$Lu = \sum_{\substack{|\rho| \le m \\ |\sigma| \le m}} (-1)^{|\rho|} D^\rho(a^{\rho\sigma} D^\sigma u)$$

be a strongly elliptic operator in a domain Ω with $a^{\rho\sigma} \in C^{|\rho|+k}(\Omega) \cap C^m(\Omega)$ for some integer $k \ge 0$ with $2m + k > [n/2]$, and let $f \in C^k(\Omega)$. If u is a weak solution of $Lu = f$ in Ω, then $u \in C^h(\Omega)$, where $h = 2m + k - [n/2] - 1$.

COROLLARY. If f and the $a^{\rho\sigma}$ belong to $C^\infty(\Omega)$, then any weak solution u of $Lu = f$ in Ω is in $C^\infty(\Omega)$.

We have proved above the regularity of weak solutions of strongly elliptic equations. In the remaining part of this section we shall give an independent proof of the regularity of weak solutions of elliptic equations. Some of the details will be left to the reader. For simplicity we shall assume that the coefficients of the elliptic operator L are C^∞ functions.

We introduce the class C_π^∞ of periodic C^∞ functions $u(x)$ with period 2π in each variable x_j. For such functions, the Fourier series

$$u(x) = \sum_\xi u_\xi e^{ix\cdot\xi} \qquad (\xi = (\xi_1, \ldots, \xi_n), \xi_i \text{ integers})$$

converges uniformly with any number of the derivatives.

Parseval's equality gives, for any integer $s \geq 0$,

$$c^{-1} \sum_\xi (1 + |\xi|^2)^s |u_\xi|^2 \leq \int_Q \sum_{|\alpha| \leq s} |D^\alpha u|^2 \, dx \leq c \sum_\xi (1 + |\xi|^2)^s |u_\xi|^2, \quad (16.7)$$

where Q is the cube $\{x; 0 \leq x_j \leq 2\pi \text{ for } j = 1, \ldots, n\}$ and c is an absolute positive constant.

We introduce the scalar product, for *any* integer s,

$$(u, v)_s = (2\pi)^n \sum_\xi (1 + |\xi|^2)^s u_\xi \bar{v}_\xi$$

and the norm $\|u\|_s = (u, u)_s^{1/2}$. We also set $(u, v) = (u, v)_0$, $\|u\| = \|u\|_0$.

PROBLEMS. (1) Prove that $\|u\|_s$ is increasing in s. Furthermore, for $t_1 < s < t_2$ and any $\varepsilon > 0$, there is a constant $C = C(\varepsilon, s, t_1, t_2)$ such that

$$\|u\|_s \leq \varepsilon \|u\|_{t_2} + C \|u\|_{t_1} \tag{16.8}$$

[*Hint:* Use Parseval's equality.]

(2) Let t, s be integers. If $u \in C_\pi^\infty$, then the function ϕ given by $\phi = \sum u_\xi (1 + |\xi|^2)^t e^{ix\cdot\xi}$ also belongs to C_π^∞. We write: $\phi = (1 - \Delta)^t u$. If $t \geq 0$, then this relation can be proved, with Δ being the Laplacian. Setting $\psi = (1 - \Delta)^t v$, prove:

$$\|u\|_s = \|\phi\|_{s-2t} = \|(1 - \Delta)^t u\|_{s-2t}, \tag{16.9}$$

$$(u, v)_s = (u, (1 - \Delta)^t v)_{s-t} = ((1 - \Delta)^t u, v)_{s-t}. \tag{16.10}$$

(3) Let $\zeta \in C_\pi^\infty$, then

$$|(\zeta u, v)_t - (u, \bar{\zeta} v)_t| \leq C(\|u\|_t \|v\|_{t-1} + \|u\|_{t-1} \|v\|_t), \tag{16.11}$$

where C is a constant depending only on ζ, t. [*Hint:* If $t \leq 0$,

$$(\zeta u, v)_t = (\zeta(1 - \Delta)^{-t}\phi, \psi) = ((1 - \Delta)^{-t}\phi, \bar{\zeta}\psi) = (\phi, (1 - \Delta)^{-t}(\bar{\zeta}\psi))$$

$$= (\phi, \bar{\zeta}(1 - \Delta)^{-t}\psi) + I = (u, \bar{\zeta}v)_t + I,$$

where (by (16.9) and integration by parts) I is bounded by the right-hand side of (16.11). For $t > 0$ the proof is similar.]

(4) Prove:

$$|(u, v)_s| \leq \|u\|_{s+t} \|v\|_{s-t},\tag{16.12}$$

$$\|u\|_{s+t} = \text{l.u.b.} \frac{(u, v)_s}{\|v\|_{s-t}}.\tag{16.13}$$

[*Hint*: To prove (16.13), take $v = (1 - \Delta)^t u$.]

DEFINITION. We denote by H_π^s the completion of C_π^∞ with respect to the norm $\| \ \|_s$.

By (16.12) and completion, the scalar product (u, v) is defined for any functions $u \in H_\pi^s$, $v \in H_\pi^{-s}$.

PROBLEMS. (5) Prove that every bounded linear functional $f(u)$ on H_π^s can be represented in the form $f(u) = (u, v)$ for some $v \in H_\pi^{-s}$.

(6) Let $s > t$. Prove that the set $\{u \in H_\pi^s; \|u\|_s \leq 1\}$ is a compact subset in H_π^t.

(7) Let $a \in C_\pi^\infty$. Prove that

$$\|au\|_s \leq ck' \|u\|_s + ck'' \|u\|_{s-1} \qquad (u \in H_\pi^s),\tag{16.14}$$

where k', k'' are bounds on $|a|$ and on $|D^\alpha a|$ ($|\alpha| \leq s$), respectively, and c is a constant depending only on s. [*Hint:* If $s < 0$, set $\phi = (1 - \Delta)^s u$, $\psi = (1 - \Delta)^s(au)$. Then

$$\|au\|_s^2 = \|\psi\|_{-s}^2 = (au, \psi) = (a(1 - \Delta)^{-s}\phi, \psi)$$

and, integrating by parts $(-s)$ times, we get the bound

$$ck\|\phi\|_{-s}\|\psi\|_{-s} + ck'\|\psi\|_{-s}\|\phi\|_{-s-1}.$$

If $s \geq 0$, $\|au\|_s^2 = (au, (1 - \Delta)^s(au))$ and we integrate by parts s times.]

From the result of Problem 7 we get:

LEMMA 16.1. *Let* L *be any differential operator of order* k *with coefficients in* C_π^∞. *Then*

$$\|Lu\|_s \leq cK\|u\|_{s+k} + cK'\|u\|_{s+k-1},\tag{16.15}$$

where $c = c(k, s)$, K *is a bound on the leading coefficients, and* K' *is a bound on all the other coefficients and on their derivatives up to order* s.

l

COROLLARY.   L *can be extended as a bounded map from all of* H_π^s *into* H_π^{s-k}.

We denote this extension also by *L*.

PROBLEMS.   (8) Let ζ be a real-valued function in C_π^∞. Prove that

$$|(L(\zeta^2 u), Lu)_s - \|L(\zeta u)\|_s^2| \le c(\|u\|_{s+k}\|u\|_{s+k-1}), \qquad (16.16)$$

where *c* depends only on *K*, *K'* (of Lemma 16.1), *s*, *k*, and ζ. [*Hint:* Evaluate

$$(L(\zeta^2 u), Lu)_s - (\zeta L(\zeta u), Lu)_s, (\zeta L(\zeta u), Lu)_s - (L(\zeta u), \zeta Lu)_s,$$

$$(L(\zeta u), \zeta Lu)_s - (L(\zeta u), L(\zeta u))_s.]$$

(9) If u^h denotes the difference quotient, verify that $\|u(x + h)\|_s = \|u(x)\|_s$ and that

$$\|u^h\|_s \le \|u\|_{s+1}. \qquad (16.17)$$

Furthermore, if $u \in H_\pi^s$, $u^h \in H_\pi^s$ and $\|u^h\|_s \le C$ for all *h*, $|h| \le 1$, then $u \in H_\pi^{s+1}$ and $\|u\|_{s+1} \le C$. [*Hint:* Prove that the partial sums of *u* form a Cauchy sequence in H_π^{s+1}.]

We shall now prove a fundamental inequality.

THEOREM 16.3.   *Let L be an elliptic operator of order* k *with coefficients in* C_π^∞. *Then, for any s, s_0*

$$\|u\|_{s+k} \le c\|Lu\|_s + c'\|u\|_{s_0} \qquad (16.18)$$

where c, c' *depend only on* L, s, s_0.

Proof.   It suffices to consider the case $s_0 < s + k$. The proof's outline is similar to that of Gårding's inequality. If *L* has constant coefficients and is homogeneous, then we use Parseval's equality and the fact that $|L(\xi)| \ge \gamma|\xi|^k$ for some $\gamma > 0$. If *L* is homogeneous with nonconstant coefficients, and if the support of *u* in the cube *Q* is a set of diameter $\le \delta$, then using Lemma 16.1 and (16.8) we find that (16.18) holds provided δ is sufficiently small. In the general case we use partition of unity and the results of Problems 3, 8.

We now state the interior differentiability theorem for the periodic case.

THEOREM 16.4.   Let L *be an elliptic operator of order* k *with coefficients in* C_π^∞. *If for some integer* s, $u \in H_\pi^s$, Lu $\in H_\pi^{s-k+1}$, *then* $u \in H_\pi^{s+1}$.

COROLLARY.   If $u \in H_\pi^s$ *and* Lu $\in H_\pi^{t-k}$, *then* $u \in H_\pi^t$.

Proof.   As easily seen, $Lu^h = (Lu)^h - L^h u(x^h)$, where the operator L^h is obtained from L by replacing each coefficient in L by its difference quotient. Using Theorem 16.3, Lemma 16.1, and Problem 9, we get $\|u^h\|_s \leq C$. Hence $u \in H_\pi^{s+1}$.

Using Theorem 16.4, we shall prove the following differentiability theorem for the nonperiodic case:

THEOREM 16.5.   Let L *be an elliptic operator of order* k *in a bounded domain* Ω, *with coefficients in* $C^\infty(\Omega)$, *and let* f $\in H^t(\Omega)$ *for some* t ≥ 0. *If* u *is a weak solution of* Lu = f *in* Ω, *then* $u \in H^{k+t}(\Omega_0)$ *for any domain* Ω_0 *with* $\overline{\Omega}_0 \subset \Omega$.

Proof.   Suppose first that $u \in H^k(\Omega_0)$ for any domain $\Omega_0 \subset \Omega$ with $\overline{\Omega}_0 \subset \Omega$. The proof is by induction on t. Suppose the theorem is true for all $t \leq p$. We shall prove that, for any point $x^0 \in \Omega$, $u \in H^{k+p+1}$ in some neighborhood V of x^0. Let Q_0 be a cube contained in Ω and containing V in its interior. For simplicity we may take $Q_0 = Q$. Take $\zeta \in C_0^\infty(Q)$ such that $\zeta = 1$ in V. Set $v = \zeta u$ in Q. Extend f, ζ, v, u outside the support of ζ so that they all become periodic in R^n with period 2π. We also extend the coefficients of L outside the support of ζ in a similar way. We have (with L extended as in the corollary to Lemma 16.1)

$$Lv = L(\zeta u) = \zeta f + g, \tag{16.19}$$

where g contains derivatives of u only up to order $k - 1$. Thus, by the inductive hypothesis and Lemma 16.1 it easily follows that $g \in H_\pi^{p+1}$. Theorem 16.4 then shows that $v \in H_\pi^{k+p+1}$. Consequently, $u \in H^{k+p+1}(V)$.

It remains to prove that if $f \in H^0(\Omega)$ and u is a weak solution of $Lu = f$, then $u \in H^k(\Omega_0)$ for any domain Ω_0, $\overline{\Omega}_0 \subset \Omega$. This we also prove by induction —that is, we first prove that $u \in H^1$, then that $u \in H^2$, and so on. The proof is similar to the previous proof by induction provided one observes the following:

If $u \in H^j(Q_0)$ $(0 \leq j < k)$, then the relation (16.19) is valid with g an element in H_π^{j-k}.

17 | REGULARITY ON THE BOUNDARY

We first consider the situation in a domain $\Omega = \Omega_R$ which is a half-ball $\{x; |x| < R, x_n > 0\}$. We take R' to be an arbitrary number satisfying $0 < R' < R$, and define $R'' = (R' + R)/2$. We also define $\Omega_{R'}$, $\Omega_{R''}$ similarly to Ω_R, and set $\Omega' = \Omega_{R'}$, $\Omega'' = \Omega_{R''}$.

LEMMA 17.1. Let the assumption (A_1), (A_2), (A_4) *hold and let* u *be in* $H^m(\Omega_R)$ *and such that* $\zeta u \in H_0^m(\Omega_R)$ *for any* $\zeta \in C_0^\infty \{x; |x| < R\}$. *Assume that* (15.6) *holds for any* $\phi \in C_0^\infty(\Omega_R)$. *Then the assertion* (15.7) *of Lemma 15.4 holds for all* $i = 1, 2, \ldots, n - 1$.

The proof is very similar to the proof of Lemma 15.4 and is left to the reader.

In order to deal with the case $i = n$, we need the following lemma.

LEMMA 17.2. Let $u \in L^2(\Omega_R)$ *and assume that, for some integer* $j \geq 1$, *the weak derivatives* $D_i^j u$ *exist and belong to* $L^2(\Omega_R)$ *for* $i = 1, \ldots, n - 1$. *Assume, further, that for some integer* $m \geq j$ *and for any* $\phi \in C_0^\infty(\Omega_R)$

$$|(D_n^m \phi, u)| \leq P \|\phi\|_{m-j}^\Omega. \qquad (17.1)$$

Then, for any $R' < R$, $u \in H^j(\Omega_{R'})$ *and*

$$\|u\|_j^{\Omega'} \leq C \left(P + \sum_{i=1}^{n-1} \|D_i^j u\|_0^\Omega + \|u\|_0^\Omega \right); \qquad (17.2)$$

C *is a constant depending only on* m, R, R'.

Proof. By completion, (17.1) holds for any $\phi \in C_0^m(\Omega_R)$. We claim that (17.1) holds also for any $\phi \in C^m(\overline{\Omega}_R)$ such that, for some $\delta > 0$, $\phi(x) = 0$ if $R - \delta < |x| < R$ and $D^\alpha \phi(x) = 0$ on $x_n = 0$ for $0 \leq |\alpha| \leq m$. Indeed, define

$$\phi_\varepsilon(x) = \begin{cases} \phi(x - \varepsilon e^n) & \text{if } x \in \Omega_R, \, x_n \geq \varepsilon, \\ 0 & \text{if } x \in \Omega_R, \, x_n < \varepsilon. \end{cases}$$

Then $\phi_\varepsilon \in C_0^m(\Omega_R)$. Since $\phi_\varepsilon \to \phi$ in $H^m(\Omega_R)$ and since (17.1) holds with ϕ replaced by ϕ_ε, (17.1) follows for ϕ.

We now extend u by reflection (compare (5.2)), setting

$$u(x', x_n) = \sum_{k=1}^{2m+1} \lambda_k u\left(x', -\frac{x_n}{k}\right) \qquad \text{for } (x', -x_n) \in \Omega_R, \qquad (17.3)$$

where $x' = (x_1, \ldots, x_{n-1})$. The constants λ_k are determined by the equations

$$\sum_{k=1}^{2m+1} \lambda_k(-k)^{-h} = 1 \qquad (-1 \le h \le 2m-1). \qquad (17.4)$$

Then, formally,

$$D_n^h u(x', x_n)|_{x_n=0+} = D_n^h u(x', x_n)|_{x_n=0-} \qquad \text{for } 0 \le h \le 2m-1.$$

(The equation in (17.4) with $h = -1$ will be needed later on.)

Denoting by S_R the ball $\{x; |x| < R\}$, it is clear that $u \in L^2(S_R)$ and

$$\|u\|_0^{S_R} \le c \|u\|_0^{\Omega_R}. \qquad (17.5)$$

We claim that if $\alpha = (\alpha_1, \ldots, \alpha_n)$ is such that $|\alpha| \le 2m-1$ and $\alpha_n = 0$, and if the weak derivative $D^\alpha u$ exists in Ω_R, then the weak derivative $D^\alpha u$ exists in S_R and it is the extension (by (17.3)) of $D^\alpha u$ on Ω_R—that is,

$$\int_{S_R} D^\alpha \phi \cdot u \, dx = (-1)^{|\alpha|} \left\{ \int_{x_n>0} \phi D^\alpha u \, dx + \int_{x_n<0} \phi \sum_{k=1}^{2m+1} \lambda_k D^\alpha u\left(x', -\frac{x_n}{k}\right) dx \right\}$$

$$(17.6)$$

for any $\phi \in C_0^\infty(S_R)$.

To prove this, we notice that

$$\int_{x_n<0} D^\alpha \phi \cdot u \, dx = \sum_{k=1}^{2m+1} \lambda_k \int_{x_n<0} D^\alpha \phi(x) \cdot u\left(x', -\frac{x_n}{k}\right) dx$$

$$= \sum_{k=1}^{2m+1} k\lambda_k \int_{x_n>0} D^\alpha \phi(x', -kx_n) \cdot u(x) \, dx.$$

Hence,

$$\int_{S_R} D^\alpha \phi \cdot u \, dx = \int_{\Omega_R} D^\alpha \phi_0 \cdot u \, dx \qquad (17.7)$$

where

$$\phi_0(x) = \phi(x) + \sum_{k=1}^{2m+1} k\lambda_k \phi(x', -kx_n).$$

Let $g(t)$ be a C^∞ function satisfying $g(t) = 0$ if $t \leq 1$, $g(t) = 1$ if $t \geq 2$ and set $\zeta_\varepsilon(x) = g(|x_n|/\varepsilon)$. Then $D_i \zeta_\varepsilon = 0$ if $1 \leq i \leq n - 1$. Hence

$$\zeta_\varepsilon D^\alpha \phi_0 = D^\alpha (\zeta_\varepsilon \phi_0).$$

Since $\zeta_\varepsilon \phi_0$ is a test function on Ω_R, the existence of the weak derivative $D^\alpha u$ in Ω_R implies that

$$\int_{\Omega_R} \zeta_\varepsilon D^\alpha \phi_0 \cdot u \, dx = (-1)^{|\alpha|} \int_{\Omega_R} \zeta_\varepsilon \phi_0 D^\alpha u \, dx.$$

Hence, from (17.7),

$$\int_{S_R} D^\alpha \phi \cdot u \, dx = (-1)^{|\alpha|} \int_{\Omega_R} \zeta_\varepsilon \phi_0 D^\alpha u \, dx + \int_{\Omega_R} (1 - \zeta_\varepsilon) D^\alpha \phi_0 \cdot u \, dx.$$

As $\varepsilon \to 0$, the second term on the right tends to 0 and the first term on the right converges to

$$(-1)^{|\alpha|} \int_{\Omega_R} \phi_0 D^\alpha u \, dx = (-1)^{|\alpha|} \int_{\Omega_R} \left[\phi(x) + \sum_{k=1}^{2m+1} k \lambda_k \phi\left(x', - k x_n \right) \right] D^\alpha u(x) \, dx,$$

—that is, to the right-hand side of (17.6). This proves (17.6).

Since (17.6) holds for $\alpha = je^i$, we get

$$\| D_i^j u \|_0^{S_R} \leq c \| D_i^j u \|_0^{\Omega_R} \qquad (1 \leq i \leq n - 1). \tag{17.8}$$

Now let $\psi \in C_0^\infty(S_R)$. Proceeding similarly to the proof of (17.7), we find that

$$(D_n^{2m} \psi, u)_0^{S_R} = (D_n^{2m} \psi^*, u)_0^{\Omega_R} \tag{17.9}$$

where

$$\psi^*(x) = \psi(x', x_n) - \sum_{k=1}^{2m+1} \lambda_k (-k)^{1 - 2m} \psi(x', - k x_n). \tag{17.10}$$

Using (17.4), we see that

$$D_n^h \psi^*(x', 0) = 0 \qquad \text{for } 0 \leq h \leq 2m.$$

By the remark made at the beginning of the proof we see that (17.1) holds for $\phi = D_n^m \psi^*$—that is,

$$|(D_n^{2m} \psi^*, u)_0^{\Omega_R}| = |(D_n^m \phi, u)_0^{\Omega_R}| \leq P \| \psi^* \|_{2m-j}^{\Omega_R}.$$

From (17.9) we then obtain

$$|(D_n^{2m}\psi, u)_0^{S_R}| \le P \|\psi^*\|_{2m-j}^{\Omega_R} \le cP \|\psi\|_{2m-j}^{S_R}. \tag{17.11}$$

If $i \ne n$, we can use (17.8) and find that

$$|(D_i^{2m}\psi, u)_0^{S_R}| = |(D_i^{2m-j}\psi, D_i^j u)_0^{S_R}| \le c \|D_i^j u\|_0^{\Omega_R} \|\psi\|_{2m-j}^{S_R}. \tag{17.12}$$

Consider now the strongly elliptic operator $A = (-1)^m \sum_{i=1}^{2m} D_i^{2m}$. From (17.11), (17.12) we see that

$$|(A\psi, u)_0^{S_R}| \le c \left(P + \sum_{i=1}^{n-1} \|D_i^j u\|_0^{\Omega_R} \right) \|\psi\|_{2m-j}^{S_R}$$

for any $\psi \in C_0^\infty(S_R)$. Applying Theorem 16.1 and (17.5), we conclude that, for any $R' < R$, $u \in H^j(\Omega_{R'})$ and the inequality (17.2) holds.

We shall need another lemma.

LEMMA 17.3.   *If* $u \in H^{m+1}(\Omega_R)$ *and if* $\zeta u \in H_0^m(\Omega_R)$ *for all* $\zeta \in C_0^\infty(S_R)$, *then* $\zeta D_i u \in H_0^m(\Omega_R)$ *for all* $\zeta \in C_0^\infty(S_R)$ *and* $i = 1, 2, \ldots, n - 1$.

Proof.   For h sufficiently small, $(\zeta u)^h \in H_0^m(\Omega)$, where $x^h = x + he^i$, $1 \le i \le n - 1$. Since $\zeta^h \in C_0^\infty(\Omega)$, also $\zeta^h(x)u(x + he^i)$ belongs to $H_0^m(\Omega)$. Using the rule (15.9), we find that $\zeta u^h \in H_0^m(\Omega)$ if h is small. Let $R' < R'' < R$, where R' is such that the support of ζ lies in $\Omega_{R''}$. Then, by Lemma 15.2,

$$\|(\zeta u)^h\|_m^\Omega \le \|\zeta u\|_{m+1}^\Omega \le c \|\zeta\|_{m+1}^\Omega \|u\|_{m+1}^{\Omega'}. \tag{17.13}$$

Using the rule (15.9), we then easily find that

$$\|\zeta u^h\|_m^\Omega \le c \|u\|_{m+1}^\Omega,$$

provided $|h|$ is sufficiently small, where c is a constant independent of h. Since $\{\zeta u^h\}$ ($|h|$ small) is a bounded set in the Hilbert space $H_0^m(\Omega)$, we can find a subsequence $\{\zeta u^{h_k}\}$ that is weakly convergent, in $H_0^m(\Omega)$ to some element, say w. As easily verified, $w = \zeta D_i u$. Hence, $\zeta D_i u \in H_0^m(\Omega)$, and the proof is complete.

With the aid of Lemmas 17.2, 17.3 we shall now prove a result that extends Lemma 15.5 to the case where $\Omega = \Omega_R$ is a half-ball.

LEMMA 17.4.   *Let the assumptions* (A_1), (A_2), *and* (B_j) *hold in* $\Omega = \Omega_R$, *for some* j, $1 \le j \le m$. *Assume that* $u \in H^m(\Omega_R)$, *that* $\zeta u \in H_0^m(\Omega_R)$ *for any* $\zeta \in C_0^\infty\{x; |x| < R\}$, *and that* (15.13) *holds for all* $\phi \in C_0^\infty(\Omega_R)$. *Then, for any* $R' < R$, $u \in H^{m+j}(\Omega_{R'})$ *and* (15.14) *holds (where* $\Omega = \Omega_R$, $\Omega' = \Omega_{R'}$).

Proof. The case $j = 0$ is trivial. We proceed by induction on j. Assuming the lemma to hold for $j - 1$, we shall prove it for j ($1 \leq j \leq m$). We set $\Omega''' = \Omega_{R'''}$, where $R''' = (R + R'')/2$. As in the proof of Lemma 15.5, the inductive hypothesis immediately shows that $u \in H^{m+j-1}(\Omega''')$. For $j = 1$, Lemma 17.1 gives $D_i u \in H^{m+j-1}(\Omega'')$ if $1 \leq i \leq n - 1$. We shall now prove the same for $j > 1$. We have: $u \in H^{m+j-1}(\Omega''') \subset H^{m+1}(\Omega''')$. Hence, by Lemma 17.3, $\zeta D_i u \in H_0^m(\Omega''')$ ($1 \leq i \leq n - 1$) for any $\zeta \in C_0^\infty \{x; |x| < R'''\}$.

Proceeding now as in (15.16), for $\phi \in C_0^\infty(\Omega''')$, we arrive at (15.17). We can then apply the inductive assumption and conclude that

$$\|D_i u\|_{m+j-1}^{\Omega''} \leq c(P + \|u\|_m^\Omega) \qquad (1 \leq i \leq n - 1). \tag{17.14}$$

We shall now use Lemma 17.2 to show that $D_n u \in H^{m+j-1}(\Omega')$. For $\phi \in C_0^\infty(\Omega'')$

$$B[\phi, u] = \left(D_n^m \phi, \sum_{|\sigma| \leq m} a^{me^n, \sigma} D^\sigma u \right) + \sum_{\substack{m-j < |\rho| \leq m \\ \rho \neq me^n \\ |\sigma| \leq m}} (D^\rho \phi, a^{\rho\sigma} D^\sigma u), \tag{17.15}$$

where $me^n = (0, 0, \ldots, 0, m)$. For any term in the last sum, $D^\rho \phi = D^{e^i} D^{\rho - e^i} \phi$ for some $i \neq n$. Hence

$$(D^\rho \phi, a^{\rho\sigma} D^\sigma u) = -(D^{\rho - e^i} \phi, D_i(a^{\rho\sigma} D^\sigma u))$$

$$= -(D^{\rho - e^i} \phi, D_i a^{\rho\sigma} \cdot D^\sigma u) - (D^{\rho - e^i} \phi, a^{\rho\sigma} D^\sigma D_i u).$$

We estimate the right side by transferring $|\rho| - m + j - 1$ differentiations from $D^{\rho - e^i} \phi$ to the second factor of the scalar product. We then get

$$|(D^{\rho - e^i} \phi, D_i a^{\rho\sigma} \cdot D^\sigma u) \leq c \|\phi\|_{m-j}^{\Omega''} \|u\|_{m+j-1}^{\Omega''},$$

$$|(D^{\rho - e^i} \phi, a^{\rho\sigma} D^\sigma D_i u)| \leq c \|\phi\|_{m-j}^{\Omega''} \|D_i u\|_{m+j-1}^{\Omega''}$$

$$\leq c \|\phi\|_{m-j}^{\Omega''} (P + \|u\|_m^\Omega)$$

by (17.14). We thus have

$$\left| \left(D_n^m \phi, \sum_{|\sigma| \leq m} a^{me^n, \sigma} D^\sigma u \right) \right| \leq |B[\phi, u]| + c \|\phi\|_{m-j} (P + \|u\|_{m+j-1}^{\Omega''})$$

for any $\phi \in C_0^\infty(\Omega')$.

Set

$$v = \sum_{|\sigma| \leq m} a^{me^n, \sigma} D^\sigma u.$$

Using the last inequality and the assumption (15.13), we get

$$|(D_n^m \phi, v)| \leq c(P + \|u\|_m^\Omega + \|u\|_{m+J-1}^{\Omega''})\|\phi\|_{m-j} \leq c(P + \|u\|_m^\Omega)\|\phi\|_{m-j}$$

for any $\phi \in C_0^\infty(\Omega'')$. Since $D_i u \in H^{m+J-1}(\Omega'')$ if $i \neq n$, we also see that $D_i^j v \in L^2(\Omega'')$. Hence we can apply Lemma 17.2 to v. It follows that $v \in H^j(\Omega')$ and

$$\|v\|_j^{\Omega'} \leq c\left(P + \|u\|_m^\Omega + \sum_{i=1}^{n-1} \|D_i^j v\|_0^{\Omega''} + \|v\|_0^{\Omega''}\right) \leq c(P + \|u\|_m^\Omega);$$

here we have used (17.14).

Writing

$$a^{me^n, me^n} D_n^m u = v - \sum_{\substack{|\sigma| \leq m \\ \sigma \neq me^n}} a^{me^n, \sigma} D^\sigma u$$

and noting, by the ellipticity condition, that $a^{me^n, me^n} \neq 0$ in $\overline{\Omega}$, we conclude that $D_n^m u \in H^j(\Omega')$ and

$$\|D_n^m u\|_j^{\Omega'} \leq c(P + \|u\|_m^\Omega).$$

This, together with (17.14), shows that $u \in H^{m+j}(\Omega')$ and

$$\|u\|_{m+j}^{\Omega'} \leq c(P + \|u\|_m^\Omega).$$

The proof of the assertion for j is thereby completed.

COROLLARY.   *The corollary to Lemma 15.5 remains true for $\Omega = \Omega_R$ a half-ball, provided $\zeta u \in H_0^m(\Omega)$ for any $\zeta \in C_0^\infty\{x; |x| < R\}$.*

Indeed, the proof is the same as the proof of that corollary.

THEOREM 17.1.   *Theorem 15.1 remains true in case $\Omega = \Omega_R$ is a half-ball, provided $\zeta u \in H_0^m(\Omega)$ for any $\zeta \in C_0^\infty\{x; |x| < R\}$.*

Proof.   The proof is by induction on k. The case $k = 0$ follows from the last corollary. Suppose the theorem holds for $k - 1$ $(k \geq 1)$. We shall prove it for k. If $i = 1, \ldots, n - 1$, then we can proceed as in the proof of Theorem 15.1 and derive (15.23), (15.24). Since, by Lemma 17.3, $\zeta D_i u \in H_0^m(\Omega)$, we can apply the inductive hypothesis to $D_i u$. In this way we obtain (15.25). It remains to consider $D_i u$ for $i = n$. Since $a^{\rho\sigma} \in C^{|\rho|+k}(\overline{\Omega})$, we can write (15.18) in the form

$$Lu \equiv \sum_{\substack{|\rho| \le m \\ |\sigma| \le m}} (-1)^{|\rho|} D^\sigma(a^{\rho\sigma}D^\sigma u) = f.$$

Hence,

$$(-1)^m a^{me^n, me^n} D_n^{2m} u = f + \sum_{|\tau| \le 2m} b^\tau D^\tau u,$$

where $b^\tau \in C^k(\overline{\Omega})$ and $b^{2me^n} = 0$. Estimating the kth derivatives of each of the terms in the last sum by using (15.25), we find that

$$\|D_n^{2m} u\|_k^{\Omega'} \le c(\|f\|_k^\Omega + \|u\|_m^\Omega).$$

Thus u has weak derivatives of any order $\le 2m + k$ in $L^2(\Omega')$. By Theorem 6.3 $u \in H^{2m+k}(\Omega')$. Furthermore,

$$\|u\|_{2m+k}^{\Omega'} \le c(\|f\|_k^\Omega + \|u\|_m^\Omega).$$

So far we have dealt with the special case of a half-ball Ω_R. We can now easily deal with the general case.

THEOREM 17.2. *Let Ω be a bounded domain with $\partial\Omega$ in C^{2m}. Let the assumptions (A_1), (A_2) hold and let (B_j) hold for some j, $1 \le j \le m$. Assume that $n \in H_0^m(\Omega)$, $f \in L^2(\Omega)$, and that*

$$B[\phi, u] = (\phi, f) \qquad \textit{for all } \phi \in C_0^\infty(\Omega). \tag{17.16}$$

Then $u \in H^{m+j}(\Omega)$ and

$$\|u\|_{m+j}^\Omega \le C(\|f\|_0^\Omega + \|u\|_0^\Omega). \tag{17.17}$$

If, for some $k \ge 0$, $\partial\Omega$ is of class C^{2m+k}, if $f \in H^k(\Omega)$, and if (B_{m+k}) holds, then $u \in H^{2m+k}(\Omega)$ and

$$\|u\|_{2m+k}^\Omega \le C(\|f\|_k^\Omega + \|u\|_0^\Omega). \tag{17.18}$$

C *is a constant independent of* u, f.

Proof. We cover $\partial\Omega$ by a finite number of domains Ω_t such that each domain $\Omega_i \cap \Omega$ can be mapped by a one-to-one map onto a half-ball, $\tilde{\Omega}_{R_t}$, and the map is of the same differentiability class as $\partial\Omega$. We take $R_i' < R_t$ such that the inverse images $\Omega_{R_i'}$ of the sets $\tilde{\Omega}_{R_i'}$ also cover $\partial\Omega$. We now apply the corollary to Lemma 17.4 to each $\tilde{\Omega}_{R_i}$ with $R' = R'_i$. We also apply the corollary to Lemma 15.5 in a subdomain Ω' of Ω containing $\Omega - (\cup_i \overline{\Omega}_{R_i})$. Combining the results thus obtained and recalling that a C^k mapping establishes a bounded mapping between H^k spaces (cf. the proof of Theorem 7.1), we conclude that if the assumption (B_j) holds with $1 \leq j \leq m$, then $u \in H^{m+j}(\Omega)$ and

$$\|u\|^{\Omega}_{m+j} \leq C(\|f\|^{\Omega}_0 + \|u\|^{\Omega}_m). \tag{17.19}$$

Using Gårding's inequality and (17.16), we get

$$\|u\|^{\Omega}_m \leq c\|f\|^{\Omega}_0 + c\|u\|^{\Omega}_0.$$

Substituting this into (17.19), (17.17) follows. The proof of the second part of the theorem is similar.

Combining Theorem 17.2 and Problem 1, Section 11, we have:

THEOREM 17.3. Let Ω be a bounded domain with boundary $\partial\Omega$ in class C^{2m+k} ($2m + k > [n/2]$), and let L be a strongly elliptic operator in $\overline{\Omega}$ satisfying (B_{m+k}). If u is a solution of the generalized Dirichlet problem $Lu = f$ in Ω, $u \in H^m_0(\Omega)$ with $f \in C^k(\overline{\Omega})$, then $u \in C^h(\overline{\Omega})$, where $h = 2m + k - [n/2] - 1$.

18 | A PRIORI INEQUALITIES

In this section we write the strongly elliptic operator in the form

$$A = \sum_{|\alpha| \leq 2m} a_\alpha D^\alpha. \tag{18.1}$$

We shall·use Theorem 17.2 to derive a priori inequalities.

THEOREM 18.1. Let Ω be a bounded domain with $\partial\Omega$ in C^{2m+k} for some $k \geq 0$ and assume that A is uniformly strongly elliptic in Ω with $a_\alpha \in C^k(\overline{\Omega})$. If $u \in H^{2m+k}(\Omega) \cap H^m_0(\Omega)$, then

$$\|u\|^{\Omega}_{2m+k} \leq C(\|Au\|^{\Omega}_k + \|u\|^{\Omega}_0), \tag{18.2}$$

where C is a constant independent of u.

Proof. Assume first that A is an operator with constant coefficients and homogeneous. Clearly $f \equiv Au \in H^k(\Omega)$ and for any $\phi \in C_0^\infty(\Omega)$

$$B[\phi, u] = (\phi, Au) = (\phi, f),$$

where $B[\phi, u]$ is a bilinear form associated to A. Since all the assumptions of the second part of Theorem 17.2 hold,

$$\|u\|_{2m+k} \le C(\|Au\|_k + \|u\|_0). \qquad (18.3)$$

This proves the theorem in case A has constant coefficients and is homogeneous.

We can now proceed analogously to the proof of Gårding's inequality. We introduce a finite open covering $\{\Omega_i\}$ of $\overline{\Omega}$ such that the diameter of each Ω_i is $\le \delta$. We also introduce a partition of unity $\{\psi_i\}$ subordinate to the covering $\{\Omega_i\}$. Let x^i be any point of Ω^i. To evaluate the norms of $A(u\psi_i)$, we write

$$A = A' + A'',$$

where A' is obtained from the principal part of A by replacing the coefficients $a_\alpha(x)$ by $a_\alpha(x^i)$. If δ is sufficiently small, then we find, upon using the previous case of constant coefficients (for A') and Theorem 8.1, that

$$\|A(u\psi_i)\|_k^2 \ge c_0 \|u\psi_i\|_{2m+k}^2 - k_0 \|u\psi_i\|_0^2,$$

where c_0, k_0 are constants and $c_0 > 0$. Using these inequalities and Theorem 8.1, one can without difficulty complete the proof of the theorem. Details are left to the reader.

Remarks. If the a_α for $0 \le |\alpha| \le 2m - 1$ are bounded measurable functions in $\overline{\Omega}$ and if $a_\alpha \in C(\overline{\Omega})$ for $|\alpha| = 2m$, then the assertion of Theorem 18.1 for $k = 0$ is still valid, without any change in the proof. From the proof of Theorem 18.1 it follows that the constant C depends only on Ω, on a module of strong ellipticity of A, on an upper bound on all the derivatives of the $a_\alpha(x)$ (in $\overline{\Omega}$) up to order k, and (if $k = 0$) on a modulus of continuity of the leading coefficients of A.

THEOREM 18.2. Let Ω be a bounded domain with $\partial\Omega$ in C^{2m} and assume that A is uniformly strongly elliptic in Ω with $a_\alpha \in C^0(\overline{\Omega})$. Then, for any $\varepsilon > 0$, there exist positive constants C, Λ_0 such that for every $u \in H^{2m}(\Omega) \cap H_0^m(\Omega)$,

$$\sum_{j=0}^{2m} |\lambda|^{(2m-j)/2m} \|u\|_j^\Omega \le C \|(A + \lambda)u\|_0^\Omega \qquad (18.4)$$

for all λ with $-\frac{1}{2}\pi + \varepsilon \le \arg\lambda \le \frac{1}{2}\pi - \varepsilon$, $|\lambda| \ge \Lambda_0$.

Proof. We introduce the operator

$$A_0 = A + (-1)^m e^{i\theta} \frac{\partial^{2m}}{\partial t^{2m}} \qquad (-\tfrac{1}{2}\pi + \varepsilon \leq \theta \leq \tfrac{1}{2}\pi - \varepsilon)$$

which is uniformly strongly elliptic in any cylinder $\Omega_T = \Omega \times \{t; |t| \leq T\}$. Since $\partial\Omega$ is of class C^{2m}, we can construct a domain Ω_* that contains Ω_2 and is contained in Ω_3, such that $\partial\Omega_*$ is of class C^{2m}. Let $\zeta(t)$ be a C^∞ function satisfying $\zeta(t) = 1$ if $|t| \leq 1$, $\zeta(t) = 0$ if $|t| \geq \tfrac{3}{2}$. The function

$$v(x, t) = u(x)e^{i\mu t}\zeta(t) \qquad (\mu \text{ real number})$$

belongs to $H^{2m}(\Omega_*)$ and to $H_0^m(\Omega_*)$. Applying Theorem 18.1 to A_0, v in Ω_*, we conclude that

$$\|v\|_{2m}^{\Omega_*} \leq c \, \|A_0 v\|_0^{\Omega_*} + c \, \|v\|_0^{\Omega_*},$$

where c is a generic constant. Hence,

$$\|ue^{i\mu t}\|_{2m}^{\Omega_1} \leq c \, \|Au + \mu^{2m}e^{i\theta}u\|_0^{\Omega} + c \, |\mu|^{2m-1} \, \|u\|_0^{\Omega} + c\|u\|_0^{\Omega}. \qquad (18.5)$$

Now,

$$(\|ue^{i\mu t}\|_{2m}^{\Omega_1})^2 = 2 \sum_{j=0}^{2m} \int_\Omega \sum_{|\alpha| \leq j} |D^\alpha u|^2 \, |\mu|^{2(2m-j)} \, dx$$

$$\geq 2 \, |\mu|^{2(2m-j)}(\|u\|_j^{\Omega})^2$$

for any j. Using this estimate in (18.5), and taking $|\mu| \geq \mu_0$ with μ_0 sufficiently large, we obtain (18.4) with $\lambda = \mu^{2m}e^{i\theta}$.

Remark. Theorem 18.2 is true also with $\varepsilon = 0$; see Section 19.

Throughout the rest of this section we write the differential operator (18.1) usually in the form $A(x, D)$. We also assume that it satisfies the conditions of Theorem 18.2. By A we shall denote an operator in the Hilbert space $L^2(\Omega)$ with domain $D_A = H^{2m}(\Omega) \cap H_0^m(\Omega)$, defined as follows:

$$(Au)(x) = A(x, D)u(x) \qquad \text{for } u \in D_A.$$

By Theorems 18.1, 18.2,

$$\|u\|_{2m} \leq C(\|Au\|_0 + \|u\|_0), \qquad (18.6)$$

$$|\lambda| \, \|u\|_0 \leq C\|(A + \lambda I)u\|_0 \qquad (18.7)$$

for all $u \in D_A$, where λ is restricted as in Theorem 18.2; I is the identity operator.

LEMMA 18.1.   Let $\partial\Omega \in C^{2m}$. Then A is a closed operator.

Proof. We have to prove: if $u_i \in D_A$, $u_i \to u$ in $L^2(\Omega)$, $Au_i \to w$ in $L^2(\Omega)$, then $u \in D_A$ and $Au = w$. Applying (18.6) to $u_i - u_j$, we see that $\|u_i - u_j\|_{2m} \to 0$ if $i, j \to \infty$. Hence $\|u_i - v\|_{2m} \to 0$ as $i \to \infty$, for some $v \in H^{2m}(\Omega)$. Clearly $u = v$ and $u \in H_0^m(\Omega)$. Thus, $u \in D_A$. It is also clear that $\|Au_i - Au\|_0 \to 0$ as $i \to \infty$. Hence $Au = w$.

LEMMA 18.2.   If $\partial\Omega \in C^{2m}$, then, for all λ with $-\frac{1}{2}\pi + \varepsilon \leq \arg \lambda \leq \frac{1}{2}\pi - \varepsilon$, $|\lambda| \geq \Lambda_0$ ($\varepsilon > 0$ arbitrarily small, Λ_0 sufficiently large) the operator $A + \lambda I$ maps D_A onto $L^2(\Omega)$.

Proof. Let

$$A_i = \sum_{|\alpha| \leq 2m} a_\alpha^i D^\alpha$$

where $a_\alpha^i \in C^\infty(\overline{\Omega})$ and

$$\sup_\Omega |a_\alpha^i(x) - a_\alpha(x)| \to 0 \qquad \text{if } i \to \infty. \tag{18.8}$$

The inequality (18.4) for $A = A_i$ gives

$$\|u\|_{2m} \leq C \|(A_i + \lambda I)u\|_0 \qquad (u \in H^{2m}(\Omega) \cap H_0^m(\Omega)) \tag{18.9}$$

for all λ as in the assertion of the Theorem 18.2. It is important to note that the constants C, Λ_0 can be taken to be *independent* of i (see the remarks following Theorem 18.1).

Theorem 14.6 applies to $L = A_i + \lambda$ since A_i can be written in the form $\sum (-1)^{|\rho|} D^\rho(b^{\rho\sigma} D^\sigma)$ with $b^{\rho\sigma}$ satisfying the assumptions (A_1)–(A_3). We claim that the first alternative holds. In view of Theorem 14.2 it suffices to show that if w is a generalized solution of $(A_i + \lambda)w = 0$, $w \in H_0^m(\Omega)$, then $w = 0$. By Theorem 17.2, $w \in H^{2m}(\Omega)$. Applying (18.9) to $u = w$, we get $w = 0$.

Having proved that the first alternative holds, it follows that for any $f \in L^2(\Omega)$ there exists a unique solution u_i of the generalized Dirichlet problem

$$A_i(x, D)u_i + \lambda u_i = f \qquad \text{in } \Omega, \quad u_i \in H_0^m(\Omega). \tag{18.10}$$

As before (in the case $f = 0$) we have $u_i \in H^{2m}(\Omega)$. Write

$$(A_i + \lambda I)(u_i - u_j) = f_{ij}$$

where $f_{ij} = [A_j(x, D) - A_i(x, D)]u_j$. Since, by (18.9),

$$\|u_j\|_{2m} \leq C\|f\|_0,$$

it follows, upon using (18.8), that $\|f_{ij}\|_0 \to 0$ if $i, j \to \infty$. Applying (18.9) to $u_i - u_j$, we then get

$$\|u_i - u_j\|_{2m} \to 0 \qquad \text{if } i, j \to \infty.$$

Hence there exists an element u in $H^{2m}(\Omega)$ such that $\|u_i - u\|_{2m} \to 0$. It follows that $u \in D_A$ and

$$\|u_i - u\|_0 \to 0, \quad \|Au_i - Au\|_0 \to 0 \qquad \text{if } i \to \infty.$$

From this and from (18.8), (18.10) we get $(A + \lambda I)u = f$. Since f is an arbitrary function in $L^2(\Omega)$, the proof is complete.

The inequality (18.7) and Lemma 18.2 show that the operator $A + \lambda I$ maps D_A in one-to-one fashion onto $L^2(\Omega)$, and its inverse $(A + \lambda I)^{-1}$ is a bounded operator. We shall now prove:

THEOREM 18.3. Let Ω be a bounded domain with $\partial\Omega$ in C^{2m} and let $A(x, D)$ be a uniformly strongly elliptic operator in Ω with $a_\alpha \in C^j(\overline{\Omega})$, $j = \max(0, |\alpha| - m)$. Then the resolvent $(\lambda I - A)^{-1}$ of A exists for all complex λ in the set $\frac{1}{2}\pi \leq \arg(\lambda + k) \leq \frac{3}{2}\pi$ and

$$\|(\lambda I - A)^{-1}\| \leq \frac{C}{|\lambda| + 1}; \tag{18.11}$$

C *and* k *are positive constants.*

The notation $\| \ \|$ in (18.11) stands for the norm of operators in $L^2(\Omega)$.

Proof. From the previous considerations we already know that $R(\lambda; A)$ exists and (18.11) holds if $\frac{1}{2}\pi + \varepsilon \leq \arg\lambda \leq \frac{3}{2}\pi - \varepsilon$, $|\lambda| \geq \Lambda_0(\varepsilon)$, where ε is any positive number arbitrarily small. We shall next show that for $u \in D_A$ and all real τ,

$$\|[A + (\sigma_0 + i\tau)I]u\|_0 \geq c|\tau| \|u\|_0, \tag{18.12}$$

where σ_0, c are some positive constants. The proof of Theorem 18.3 then follows by using the assertions of Problems 3, 4 below.

To prove (18.12), recall, by Gårding's inequality, that

$$\text{Re}\,([A + (\sigma_0 + i\tau)I]u, u) \geq \sigma_0\|u\|_0^2 + c_0\|u\|_m^2 - k_0\|u\|_0^2 = c_0\|u\|_m^2$$

if we take $\sigma_0 = k_0$. Also, $|(Au, u)| \le C\|u\|_m^2$, so that

$$|\text{Im} ([A + (\sigma_0 + i\tau)I]u, u)| + \frac{C}{c_0} \text{Re} ([A + (\sigma_0 + i\tau)I]u, u) \ge |\tau| \|u\|_0^2.$$

Consequently

$$|([A + (\sigma_0 + i\tau)I]u, u)| \ge c|\tau| \|u\|_0 \qquad (c > 0),$$

from which (18.12) follows.

PROBLEMS. (1) Let X be a Banach space and let B be a bounded linear operator in X with $\|B\| < 1$. Prove that $(I + B)^{-1}$ is a bounded operator and $\|(I + B)^{-1}\| \le (1 - \|B\|)^{-1}$.

(2) Let X be a Banach space and let B, E be bounded operators in X. Assume that B^{-1} is a bounded operator and $\|E\| < 1/\|B^{-1}\|$. Show that $B + E$ has a bounded inverse and

$$\|(B + E)^{-1} - B^{-1}\| \le \frac{\|B^{-1}\|}{1 - \|B^{-1}E\|}.$$

(3) Prove that if the assertion (18.11) of Theorem 18.3 holds for all λ in a set $\arg \lambda = \theta_0$, $|\lambda| \ge \Lambda_0$, then it also holds for all λ in the set $|\theta_0 - \arg \lambda| < \delta$, $|\lambda| \ge \Lambda_0$, for some sufficiently small $\delta > 0$.

(4) Let X be a Banach space and let B be a closed operator in X with domain D_B. Assume that for any λ in a domain W of the complex plane,

$$\|(B - \lambda I)x\| \ge c\|x\| \qquad \text{for all } x \in D_B,$$

where c is a positive constant. Prove that if the resolvent $(\lambda I - B)^{-1}$ exists for some $\lambda \in W$, then it exists for all $\lambda \in W$.

(5) If in Theorem 18.2 the coefficients of $A(x, D)$ are real, then, for any $\varepsilon > 0$, the inequality (18.4) holds if $|\pi - \arg \lambda| \ge \varepsilon$, $|\lambda| \ge \Lambda_0$, and the inequality (18.11) holds if $|\arg (\lambda + k)| \ge \varepsilon$.

19 | GENERAL BOUNDARY CONDITIONS

In this section we describe general boundary conditions for which one can derive existence and regularity theorems. We also state the corresponding a priori inequalities, not just in the L^2-norm but also in the L^p-norm. The proofs of these results are quite difficult and lengthy, and will not be given here.

We denote by $A(x, D)$ an elliptic operator given by (18.1). Its principal part will be denoted by $A_0(x, D)$. If $n = 2$, then we assume:

ROOT CONDITION. For every pair of linearly independent real vectors ξ, η and $x \in \overline{\Omega}$ the polynomial in z: $A_0(x, \xi + z\eta)$ has exactly m roots with positive imaginary parts.

PROBLEM. (1) Prove that if $n \geq 3$ or if $n = 2$ and the coefficients of A are real, then the root condition is satisfied.

We introduce m differential boundary operators $\{B_j\}_{j=1}^m$ of respective orders m_j, given by

$$B_j(x, D) = \sum_{|\alpha| \leq m_j} b_\alpha^j(x) D^\alpha. \tag{19.1}$$

The boundary value problem we consider is the following:

$$\begin{cases} A(x, D)u = f & \text{in } \Omega, \\ B_j(x, D)u = 0 & \text{on } \partial\Omega \end{cases} \quad (1 \leq j \leq m). \tag{19.2}$$

The following conditions are assumed:

SMOOTHNESS CONDITION. Ω is a bounded domain and $\partial\Omega$ is of class C^{2m}. The leading coefficients of $A(x, D)$ are continuous in $\overline{\Omega}$, the other coefficients being measurable and bounded. $2m > m_j$ for $1 \leq j \leq m$ and the coefficients of the B_j belong to $C^{2m - m_j}$ on $\partial\Omega$.

COMPLEMENTARY CONDITION. At any point $x \in \partial\Omega$ let v denote the outward normal to $\partial\Omega$ at x and let $\xi \neq 0$ be a real vector in the tangent hyperplane to $\partial\Omega$ at x. Then the polynomials in z: $B_{j0}(x, \xi + zv)$ (B_{j0} denoting the principal part of B_j) are linearly independent modulo the polynomial

$$\prod_{k=1}^m (z - z_k^+(\xi)),$$

where $z_k^+(\xi)$ are the roots of $A_0(x, \xi + zv)$ with positive imaginary parts.

THEOREM 19.1. Consider the class of functions u in $C^{2m}(\overline{\Omega})$ satisfying the boundary conditions

$$B_j u = 0 \quad \text{on } \partial\Omega, \quad 1 \leq j \leq m,$$

and let $1 < p < \infty$. *Assume that the root condition, the smoothness condition, and the complementary condition hold. Then*

$$|u|_{2m,p}^{\Omega} \leq C(|A(x, D)u|_{0,p}^{\Omega} + |u|_{0,p}^{\Omega}), \tag{19.3}$$

where C *is a constant independent of* u.

We now define an operator A in $L^2(\Omega)$. The domain D_A of A consists of the closure in $H^{2m,p}(\Omega)$ of the set of functions u in $C^{2m}(\overline{\Omega})$ that satisfy the boundary conditions $B_j(x, D)u = 0$ on $\partial\Omega$ $(1 \leq j \leq m)$. For every $u \in D_A$ we define

$$(Au)(x) = A(x, D)u(x).$$

The inequality (19.3) is easily seen to hold for every $u \in D_A$. We can now proceed as in the proof of Lemma 18.1 and show that A is a closed operator.

Since the domain of A is independent of the particular elliptic operator $A(x, D)$, we write

$$D_A = H^{2m,p}(\Omega; \{B_j\}).$$

It is not obvious that the present definition of D_A in case $B_j = \partial^{j-1}/\partial v^{j-1}$ is the same as the definition of D_A given in Section 18. We shall therefore prove:

LEMMA 19.1. Let $\partial\Omega \in C^{2m+[n/2]+1}$. *Then*

$$H^{2m, 2}\left(\Omega; \left\{\frac{\partial^{j-1}}{\partial v^{j-1}}\right\}\right) = H^{2m}(\Omega) \cap H_0^m(\Omega).$$

Proof. Denote the left side by D_A, where $A(x, D)$ is any strongly elliptic operator with coefficients in $C^\infty(\overline{\Omega})$. If $u \in D_A$, then there exists a sequence $\{u_i\}$ with the following properties: $u_i \in C^{2m}(\overline{\Omega})$, $\partial^j u_i/\partial v^j = 0$ on $\partial\Omega$ $(0 \leq j \leq m-1)$, $\|u_i - u\|_{2m} \to 0$ as $i \to \infty$. From Lemma 13.3 it follows that $u_i \in H_0^m(\Omega)$. Hence $u \in H_0^m(\Omega)$. We have thus proved that $D_A \subset H^{2m}(\Omega) \cap H_0^m(\Omega)$.

Suppose conversely that $u \in H^{2m}(\Omega) \cap H_0^m(\Omega)$ and set $f = A(x, D)u + \lambda u$, λ being as in Theorem 18.2. Let $\{f_i\}$ be a sequence of functions in $C^\infty(\overline{\Omega})$ with $\|f_i - f\|_0 \to 0$ as $i \to \infty$. Consider the generalized Dirichlet problem

$$A(x, D)u_i + \lambda u_i = f_i \text{ in } \Omega, \qquad u_i \in H_0^m(\Omega).$$

As in the proof of Lemma 18.2 we find that there exists a unique solution u_i and it belongs to $H^{2m+[n/2]+1}(\Omega)$. By Problem 1, Section 11, $u_i \in C^{2m}(\overline{\Omega})$. By Lemma 13.2 we further see that $\partial^j u_i/\partial v^j = 0$ on $\partial\Omega$, if $0 \leq j \leq m-1$.

As in the proof of Lemma 18.2 we find that for some $w \in H^{2m}(\Omega)$, $\|u_i - w\|_{2m} \to 0$ as $i \to \infty$. Clearly, $w \in D_A$. Since $v \equiv w - u$ satisfies

$$A(x, D)v + \lambda v = 0 \quad \text{in } \Omega, \quad v \in H^{2m}(\Omega) \cap H_0^m(\Omega),$$

Theorem 18.2 shows that $v = 0$. Hence $u = w \in D_A$.

PROBLEMS.   (2) Prove that the null space $N(A)$ of A is finite-dimensional.
(3) Prove that if $N(A) = 0$, then

$$|u|_{2m, p}^{\Omega} \le C |Au|_{0, p}^{\Omega} \quad \text{for all} \quad u \in H^{2m, p}(\Omega; \{B_j\}).$$

[*Hint:* Prove first that $|u|_{0,p} \le c |Au|_{0,p}$.]

DEFINITION.   The system $\{B_j\}_1^m$ is called *normal* if $m_i \ne m_j$ when $i \ne j$, and if $\partial\Omega$ is *noncharacteristic* to the B_j at each point. The latter property means that for any $x \in \partial\Omega$, if we denote the normal to $\partial\Omega$ at x by v and any tangent vector by ξ, then in the polynomial (in z)

$$B_j(x, \xi + zv) = c_j(x)z^{m_j} + \sum_{k=1}^{m_j} c_{jk} z^{m_j - k},$$

the leading coefficient $c_j(x)$ is always $\ne 0$.

We shall call the triple $(A, \{B_j\}, \Omega)$ an *elliptic boundary value problem*. (Here A indicates the differential operator $A(x, D)$.)

DEFINITION.   An elliptic boundary value problem $(A, \{B_j\}, \Omega)$ is called *regular* if:

(i) It satisfies the root condition, the smoothness condition, and the complementary condition.

(ii) The system $\{B_j\}$ is normal.

PROBLEM.   (4) Let μ be a nontangential smoothly varying direction on $\partial\Omega$ and let

$$B_j = \frac{\partial^{s+j-1}}{\partial\mu^{s+j-1}} + \text{lower-order differential boundary operator} \qquad (19.4)$$

for some s, $0 \le s \le m$. Prove that the $\{B_j\}$ form a normal system and that if the root condition is satisfied then also the complementary condition is satisfied.

We now impose a condition on $A(x, D)$ that is stronger than the ellipticity condition: For some real θ,

$$(-1)^m \frac{A_0(x, \xi)}{|A_0(x, \xi)|} \neq e^{i\theta} \qquad \text{for all real } \xi \text{ and } x \in \bar{\Omega}. \tag{19.5}$$

Note that $A(x, D)$ is strongly elliptic if and only if (19.5) holds for all θ in the interval $\frac{1}{2}\pi \leq \theta \leq \frac{3}{2}\pi$.

We also introduce a stronger condition than the complementary condition:

STRONG COMPLEMENTARY CONDITION. With the notation of the complementary condition, for any λ on the ray $\arg \lambda = \theta$, the polynomials $B_{0j}(x, \xi + zv)$ are linearly independent modulo the polynomial

$$\prod_{k=1}^{m} (z - z_k^*(\xi, \lambda)),$$

where $z_k^*(\xi, \lambda)$ are the m roots of $(-1)^m A_0(x, \xi + zv) - \lambda$ with positive imaginary parts. (It is assumed that there are precisely m such roots.)

By Problem 4, this condition holds for B_j having the form (19.4), provided there are precisely m roots of $(-1)^m A_0(x, \xi + zv) - \lambda$ with positive imaginary parts.

THEOREM 19.2. Let $(A, \{B_j\}, \Omega)$ be a regular elliptic boundary value problem satisfying (19.5) and the strong complementary condition. Then there exist positive constants C, Λ_0 such that for any u in $H^{2m,p}(\Omega; \{B_j\})$,

$$\sum_{j=0}^{2m} |\lambda|^{(2m-j)/m} |u|_{j, p}^{\Omega} \leq C |(A - \lambda I)u|_{0, p}^{\Omega} \qquad \text{if } \arg \lambda = \theta, |\lambda| \geq \Lambda_0. \tag{19.6}$$

The proof is similar to the proof of Theorem 18.2 and is left to the reader.

THEOREM 19.3. Let $(A, \{B_j\}, \Omega)$ be a regular elliptic boundary value problem satisfying (19.5) and the strong complementary condition. Then for any λ with $\arg \lambda = \theta, |\lambda| \geq \Lambda_0$ (Λ_0 some constant) the range of $A - \lambda I$ is the whole space $L^p(\Omega)$.

Assuming Theorem 19.3, and making use also of Theorem 19.2, we can now conclude (cf. Theorem 18.3):

THEOREM 19.4.   Let $(A, \{B_j\}, \Omega)$ be a regular elliptic boundary value problem satisfying (19.5) and the strong complementary condition. Then the resolvent $(\lambda I - A)^{-1}$ exists for all λ in the set $\arg \lambda = \theta$, $|\lambda| \geq \Lambda_0$ and

$$\|(A - \lambda I)^{-1}\|_p \leq C/|\lambda|; \tag{19.7}$$

C and Λ_0 are positive constants.

The notation $\| \ \|_p$ stands for the norm of operators in $L^p(\Omega)$.

PROBLEMS.   (5) Let $(A, \{B_j\}, \Omega)$ be a regular elliptic boundary value problem. Assume that $A(x, D)$ is strongly elliptic and that the strong complementary condition holds for all θ with $\frac{1}{2}\pi \leq \theta \leq \frac{3}{2}\pi$. Prove that (19.7) holds for all λ with $\text{Re } \lambda \leq 0$, $|\lambda| \geq \Lambda_0$, where Λ_0, C are constants (independent of $\arg \lambda$).

(6) Using Theorem 19.3, prove that Lemma 19.1 holds, assuming that $\partial\Omega$ is only in C^{2m}.

We shall now make a few remarks concerning the proofs of Theorems 19.1 and 19.3.

Theorem 19.1 is due to Agmon, Douglis, and Nirenberg [1]. Browder [5] has proved it for the Dirichlet boundary conditions. The general structure of the proof is as follows:

First, it suffices to obtain the estimate in a neighborhood of the boundary. This situation can further be transformed into the case of a half-space $x_n > 0$, and $A(x, D)$ being homogeneous with constant coefficients. Denote by $M(x)$ a fundamental solution (see Section 4). We extend u by reflection into $x_n < 0$ (cf. Section 5) and set $f = A(D)u$. The function

$$v(x) = \int_{R^n} M(x - y)f(y) \, dy$$

satisfies $A(D)v = f$. Set $\Phi_j = -\partial^{j-1}v/\partial v^{j-1}$ on $x_n = 0$. It will be enough to estimate $w = u - v$. w satisfies

$$A(D)w = 0 \quad \text{in } \Omega, \qquad \frac{\partial^j w}{\partial v^j} = \Phi_{j+1} \quad \text{on } x_n = 0 \quad (0 \leq j \leq m - 1).$$

Set $x = (x', t)$, where $x' = (x_1, \ldots, x_{n-1})$, $t = x_n$. One constructs kernels $K_j(x', t)$ $(j = 1, \ldots, m)$ such that the solution w can be represented in the form

$$w(x', t) = \sum_{j=1}^{m} \int K_j(x' - y', t)\Phi_j(y') \, dy'. \tag{19.8}$$

We shall estimate here just the $(2m)$th derivatives $D^{2m}w$ of w. The formulas for the K_j show that $K(x', t) \equiv D^{2m}K_j$ has the form

$$K(x', t) = G\left(\frac{x'}{|x|}, \frac{t}{|x|}\right)\bigg/ |x|^n,$$

where G is continuous on the hemisphere $|x| = 1$, $t \geq 0$ together with its first derivatives. Also,

$$\int_{|x'|=1} G(x', 0) \, dS_{x'} = 0.$$

For such a kernel G the following is true: Set

$$w_j(x', t) = \int_{R^{n-1}} K(x' - y', t)\Phi_j(y') \, dy' \qquad \text{for } t > 0.$$

Then for any $t > 0$, $1 < p < \infty$

$$\left(\int_{R^{n-1}} |w_j(x', t)|^p \, dx'\right)^{1/p} \leq C\left(\int_{R^{n-1}} |\Phi_j(y')|^p \, dy'\right)^{1/p}, \tag{19.9}$$

where $C = c\gamma$, c depending only on p, and γ is a bound on G and on its first derivatives for $|x| = 1$, $t \geq 0$.

The result (19.9) (for $t = 0$) is due to Calderon and Zygmund [1].

Theorem 19.3 was proved by Agmon [4]. We shall outline a shorter proof under additional assumptions. An elliptic boundary value problem $(A, \{B_j\}, \Omega)$ is said to be *formally adjoint* to the elliptic boundary value problem $(A, \{B_j\}, \Omega)$ if for any $v \in H^{2m,q}(\Omega)$ $(1/p + 1/q = 1)$ the following is true: the relation $(A(x, D)u, v) = (u, \hat{A}(x, D)v)$ holds for all $u \in H^{2m,p}(\Omega; \{B_j\})$ if and only if $v \in H^{2m,q}(\Omega; \{\hat{B}_j\})$. This implies, of course, that $\hat{A}(x, D) = A^*(x, D)$. If $B_j = \partial^{j-1}/\partial v^{j-1}$ (v normal to $\partial\Omega$), then $(A^*, \{\partial^{j-1}/\partial v^{j-1}\}, \Omega)$ is formally adjoint to $(A, \{\partial^{j-1}/\partial v^{j-1}\}, \Omega)$.

We now assume that there exists a formally adjoint elliptic boundary value problem $(\hat{A}, \{\hat{B}_j\}, \Omega)$ satisfying the same assumptions as $(A, \{B_j\}, \Omega)$ in Theorem 19.3 with θ replaced by $-\theta$, and that $\partial\Omega$ and the coefficients of A, B_j, $\hat{B}_j$ are sufficiently smooth. (For the Dirichlet problem, for instance, this assumption holds if only $\partial\Omega$ and the coefficients of $A(x, D)$ are sufficiently smooth.) Then it is not difficult to show (see Browder [3]) that $A^* = \hat{A}$ (considered as operators in $L^q(\Omega)$). It follows that

$$(A - \lambda I)^* = \hat{A} - \bar{\lambda}I \qquad (\arg \lambda = \theta).$$

Set $T = A - \lambda I$ (arg $\lambda = \theta$). Since T is closed and densely defined, we have (by a well-known theorem)

$$R(T)^{\perp} = N(T^*), \tag{19.10}$$

where " $\perp$ " indicates the orthogonal complement and $N(T^*) =$ null space of T^*. By Theorem 19.1, $N(T^*) = 0$ if arg $\lambda = \theta$ and $|\lambda| \geq \Lambda_0$, and also (as easily verified) $R(T)$ is a closed set. It follows from (19.10) that $R(T) = L^p(\Omega)$, which is precisely the assertion of Theorem 19.3.

Theorem 19.1 has important applications. We shall give here an application to the question of regularity of solutions.

THEOREM 19.5. Let A *be an elliptic operator with coefficients in* $C^k(\Omega)$, *where* Ω *is a bounded domain, and let* $u \in H^{2m,p}(\Omega_0)$ *for every domain* Ω_0 *with* $\bar{\Omega}_0 \subset \Omega$. *Let* $f \in H^{k,p}(\Omega)$ *and assume that* $A(x, D)u = f$ *in* Ω. *Then* $u \in H^{2m+k,p}(\Omega_0)$ *for every domain* Ω_0 *with* $\bar{\Omega}_0 \subset \Omega$.

We only sketch the proof, leaving some details to the reader. If $k = 1$, then we introduce $v = \zeta u$ for a fixed $\zeta \in C_0^{\infty}(\Omega)$. We consider the elliptic equation for v^h. It has the form

$$Av^h = f_h,$$

where $|f_h|_{0,p} \leq C$, C independent of h. Since the $v^h \in C_0^{\infty}(\Omega)$, we can apply Theorem 19.1 and get

$$|v^h|_{2m,p} \leq C.$$

Now use the remark following Lemma 15.3.

If $k > 1$, we use induction. In fact we apply the inductive assumption to $D_i v$, which is easily seen to satisfy $A(D_i v) = \tilde{f}_i$ with $\tilde{f}_i \in H^{2m+k-1,p}(\Omega)$.

COROLLARY. Assume that $a_{\alpha} \in C^k(\Omega)$, $f \in C^k(\Omega)$. *If* $u \in C^{2m}(\Omega)$ *and* $Au = f$ *in* Ω, *then* $u \in C^{2m+k-1}(\Omega)$.

Indeed, use Theorem 19.1 with $p > n$ and apply Theorem 11.1 with $j = 1$, $m = 0$.

One can also combine arguments of Section 17 together with Theorem 19.1 and thus prove a regularity theorem on the boundary analogous to Theorem 17.3. For details, see Agmon [1].

20 | PROBLEMS

(1) Green's function for the Laplace equation in a bounded domain Ω is, by definition, a function $G(x, y)$ defined for all $y \in \Omega$ and $x \in \overline{\Omega}$, $x \neq y$, and having the form

$$G(x, y) = K(x - y) + H(x, y), \qquad K \text{ as in } (3.6);$$

$H(x, y)$, for fixed y, is a solution of the Laplace equation in Ω, continuous in $\overline{\Omega}$, and $K(x - y) + H(x, y) = 0$ for $x \in \partial\Omega$.

Prove: if u is harmonic in Ω, and if u and $G(x, y)$ are continuously differentiable in $\overline{\Omega}$ and in $\overline{\Omega} - \{y\}$ respectively, then

$$u(x) = -\int_{\partial\Omega} u(\xi) \frac{\partial G(\xi, x)}{\partial v} \, dS_\xi \qquad \text{for any } x \in \Omega, \tag{20.1}$$

where v is the outward normal to $\partial\Omega$ at ξ.

(2) The analog of (20.1) for the half-space $\Omega = \{(x_1, \ldots, x_n); x_n > 0\}$ is

$$u(x_1, x_2, \ldots, x_n)$$
$$= \frac{2}{\Omega_n} \int_{-\infty}^{\infty} \cdots \int_{-\infty}^{\infty} \frac{x_n f(\xi_1, \ldots, \xi_{n-1}) \, d\xi_1 \cdots d\xi_{n-1}}{[(x_1 - \xi_1)^2 + \cdots + (x_{n-1} - \xi_{n-1})^2 + x_n^2]^{n/2}}, \tag{20.2}$$

where Ω_n is the surface area of the unit sphere in R^n.

Prove: If f is continuous and bounded in R^{n-1}, then the right-hand side of (20.2) is a solution of $\Delta u = 0$ in Ω, continuous in $\overline{\Omega}$, and $u = f$ on $x_n = 0$.

(3) Let $u(x)$ be a harmonic function in a domain Ω of R^n and consider the function

$$v(x) = \frac{1}{r^{n-2}} u\left(\frac{x_1}{r^2}, \frac{x_2}{r^2}, \ldots, \frac{x_n}{r^2}\right) \qquad \left(r = \left(\sum_{i=1}^{n} x_i^2\right)^{1/2}\right)$$

for x/r^2 in Ω. Prove: $v(x)$ is harmonic.

(4) Let Ω be the n-ball $\{x; |x| < R\}$. Consider the integral

$$u(x) = \frac{R^{n-2}(R^2 - |x|^2)}{\Omega_n} \int_{|\xi| = R} \frac{f(\xi) \, dS_\xi}{(|x|^2 + R^2 - 2|x| R \cos \theta)^{n/2}}. \tag{20.3}$$

where θ is the angle between the vector x and the radius drawn to a variable point of integration ξ. Prove: If $f(\xi)$ is continuous, then $u(x)$ is harmonic in Ω, continuous in $\overline{\Omega}$, and $u = f$ on $\partial\Omega$.

The integral in (20.3) is called the *Poisson integral* and

$$R^{n-2}(R^2 - |x|^2)(\Omega_n)^{-1}(|x|^2 + R^2 - 2|x|R\cos\theta)^{-n/2}$$

is called the *Poisson kernel*.

[*Hint:* Writing the Poisson kernel in the form

$$c\left\{-|x - \xi|^{2-n} + \frac{2}{n-2}\sum \xi_i \frac{\partial}{\partial \xi_i}|x - \xi|^{2-n}\right\} \qquad \text{(if } n \geq 3\text{)},$$

it follows that $\Delta u = 0$.]

(5) If u is harmonic in a ball B and if B' is a closed ball lying in the interior of B, then, for any integer $k \geq 0$,

$$\sup_{0 \leq |\alpha| \leq k} \sup_{x \in B'} |D^\alpha u(x)| \leq C \sup_B |u(x)|,$$

where C is a constant depending only on B, B', k.

(6) Extend the result of the previous problem to the case where B is any domain and B' is any bounded domain with $\bar{B}' \subset B$.

(7) Let $\{u_m\}$ be a sequence of harmonic functions in a domain Ω, and suppose that $|u_m(x)| \leq C$ for all $x \in \Omega$, where C is a constant independent of m. Prove: There exists a subsequence of $\{u_m\}$ that is uniformly convergent in every compact subdomain of Ω.

(8) Show that if u, v are in $C^2(\Omega_0)$ and if Ω is any bounded subdomain of Ω_0 with smooth boundary and with $\bar{\Omega} \subset \Omega_0$, then

$$\int_\Omega v\Delta^2 u \, dx + \int_\Omega \nabla v \cdot \nabla u \, dx = \int_{\partial\Omega} v \frac{\partial u}{\partial v} \, dS, \qquad (20.4)$$

where v is the outward normal to $\partial\Omega$. Use (20.4) to prove:

(a) If u is harmonic in Ω_0, then

$$\int_{\partial\Omega} \frac{\partial u}{\partial v} \, dS = 0.$$

(b) If u is harmonic in Ω_0, then

$$\int_{\partial\Omega} u \frac{\partial u}{\partial v} \, dS \geq 0.$$

(c) If u, v are harmonic in Ω_0, then

$$\int_{\partial\Omega} \left(u \frac{\partial v}{\partial v} - v \frac{\partial u}{\partial v}\right) dS = 0.$$

(9) If u is harmonic in a domain Ω_0 and if Ω is any domain with smooth boundary such that $\overline{\Omega} \subset \Omega_0$, then for any harmonic function u in Ω_0,

$$u(x) = \frac{1}{\Omega_n} \int_{\partial\Omega} \frac{\partial u}{\partial v} \frac{1}{r^{n-2}} \, dS - \frac{1}{\Omega_n} \int_{\partial\Omega} u \frac{\partial}{\partial v} \frac{1}{r^{n-2}} \, dS$$

for all $x \in \Omega$.

(10) If u is harmonic in a domain Ω, then, for any closed ball B in Ω with center x^0,

$$u(x^0) = \frac{1}{|S|} \int_S u \, dS, \tag{20.5}$$

where S is the boundary of B and $|S|$ is the surface area of S. This assertion is known as the *mean value theorem* (or *Gauss theorem*).

(11) Prove the *maximum principle* for harmonic functions—that is, a nonconstant harmonic function u in a domain Ω cannot assume its supremum at any point of Ω.

[*Hint:* If $u(x^0) = M = \sup_\Omega u$, then use (20.5) to show that for any sphere $S_\varepsilon = \{x; |x - x^0| = \varepsilon\}$ with ε sufficiently small, $u \equiv M$ on S_ε.]

(12) Prove *Harnack's first theorem*: If $\{u_m\}$ is a sequence of harmonic functions in a bounded domain Ω, continuous in $\overline{\Omega}$, and if the sequence is uniformly convergent to a function v on $\partial\Omega$, then there is a harmonic function u in Ω, continuous in $\overline{\Omega}$ and with $u = v$ on $\partial\Omega$, such that, for any α, $D^\alpha u_m \to D^\alpha u$ on compact subsets of Ω.

(13) The maximum principle implies uniqueness for the Dirichlet problem. Hence, the results of previous sections show that there exists a unique classical solution to the problem $\Delta u = g$ in Ω, $u = f$ on $\partial\Omega$ provided g, f, and $\partial\Omega$ are sufficiently smooth. In particular, Green's function exists. Prove that $G(x, \xi) > 0$ for all $x \in \Omega$, $x \neq \xi$.

(14) Prove the converse of Gauss' theorem, namely, if $u(x)$ is continuous in Ω and if (20.5) holds for all balls B in Ω, then u is harmonic in Ω.

[*Hint:* For a fixed ball B_0 in Ω, let $v(x)$ be the solution of $\Delta v = 0$ in B_0, $v = u$ on ∂B_0. Show that $u \equiv v$.]

(15) If u is harmonic in a ball $B = \{x; |x| < R\}$ and if it is continuous and nonnegative in $\overline{B}$, then

$$\left(\frac{R}{R+\rho}\right)^{n-2} \frac{R-\rho}{R+\rho} u(x) \leq u(0) \leq \left(\frac{R}{R-\rho}\right)^{n-2} \frac{R+\rho}{R-\rho} u(x) \qquad \text{for all } x \in B,$$
$$\tag{20.6}$$

where $\rho = |x|$.

[*Hint:* Use the Poisson integral formula.]

(16) Let Ω_0 be a bounded domain and let Ω be any subdomain with $\bar{\Omega} \subset \Omega_0$. Then there exists a positive constant γ, depending only on Ω_0, Ω, such that for any nonnegative harmonic function u in Ω_0,

$$\gamma u(y) \le u(x) \le \frac{1}{\gamma} u(y) \qquad \text{for all } x, y \text{ in } \Omega.$$

This inequality is called the *Harnack inequality*.

(17) Prove *Harnack's second theorem*: Let $\{u_m\}$ be a sequence of monotone nondecreasing harmonic functions in a domain Ω. If, for some point $x^0 \in \Omega$, the sequence $\{u_m(x^0)\}$ is bounded, then the sequence $\{u_m\}$ is uniformly convergent in any compact subset of Ω.

(18) Prove *Liouville's theorem*: Any harmonic function in R^n that is bounded in R^n is necessarily a constant.

[*Hint:* Use (20.6).]

(19) If u is harmonic in $B - \{0\}$, where B is the n-dimensional unit ball, $n \ge 3$, and if $u(x) = o(|x|^{2-n})$ as $x \to 0$, then $u(x)$ is harmonic in B when suitably defined at $x = 0$.

[*Hint:* Let $\Delta v = 0$ in B, $v = u$ on ∂B. Show that $u - v \equiv 0$.]

(20) Let B_R be the ball $\{x; |x| < R\}$ in R^3. Prove that for any function u in $C^2(\bar{B}_R)$,

$$\frac{1}{4\pi R^2} \int_{\partial B_R} u \, dS = u(0) + \frac{1}{4\pi} \int_{B_R} \left(\frac{1}{r} - \frac{1}{R}\right) \Delta u \, dx.$$

(21) Set

$$v_m = \frac{1}{4\pi(2m+1)!} \frac{(R-r)^{2m+1}}{Rr} \qquad (m = 0, 1, 2, \ldots),$$

where $r = |x|$. Show that

$$\Delta v_{m+1} = v_m,$$

$$v_m = \frac{\partial v_m}{\partial r} = 0 \qquad \text{on } r = R,$$

where Δ is the Laplacian in R^3.

(22) If $u \in C^{2m}(\bar{B}_R)$, where B_R is as in Problem 20, then

$$C_m \Delta^m u(0) = \int_{B_R} (v_{m-1} \Delta^m u - v_m \Delta^{m+1} u) \, dx,$$

where $C_m = R^{2m}/(2m+1)!$.

(23) If $u \in C^{2m+2}(\bar{B}_R)$, B_R as in Problem 20, then

$$\frac{1}{4\pi R^2} \int_{\partial B_R} u \, dS = \sum_{j=0}^{m} \frac{R^{2j}}{(2j+1)!} \Delta^j u(0)$$

$$+ \frac{1}{4\pi(2m+1)!} \int_{B_R} \frac{(R-r)^{2m+1}}{Rr} \Delta^{2m+1} u \, dx. \qquad (20.7)$$

(24) If u satisfies the equation $\Delta u + cu = 0$ in B_R (B_R as in Problem 20), where c is a constant, and if it is continuous in $\bar{B}_R$, then

$$\frac{1}{4\pi R^2} \int_{\partial B_R} u \, dS = u(0) \frac{\sin R\sqrt{c}}{R\sqrt{c}}. \qquad (20.8)$$

(25) Extend the results of Problems 20–24 to a ball B_R of dimension $n \geq 3$, taking

$$v_{m+1} = \frac{1}{(n-2)r^{n-2}} \int_r^R \rho v_m(\rho)(\rho^{n-2} - r^{n-2}) \, d\rho,$$

$$v_0 = \frac{1}{(n-2)\Omega_n} \left(\frac{1}{r^{n-2}} - \frac{1}{R^{n-2}} \right)$$

and proving that

$$\frac{1}{|\partial B_R|} \int_{\partial B_R} u \, dS = \Gamma\left(\frac{n}{2}\right) \sum_{j=0}^{m} \left(\frac{R}{2}\right)^{2j} \frac{\Delta^j u(0)}{j! \, \Gamma(j+n/2)} + \int_{B_R} v_m \Delta^{m+1} u \, dx, \qquad (20.9)$$

where $|\partial B_R|$ is the surface area of ∂B_R. The analog of (20.8) is

$$\frac{1}{|\partial B_R|} \int_{\partial B_R} u \, dS = u(0)\Gamma\left(\frac{n}{2}\right) \left(\frac{R\sqrt{c}}{2}\right)^{(2-n)/2} J_{(n-2)/2}(R\sqrt{c}).$$

where $J_k(x)$ is the kth Bessel function.

(26) Extend the result of Problem 25 to $n = 2$, taking

$$v_{m+1} = \int_r^R \rho v_m(\rho) \log \frac{\rho}{r} \, d\rho,$$

$$v_0 = \frac{1}{2\pi} \log \frac{R}{r},$$

and proving that

$$\frac{1}{2\pi R} \int_{\partial B_R} u \, dS = \sum_{j=0}^{m} \left(\frac{R}{2}\right)^{2j} \frac{\Delta^j u(0)}{(j!)^2} + \int_{B_R} v_m \Delta^{m+1} u \, dx, \qquad (20.10)$$

and, if $\Delta u + cu = 0$,

$$\frac{1}{2\pi R} \int_{\partial B_R} u \, dS = u(0) R J_0(R\sqrt{c}).$$

(27) Show that if u is a solution of $\Delta u - \alpha u = 0$ in R^n, where α is a positive constant, and if

$$u(x) = 0(|x|^{(1-n)/2} e^{\gamma|x|}) \qquad \text{as } |x| \to \infty \qquad (\gamma = \sqrt{\alpha}), \qquad (20.11)$$

then $u(x) \equiv 0$.

(28) Let u be a solution of $\Delta u - \alpha u = 0$ in R^n, where α is a negative constant. Assume that $u(\infty) = \lim_{|x| \to \infty} u(x)$ exists and that

$$\lim_{R \to \infty} R^{(n-1)/2} \left| \frac{1}{|S_R|} \int_{S_R} u(x) \, dS - u(\infty) \right|$$

exists, where S_R is a sphere of radius R about any point $x^0 \in R^n$ and $|S_R|$ is the surface area of S_R. Prove that $u \equiv \text{const.}$

(29) Show that any solution of $\Delta^m u = 0$ in R^n, which is a bounded function in R^n, is necessarily a constant.

(30) Let $u \in C^\infty(R^n)$ and suppose that, for any $x \in R^n$,

$$\lim_{n \to \infty} |\Delta^n u(x)|^{1/n} = 0.$$

Suppose also that, as $n \to \infty$,

$$\frac{1}{n^2} |\Delta^n u(x)|^{1/n} \to 0 \qquad \text{uniformly in } x \text{ in bounded sets of } R^n.$$

Prove that if $u(x)$ is a bounded function in R^n, then $u(x) \equiv \text{const.}$

[*Hint:* For $n = 3$, consider $g(z) \equiv \sum a_{2n} z^{2n}$, where $a_{2n} = \Delta^n u(x^0)/(2n + 1)!$. It is an entire function of either order <1 or of order 1 and minimal type. Use a theorem of Phragmén-Lindelöf.]

(31) If w is a solution of $(\Delta - \alpha)w = 0$ in a domain Ω, where α is a constant, then

$$(\Delta - \alpha)^m (x_1^m v(x)) = m! \frac{\partial^m v(x)}{\partial x_1^m}.$$

(32) If u_i $(0 \le i \le m - 1)$ are solutions of $(\Delta - \alpha)u_i = 0$ in a domain Ω, where α is a constant, then the function

$$u(x) = u_0(x) + x_1 u(x) + x_1^2 u_2(x) + \cdots + x_1^{m-1} u_{m-1}(x) \qquad (20.12)$$

is a solution of $(\Delta - \alpha)^m u = 0$.

(33) Prove the converse of Problem 32 where Ω is a ball—that is, if $(\Delta - \alpha)^m u = 0$ in Ω, then for every closed ball Ω_0 lying in the interior of Ω there exist functions u_i $(0 \le i \le m - 1)$ satisfying $(\Delta - \alpha)u_i = 0$ in Ω_0 such that (20.12) holds in Ω_0.

[*Hint:* Let $y = (x_2, \ldots, x_n)$,

$$v_0(x, y) = \frac{1}{(m-1)!} \int_{x_1^0}^{x_1} \int_{x_1^0}^{\xi_{n-1}} \cdots \int_{x_1^0}^{\xi_2} (\Delta - \alpha)^{m-1} u(\xi_1, y) \, d\xi_1 \cdots d\xi_{n-2} \, d\xi_{n-1}.$$

Then

$$\frac{\partial^{m-1}}{\partial x_1^{m-1}} (\Delta - \alpha)v_0 = 0, \quad \text{so that} \quad (\Delta - \alpha)v_0 = \sum_{j=0}^{m-2} x_1^j a_j(y).$$

Construct

$$v_1(x, y) = \sum_{j=0}^{m-2} x_1^j b_j(y)$$

such that $(\Delta - \alpha)v_1 = -(\Delta - \alpha)v_0$. Then $v \equiv v_0 + v_1$ satisfies

$$(\Delta - \alpha)v = 0, \qquad (\Delta - \alpha)^{m-1}(u - x_1^{m-1}v) = 0.]$$

(34) Let u be a solution in R^3 of $(\Delta - \alpha)^m u = 0$, where α is a constant. Then

$$\frac{1}{4\pi R^2} \int_{|x| = R} u \, dS = \sum_{j=0}^{m-1} c_j A_j(R\sqrt{\alpha}), \qquad A_j(\sqrt{t}) = t^j \frac{d^j}{dt^j} \frac{\sinh \sqrt{t}}{\sqrt{t}}.$$

(35) If u is a solution in R^3 of $(\Delta - \alpha)^m u = 0$, where $\alpha > 0$, and if it satisfies (20.11) (with $n = 3$), then $u \equiv 0$.

(36) If u_i $(0 \le i \le m - 1)$ are harmonic functions, then the function

$$u(x) = u_0(x) + |x|^2 u_1(x) + \cdots + |x|^{2m-2} u_{m-1}(x)$$

is a solution of $\Delta^m u = 0$.

(37) Let u be a harmonic function in the three-dimensional ball $B = \{x; |x - x^0| < R\}$. Assume that u is continuous in $\bar{B}$ and $|u| \leq M$ on ∂B. Then, for any $i = 1, 2, \ldots, n$,

$$\left| \frac{\partial}{\partial x_i} u(x^0) \right| \leq \frac{3M}{R}.$$

[*Hint:* Use the Gauss theorem for $\partial u / \partial x_i$.]

(38) Let Ω be a bounded domain and let x^0 be a point on its boundary $\partial \Omega$. A *barrier* at x^0 (for the Laplace operator) is a function w satisfying the following properties: (i) w is continuous in $\bar{\Omega}$; (ii) $w > 0$ in $\bar{\Omega} - \{x^0\}$ and $w(x^0) = 0$; (iii) $\Delta w \leq 0$ in Ω. Assume that there exists a closed ball B with center y and radius R such that $B \cap \bar{\Omega} = \{x^0\}$. Prove that the function

$$w(x) = \frac{1}{R^{n-2}} - \frac{1}{|x - y|^{n-2}} \qquad (n \geq 3) \tag{20.13}$$

is a barrier at x^0.

(39) Under the same assumptions as in Problem 38, prove that if

$$\Delta u = f \quad \text{in } \Omega, \qquad u = 0 \quad \text{on } \partial \Omega,$$

where f is continuous in $\bar{\Omega}$ and u is $C^1(\bar{\Omega})$, then

$$\left| \frac{\partial u(x^0)}{\partial x_i} \right| \leq C \sup_{\Omega} |f| \qquad (1 \leq i \leq n),$$

where C is a constant independent of f.

[*Hint:* Note that $\Delta w \leq -\gamma < 0$, where w is defined in (20.13) and γ a constant. Consider $c(\sup_{\Omega} |f|)w - u$.]

(40) Let

$$Lu \equiv \sum_{i, j=1}^{n} a_{ij}(x) \frac{\partial^2 u}{\partial x_i \partial x_j} + \sum_{i=1}^{n} b_i(x) \frac{\partial u}{\partial x_i} + c(x)u \tag{20.14}$$

be a second-order elliptic operator with real continuous coefficients in a domain Ω. (We normalize L so that (a_{ij}) is positive definite.) Prove: If $u \in C^2(\Omega)$ and if u takes its supremum in Ω at some point $x^0 \in \Omega$, then $Lu(x^0) \leq c(x^0)u(x^0)$.

It follows that if at each point of Ω either $Lu \geq 0$, $c < 0$, or $Lu > 0$, $c \leq 0$, then u cannot take a positive maximum in Ω.

The last assertion can be extended. In fact the following *strong maximum principle* holds: If L is elliptic with real continuous coefficients in a domain Ω, and if $c \leq 0$ in Ω, then a nonconstant function u in $C^2(\Omega)$ satisfying $Lu \geq 0$ in Ω cannot assume positive maximum in Ω. This theorem will be proved in the following three problems.

(41) Let $B = \{x; |x - y_0| < r_0\}$, $C = \{x; |x - x_0| < \rho\}$, where $|x_0 - y_0| = r_0$, $\rho < r_0$. Assume that B and C lie in Ω. Prove that the function

$$h(x) = e^{-\alpha|x - y_0|^2} - e^{-\alpha r_0^2}$$

satisfies $Lh > 0$ in C if α is a sufficiently large constant.

(42) Suppose that $Lu \geq 0$ in Ω and that $M = \sup_\Omega u > 0$. Prove that the following situation is impossible: $u(x_0) = M$ and $u(x_0) > u(x)$ for all $x \in \bar{B} - \{x_0\}$.

[*Hint:* For a sufficiently small $\varepsilon > 0$, $L(u + \varepsilon h) > 0$ in $B \cap C$ and $u(x_0) + \varepsilon h(x_0) > u(x) + \varepsilon h(x)$ for all $x \in \partial(B \cap C) - \{x_0\}$. Use Problem 40 to conclude that $\partial(u + \varepsilon h)/\partial v \leq 0$ at x_0.]

(43) Prove the strong maximum principle.

[*Hint:* If the assertion is false, the set Σ where u takes its positive maximum M is nonempty, closed, and $\Sigma \neq \Omega$. Show that there exists a point $z_0 \notin \Sigma$ and a ball $B = \{x; |x - z_0| < \varepsilon\}$ lying in $\Omega - \Sigma$ such that $\bar{B}$ intersects Σ.]

(44) Use the maximum principle to prove the following assertions: (a) If $c(x) \leq 0$, then there exists at most one solution to the Dirichlet problem. (Hence, by results of the previous sections, there exists a unique solution to the Dirichlet problem provided $\partial\Omega$, the coefficients of the equation, and the data are sufficiently smooth.) (b) If $c(x) \leq 0$, $Lu = 0$ in Ω, $u \in C^1(\bar{\Omega})$, and $\partial u/\partial v + hu = 0$ on $\partial\Omega$, where v is the outward normal and $h > 0$, then $u \equiv 0$ in Ω.

(45) Use the maximum principle to prove the following: If $c(x) \leq 0$, $Lu = f$ in Ω, $u = 0$ on $\partial\Omega$, then there exists a constant C, independent of f, such that

$$\sup_\Omega |u| \leq C \sup_\Omega |f|.$$

(46) With L as in the assumptions of the strong maximum principle and with Ω, B, x^0 as in Problem 38, prove that the function

$$w(x) = k(R^{-p} - |x - y|^{-p}) \qquad (p, k \text{ sufficiently large constants})$$

is a *strong barrier* of L at x^0—that is, w is continuous in $\bar{\Omega}$, $w > 0$ in $\bar{\Omega} - \{x^0\}$, $w(x^0) = 0$, and $Lw < -1$ in Ω. Generalize the result of Problem 39, replacing Δ by L.

(47) Let A be a linear operator (not necessarily bounded) in a Banach space X. The set $\rho(A)$ of complex numbers λ for which $(\lambda I - A)^{-1}$ is a bounded operator is called the *resolvent set* of A. Its complement $\sigma(A)$ is called the *spectrum* of A. Prove: if $\lambda_0 \in \rho(A)$ and if $|\lambda_0 - \lambda| < \|(\lambda_0 I - A)^{-1}\|^{-1}$, then $\lambda \in \rho(A)$ and

$$\|(\lambda I - A)^{-1}\| \leq \frac{\|(\lambda_0 I - A)^{-1}\|}{1 - |\lambda_0 - \lambda| \|(\lambda_0 I - A)^{-1}\|}. \tag{20.15}$$

This implies, in particular, that $\rho(A)$ is an open set.

[*Hint:* Use Problem 3, Section 18.]

(48) Use the resolvent equation (1.19) of Part II to prove that $(\lambda I - A)^{-1}$ is a complex analytic function in $\rho(A)$—that is, prove that $d(\lambda I - A)^{-1}/d\lambda$ exists in the uniform topology.

(49) Prove that there cannot exist a sequence $\{\lambda_n\}$, $\lambda_n \to \infty$, such that $\lambda_n \in \rho(A)$ and

$$\|(\lambda_n I - A)^{-1}\| \le \frac{\theta}{|\lambda_n|} \qquad (n = 1, 2, \ldots),$$

where θ is a constant <1. (This implies that inequalities of the form (18.11) cannot be improved.)

[*Hint:* By (20.15), if $|\lambda - \lambda_n| < \delta|\lambda_n|$, where $1 < \delta, \theta\delta < 1$, then $\|(\lambda I - A)^{-1}\| \le c/|\lambda_n|$.]

(50) If λ is an eigenvalue of A, we denote by N_j the space of solutions of $(\lambda I - A)^j u = 0$. Either the sequence $\{N_1, N_2, \ldots\}$ is strictly increasing or there is a number $k \ge 1$ such that the finite sequence $\{N_1, \ldots, N_k\}$ is strictly increasing and $N_k = N_{k+1} = N_{k+2} = \cdots$. The dimension of $N \equiv \cup_{j=1}^{\infty} N_j$ is called the *multiplicity* of λ and the elements of N are called *generalized eigenvectors* (corresponding to the eigenvalue λ). Let $1/\lambda_0 \in \rho(A)$ and set $A_{\lambda_0} = A(I - \lambda_0 A)^{-1}$. Prove: If $\lambda\lambda_0 \ne 1$, then λ is an eigenvalue of A if and only if $\lambda/(1 - \lambda\lambda_0)$ is an eigenvalue of A_{λ_0}. Furthermore, the multiplicity of λ as an eigenvalue of A is the same as the multiplicity of $\lambda/(1 - \lambda\lambda_0)$ as an eigenvalue of A_{λ_0}. (In fact, the generalized eigenvectors are the same.)

Part 2 | EVOLUTION EQUATIONS

In this part we shall refer to Section m (and to Theorem $m . n$, and so on) of Part I by writing Section I.m (and Theorem I.$m.n$, and so on).

We shall consider equations of the form

$$\frac{du}{dt} + A(t)u = f(t),$$

where $u(t)$, $f(t)$ are functions from a real interval into a Banach space X, and $A(t)$ is (in general) an unbounded linear operator in X. We call such equations *evolution equations*. We shall also consider some nonlinear equations.

We denote by $B(X)$ the Banach space whose elements T are the bounded linear operators in X with norm

$$\|T\| = \sup_{x \neq 0} \frac{\|Tx\|}{\|x\|}.$$

1 | STRONGLY CONTINUOUS SEMIGROUPS

Let A be a bounded operator in a Banach space X. Then the series

$$\sum_{n=0}^{\infty} \frac{A^n}{n!} t^n$$

is easily seen to converge in the uniform operator topology (that is, in the norm of $B(X)$), for any real number t. Denoting its sum by $\exp(tA)$, one easily verifies that

$$\frac{d}{dt}\exp(tA) = A\exp(tA), \tag{1.1}$$

$$\exp(tA)\exp(sA) = \exp((t+s)A), \tag{1.2}$$

$$\exp(0A) = I \qquad (I = \text{identity operator}). \tag{1.3}$$

We would like to generalize the concept of "exp" to the case where A is not a bounded operator. If X is a Hilbert space and A is self-adjoint, this can be done very easily. Indeed, recall that there exists then a family of projections $\{E_\lambda\}$ for $-\infty < \lambda < \infty$ (a projection P is a bounded self-adjoint operator P satisfying $P^2 = P$) such that

$$E_\lambda E_\mu = E_\mu \qquad \text{if } \lambda \leq \mu;$$

$$\lim_{\lambda \uparrow \mu} E_\lambda x = E_\mu x \qquad \text{for any } x \in X;$$

$$\lim_{\lambda \to -\infty} E_\lambda x = 0, \qquad \lim_{\lambda \to \infty} E_\lambda x = x;$$

$$Ax = \int_{-\infty}^{\infty} \lambda\, dE_\lambda x \qquad \text{for all } x \in D_A,$$

where D_A is the domain of A.

Suppose now that A is negative—that is, $(Ax, x) \leq 0$ for all $x \in D_A$. This is equivalent to the condition that $E_\lambda = I$ for all $\lambda > 0$. Introduce operator $T(t)$ $(0 \leq t < \infty)$ by

$$T(t)x = \int_{-\infty}^{0} e^{\lambda t}\, dE_\lambda x.$$

The reader may easily verify that $T(t)$ is a bounded operator for any $t \geq 0$ and that the equations (1.1)–(1.3) for $\exp(tA)$ hold also for $T(t)$.

DEFINITION. A family $\{T(t)\}$ $(0 \leq t < \infty)$ of bounded linear operators in X is called a *strongly continuous semigroup* if
 (i) $T(s + t) = T(s)T(t)$, for all $s, t \geq 0$;
 (ii) $T(0) = I$;
 (iii) For each $x \in X$, $T(t)x$ is continuous in t on $[0, \infty)$.
 If, in addition, the map $t \to T(t)$ is continuous in the uniform operator topology, then we say that the family $\{T(t)\}$ is a *uniformly continuous semigroup*. In that case there exists (see, for instance, Dunford-Schwartz [1]) a bounded operator A such that $T(t) = \exp(tA)$.

DEFINITION.   For $h > 0$, introduce

$$A_h = \frac{T(h)x - x}{h} \qquad (x \in X),$$

and denote by D_A the set of all $x \in X$ for which $\lim_{h \to 0} A_n x$ exists. Define the operator A with domain D_A by

$$Ax = \lim_{h \to 0} A_h x.$$

We call A the *infinitesimal generator* of the semigroup $\{T(t)\}$.

Given an operator A, we say that it *generates a strongly continuous semigroup* $\{T(t)\}$ if A coincides with the infinitesimal generator of $\{T(t)\}$.

Examples.   If A is a bounded operator, then it generates exp (tA). If A is the negative self-adjoint operator considered above, then it is the infinitesimal generator of the operators $T(t)$ introduced above. It is well known (see, for instance, Dunford-Schwartz [1]) that a strongly continuous semigroup $\{T(t)\}$ is uniformly continuous if and only if it has a bounded infinitesimal operator.

Denote by $\rho(A)$ the resolvent set of an operator A and by $\sigma(A)$ the spectrum of A (that is, the complement of $\rho(A)$ in the complex plane). Set $R(\lambda; A) = (\lambda I - A)^{-1}$ (the resolvent of A) for $\lambda \in \rho(A)$.

We shall need the following lemma.

LEMMA 1.1.   Let $\{T(t)\}$ *be a strongly continuous semigroup generated by* B. *If* $x \in D_B$, *then* $T(t)x \in D_B$ *for* $0 \leq t < \infty$ *and* $BT(t)x = T(t)Bx$.

Proof.   If $x \in D_B$, then from $T(t)B_h x = B_h T(t)x$ we get

$$\lim_{h \to 0} B_h T(t)x = \lim_{h \to 0} T(t)B_h x = T(t)\left(\lim_{h \to 0} B_h x\right) = T(t)Bx.$$

Hence $T(t)x \in D_B$ and $B(T(t)x) = T(t)Bx$.

For a continuous function $v(s)$, on a bounded closed interval $[a, b]$, with values in X, the integral $\int_a^b v(s) \, ds$ is defined (and the integral exists) as a limit of sums $\sum v(\hat{s}_i)(s_{i+1} - s_i)$, where $s_i < \hat{s}_i < s_{i+1}$ and $\max_i (s_{i+1} - s_i)$ goes to zero. The existence of this limit follows by using the fact that a continuous function $v(s)$ in a bounded closed interval is necessarily uniformly continuous. The standard rules for integrals of real- or complex-valued functions are still valid. Their verification can be achieved either directly or by applying any bounded linear functional to both sides of the equation under consideration and then using the rules for integrals of real- or complex-valued functions. If $v(s)$ is not continuous at b then the (improper) integral $\int_a^b v(s)\,ds$ is defined as $\lim_{\varepsilon \searrow 0} \int_a^{b-\varepsilon} v(s)\,ds$. If $b = \infty$, the (improper) integral $\int_a^\infty v(s)\,ds$ is defined as $\lim_{c \to \infty} \int_a^c v(s)\,ds$.

We shall also need another simple lemma, the proof of which we leave for the reader.

LEMMA 1.2. *Let* B *be a closed linear operator in* X *and let* u(t) *be a continuous function from* t ∈ [0, T) *(for some* T ≤ ∞) *into* X *such that* u(t) ∈ D$_B$ *and* Bu(t) *is continuous for* 0 ≤ t < T *and such that the improper integrals* $\int_0^T$ u(t) dt, $\int_0^T$ Bu (t) *exist. Then* $\int_0^T$ u(t) dt *belongs to* D$_B$ *and*

$$B \int_0^T u(t)\,dt = \int_0^T Bu(t)\,dt.$$

We shall now establish the property (1.1) for strongly continuous semigroups.

THEOREM 1.1. *Let* {T(t)} *be a strongly continuous semigroup with the infinitesimal generator* A. *Then, for any* x ∈ D$_A$,

$$\frac{d}{dt} T(t)x = AT(t)x = T(t)Ax. \tag{1.4}$$

Proof. By Lemma 1.1, $T(t)x \in D_A$ and $AT(t)x = T(t)Ax$. If $t > 0$, $h > 0$, then

$$\lim_{h \to 0} \left\{ \frac{T(t)x - T(t-h)x}{h} - T(t)Ax \right\}$$

$$= \lim_{h \to 0} T(t-h)(A_h x - Ax) + \lim_{h \to 0} [T(t-h) - T(t)]Ax = 0,$$

as easily verified. Also

$$\lim_{h \to 0} \left\{ \frac{T(t+h) - T(t)x}{h} - T(t)Ax \right\} = \lim_{h \to 0} T(t)[A_h x - Ax] = 0.$$

From these two facts it follows that $dT(t)x/dt$ exists and equals $T(t)Ax$.

A family $\{T(t)\}$ of bounded operators, defined for $-\infty < t < \infty$, is called a *strongly continuous group* if (i) $T(s + t) = T(s)T(t)$ for $-\infty < t, s < \infty$; (ii) $T(0) = I$; and (iii) for each $x \in X$, $T(t)x$ is continuous in t for $-\infty < t < \infty$. The concept of an infinitesimal operator of $\{T(t)\}$ is defined in the obvious manner (by considering the limit of $(T(h)x - x)/h$ as $h \to 0$, $h \lessgtr 0$).

PROBLEMS. (1) Let A be a self-adjoint operator in a complex Hilbert space. Construct a strongly continuous group having the infinitesimal generator iA.

(2) Let $\{T(t)\}$ be a strongly continuous semigroup. Prove that the set D_A of points x for which $Ax \equiv \lim_{h \to 0}(T(h)x - x)/h$ exists is dense in X. [*Hint:* Show that $\int_0^t T(s)x \, ds$ is in D_A for any $x \in X$.]

(3) Let $\{T(t)\}$, D_A be as in Problem 2. Prove that for any $x \in D_A$, $[T(t) - T(s)]x = \int_s^t T(\sigma)Ax \, d\sigma$.

(4) Let $\{T(t)\}$ be a strongly continuous semigroup. Prove that the infinitesimal generator A is a closed operator. [*Hint:* Show that if $x_n \in D_A$, $x_n \to x_0$, $Ax_n \to y_0$, then

$$T(t)x_0 - x_0 = \int_0^t T(s)y_0 \, ds.]$$

We now state the fundamental theorem of Hille–Yosida–Phillips:

THEOREM 1.2. *A necessary and sufficient condition that a closed linear operator* A *with dense domain* D_A *be the infinitesimal generator of a strongly continuous semigroup is that there exist real numbers* M *and* ω *such that for every real* $\lambda > \omega$, λ *is in* $\rho(A)$ *and*

$$\|R(\lambda; A)^n\| \le \frac{M}{(\lambda - \omega)^n} \qquad (n = 1, 2, \ldots). \tag{1.5}$$

Proof. We first prove the sufficiency. We introduce bounded operators $B_\lambda = -\lambda[I - \lambda R(\lambda; A)]$ for $\lambda > \omega$. One easily verifies that

$$\exp(tB_\lambda) = e^{-\lambda t} \sum_{n=0}^{\infty} \frac{(\lambda^2 t)^n}{n!} (R(\lambda; A))^n. \tag{1.6}$$

It follows that

$$\|\exp(tB_\lambda)\| \le M e^{-\lambda t} \sum_{n=0}^{\infty} \frac{(\lambda^2 t)^n}{n!(\lambda - \omega)^n}$$

$$= M \exp[t\omega\lambda(\lambda - \omega)^{-1}].$$

Hence, for any $\omega_1 > \omega$,

$$\|\exp(tB_\lambda)\| < M e^{t\omega_1} \tag{1.7}$$

if λ is sufficiently large.

We next show that

$$\lim_{\lambda \to \infty} B_\lambda x = Ax \qquad \text{for any } x \in D_A. \tag{1.8}$$

Indeed, if $x \in D_A$, then

$$\|\lambda R(\lambda; A)x - x\| = \|R(\lambda; A)Ax\| \leq M\|Ax\|(\lambda - \omega)^{-1} \to 0 \qquad \text{if } \lambda \to \infty.$$

Also, $\|\lambda R(\lambda; A)\| \leq M\lambda(\lambda - \omega)^{-1} < 2M$ for large λ. Combining these two facts and recalling that D_A is dense in X, it follows (by a standard theorem) that $\lambda R(\lambda; A)x \to x$ as $\lambda \to \infty$, for any $x \in X$. Since, finally, $B_\lambda x = \lambda R(\lambda; A)Ax$, the assertion (1.8) follows.

Set $S_\lambda(t) = \exp(tB_\lambda)$. Using the well-known rule $R_\lambda R_\mu = R_\mu R_\lambda$, we see that $B_\lambda B_\mu = B_\mu B_\lambda$. This implies that $B_\mu S_\lambda(t) = S_\lambda(t)B_\mu$. Hence, if $x \in D_A$,

$$S_\lambda(t)x - S_\mu(t)x = \int_0^t \frac{d}{ds} [S_\mu(t - s)S_\lambda(s)x]\, ds$$

$$= \int_0^t S_\mu(t - s)S_\lambda(s)(B_\lambda - B_\mu)x\, ds; \tag{1.9}$$

here we have used the rule (1.1). Note that the integrands are continuous functions.

From (1.7), (1.9) we get

$$\|S_\lambda(t)x - S_\mu(t)x\| \leq M^2 t e^{t\omega_1} \|B_\lambda x - B_\mu x\| \tag{1.10}$$

for all large λ and μ. Using (1.8), we conclude that $\|S_\lambda(t)x - S_\mu(t)x\| \to 0$ if $\lambda, \mu \to \infty$ and $x \in D_A$. Since, by (1.7), $\|S_\lambda(t) - S_\mu(t)\| \leq 2Me^{t\omega_1}$, we conclude that $\|S_\lambda(t)x - S_\mu(t)x\| \to 0$ as $\lambda, \mu \to \infty$, for any $x \in X$. By a well-known theorem, there exists a bounded operator $T(t)$ such that

$$\lim_{\lambda \to \infty} S_\lambda(t)x = T(t)x \qquad \text{for all } x \in X.$$

Clearly,

$$\|T(t)\| \leq Me^{t\omega_1}. \tag{1.11}$$

From (1.10) we also see that if $x \in D_A$, then $S_\lambda(t)x \to T(t)x$ uniformly in t in bounded intervals of $[0, \infty)$. The same is easily verified for any $x \in X$ (using the uniform boundedness of the $\|S_\lambda(t)\|$ in (λ, t), t in bounded intervals). Since $S_\lambda(t)x$ is continuous in t, it thereby follows that also $T(t)x$ is a continuous function of t.

The semigroup property—that is, $T(s + t) = T(s)T(t)$—follows from the semigroup property of $S_\lambda(t)$ applied to any $x \in X$, upon taking $\lambda \to \infty$. Since also $T(0) = \lim_{\lambda \to \infty} S_\lambda(0) = I$, $\{T(t)\}$ is a strongly continuous semigroup. It remains to show that its infinitesimal generator, denoted by B, coincides with A.

We have, for $x \in X$,

$$S_\lambda(t)x - x = \int_0^t \frac{d}{ds} S_\lambda(s)x \, ds = \int_0^t S_\lambda(s)B_\lambda x \, ds, \qquad (1.12)$$

and, for $x \in D_A$,

$$\|S_\lambda(t)B_\lambda x - T(t)Ax\| = \|S_\lambda(t)\| \, \|B_\lambda x - Ax\| + \|(S_\lambda(t) - T(t))Ax\|.$$

Thus, as $\lambda \to \infty$, the integrand on the right of (1.12) converges to $T(s)Ax$ uniformly in s, provided $x \in D_A$. We conclude that

$$T(t)x - x = \int_0^t T(s)Ax \, ds \qquad \text{if } x \in D_A.$$

Consequently,

$$Bx = \lim_{t \to 0} \frac{1}{t} \int_0^t T(s)Ax \, ds = Ax.$$

This shows that $D_B \supseteq D_A$ and $Bx = Ax$ on D_A. It remains to show that $D_B = D_A$. If we prove that

$$\lambda \in \rho(B) \qquad \text{for all real and large } \lambda, \qquad (1.13)$$

then, the set $\rho(A) \cap \rho(B)$ contains a point λ. With this λ we have $(\lambda I - B)D_A = (\lambda I - A)D_A = X$ and $(\lambda I - B)D_B = X$. Since $\lambda I - B$ is one-to-one and $D_B \supseteq D_A$, we conclude that $D_B = D_A$. Thus it remains to prove (1.13).

Consider the integral

$$R(\lambda)x \equiv \int_0^\infty e^{-\lambda t}T(t)x \, dt. \qquad (1.14)$$

In view of (1.11), the integral is absolutely convergent if $x \in X$ and $\lambda > \omega_1$. Furthermore, $R(\lambda)$ is a bounded operator. Now,

$$B_h R(\lambda)x = \frac{T(h)R(\lambda)x - R(\lambda)x}{h}$$

$$= \frac{1}{h} \int_0^\infty e^{-\lambda t}T(t + h)x \, dt - \frac{1}{h} \int_0^\infty e^{-\lambda t}T(t)x \, dt$$

$$= \frac{e^{\lambda h} - 1}{h} \int_0^\infty e^{-\lambda t}T(t)x \, dt - \frac{e^{\lambda h}}{h} \int_0^h e^{-\lambda t}T(t)x \, dt \to \lambda R(\lambda)x - x$$

as $h \to 0$. Thus $R(\lambda)x \subset D_A$ and

$$(\lambda I - B)R(\lambda)x = x \qquad (x \in X). \qquad (1.15)$$

Using Lemmas 1.1 and 1.2 and Problem 4, we find that if $x \in D_B$, then $R(\lambda)x \in D_B$ and

$$B(R(\lambda)x) = B \int_0^\infty e^{-\lambda t} T(t)x \, dt = \int_0^\infty e^{-\lambda t} T(t)Bx \, dt = R(\lambda)(Bx).$$

From this and (1.15) we get

$$R(\lambda)(\lambda I - B)x = x \qquad (x \in D_B). \tag{1.16}$$

From (1.15), (1.16) it follows that $(\lambda I - B)^{-1}$ exists and is the bounded operator $R(\lambda)$—that is, $R(\lambda) = R(\lambda; B)$. This completes the proof of (1.13).

We shall now prove the necessity part of Theorem 1.2. The semigroup property of $\{T(t)\}$ implies that the function $g(t) = \log \|T(t)\|$ is subadditive—that is, $g(t + s) \leq g(t) + g(s)$.

Take any t_0, $0 < t_0 < \infty$. For any $t \geq 0$, write $t = nt_0 + s$, where $n = n(t)$ is an integer and $0 \leq s < t_0$. The subadditivity of g implies

$$\frac{g(t)}{t} \leq \frac{ng(t_0)}{t} + \frac{g(s)}{t} \to \frac{g(t_0)}{t_0} \qquad \text{as } t \to \infty.$$

Hence,

$$\|T(t)\| \leq M e^{\delta t} \qquad \text{for all } t \geq 0, \tag{1.17}$$

where δ is any constant larger than $\inf_{t_0 > 0} g(t_0)/t_0$, and M is a constant (depending on δ).

Now form the integral

$$R(\lambda)x = \int_0^\infty e^{-\lambda t} T(t)x \, dt \qquad (x \in X)$$

for $\lambda > \delta$. Since $\{T(t)\}$ is a semigroup generated by A, we can apply the analysis following (1.14) with B replaced by A and ω_1 replaced by δ. We conclude that if $\lambda > \delta$, then $\lambda \in \rho(A)$ and

$$R(\lambda; A)x = \int_0^\infty e^{-\lambda t} T(t)x \, dt. \tag{1.18}$$

We now use the resolvent formula

$$R(\lambda; A) - R(\mu; A) = (\mu - \lambda)R(\lambda; A)R(\mu; A), \tag{1.19}$$

which is proved by multiplying the right and left sides of the equations

$$(\mu I - A)(\lambda I - A)(R(\lambda; A) - R(\mu; A)) = (\mu I - A) - (\lambda I - A) = (\mu - \lambda)I$$

by $R(\lambda; A)R(\mu; A)$.

Dividing both sides of (1.19) by $\lambda - \mu$ and taking $\mu \to \lambda$, we get

$$\frac{d}{d\lambda} R(\lambda; A) = -(R(\lambda; A))^2.$$

We can now prove by induction that

$$\frac{d^n}{d\lambda^n} R(\lambda; A) = (-1)^n n!(R(\lambda; A))^{n+1}. \tag{1.20}$$

Differentiating both sides of (1.18) $(n - 1)$ times with respect to λ and using (1.20), we get

$$(R(\lambda; A))^n = \frac{1}{(n-1)!} \int_0^\infty e^{-\lambda t} t^{n-1} T(t)x \, dt. \tag{1.21}$$

The differentiation under the integral sign on the right can be justified by standard arguments (or by first applying any bounded linear functional and then using familiar theorems for integrals of complex-valued integrands).

From (1.21) and (1.17) we get

$$\|(R(\lambda; A))^n\| \le \frac{M}{(n-1)!} \int_0^\infty t^{n-1} e^{-(\lambda - \delta)} \, dt = \frac{M}{(\lambda - \delta)^n}. \tag{1.22}$$

Thus the conditions of the theorem follow with $\omega = \delta$.

Since (1.11) holds for any $\omega_1 > \omega$, it also holds for $\omega_1 = \omega$. Thus:

COROLLARY 1.   *If* (1.5) *holds for all* $\lambda > \omega$, *then*

$$\|T(t)\| \le Me^{t\omega} \qquad (0 \le t < \infty). \tag{1.23}$$

From the proof of the second part of the theorem we get:

COROLLARY 2.   *Let a closed, densely defined linear operator* A *be the infinitesimal generator of a strongly continuous semigroup* {T(t)} *satisfying* (1.23). *Then* (1.5) *holds (with the same constants* M, ω).

In the first part of the proof of Theorem 1.2 we have shown that the resolvent $R(\lambda; A)$ is given by the integral in (1.14). The proof of this fact is valid also for any complex λ with Re $\lambda > \omega$. Hence:

COROLLARY 3.   If (1.5) *holds for all* $\lambda > \omega$, *then the resolvent* $R(\lambda; A)$
exists for all complex λ *with* $\text{Re } \lambda > \omega$, *and it is given by*

$$R(\lambda; A)x = \int_0^\infty e^{-\lambda t} T(t)x \, dt \qquad (\text{Re } \lambda > \omega). \qquad (1.24)$$

Furthermore,

$$\|(R(\lambda; A))^n\| \le \frac{M}{(\text{Re } \lambda - \omega)^n} \qquad (n = 1, 2, \ldots). \qquad (1.25)$$

(1.25) follows from (1.21) and (1.23).

PROBLEMS.   (5) Let $\{T(t)\}$ be a strongly continuous semigroup generated
by A. Prove that for any $x \in X$, $\lambda R(\lambda; A)x \to x$ if $|\lambda| \to \infty$, $|\arg \lambda| \le \theta$ for some
constant $\theta < \pi/2$.

(6) Let $X = L^p(0, \infty)$ for some $1 \le p < \infty$ and let $(T(t)x)(s) = x(t + s)$
for $x \in X$. Show that $\{T(t)\}$ is a strongly continuous semigroup whose infinitesimal generator is $A = d/ds$. Show that $\sigma(A)$ is the half-plane $\text{Re } \lambda \le 0$.

(7) Let A be a closed linear operator in a Banach space X and let $\lambda_0 \in \rho(A)$.
Prove: If $R(\lambda_0; A)$ is completely continuous, then $R(\lambda; A)$ is completely continuous for each $\lambda \in \rho(A)$.

We shall now prove a uniqueness theorem.

THEOREM 1.3.   Let $\{S(t)\}, \{T(t)\}$ *be two strongly continuous semigroups having the same infinitesimal generator* A. *Then* $S(t) = T(t)$ *for all* $t \ge 0$.

Proof.   From Problems 2, 4 we know that A is closed and its domain
D_A is dense in X. As in the second part of the proof of Theorem 1.2 we find
that

$$R(\lambda; A)x = \int_0^\infty e^{-\lambda t} T(t)x \, dt,$$

$$R(\lambda; A)x = \int_0^\infty e^{-\lambda t} S(t)x \, dt$$

for all $x \in X$ and for all λ with $\text{Re } \lambda$ sufficiently large, say $\text{Re } \lambda > \omega_0$. Hence for
any bounded linear functional f in X, the Laplace transforms of the functions
$f(T(t)x)$ and $f(S(t))x$ coincide for $\text{Re } \lambda > \omega_0$. Since these functions are $0(e^{\delta t})$
for some finite δ, they must coincide (by a well-known theorem). Recalling that
f is arbitrary, the assertion of the theorem follows.

2 | ANALYTIC SEMIGROUPS

In what follows we shall often denote by e^{tA} the strongly continuous semi-group generated by A.

Let ϕ be any number satisfying $0 < \phi < \frac{1}{2}\pi$ and let M be any positive constant. A linear operator A in a Banach space X is said to be of type (ϕ, M) if:

(i) A is a closed operator with a domain D_A dense in X.

(ii) The resolvent of A contains the sector $S_\phi = \{\lambda; \lambda \neq 0, \frac{1}{2}\pi - \phi < \arg \lambda < \frac{1}{2}\pi + \phi\}$, and

$$\|R(\lambda; A)\| \leq \frac{M}{|\lambda|} \qquad \text{if } \lambda \in S_\phi. \tag{2.1}$$

Example. Let $A(x, D)$ be a uniformly strongly elliptic operator in a bounded domain Ω with coefficients continuous in $\overline{\Omega}$, and let $\partial\Omega$ be of class C^{2m}. Then the operator $A + kI$ in $L^2(\Omega)$ with domain $H^{2m}(\Omega) \cap H_0^m(\Omega)$, defined in Section I.18, is of type (ϕ, M) for some $k > 0$. The same is true for the operators $A + kI$ in $L^p(\Omega)$ with domain $H^{2m, \, p}(\Omega; \{B_j\})$ (defined in Section I.19) provided $(A, \{B_j\}, \Omega)$ is a regular elliptic boundary value problem satisfying all the conditions of Theorem I.19.4 for each θ in the interval $\frac{1}{2}\pi \leq \theta \leq \frac{3}{2}\pi$.

THEOREM 2.1.   If A *is of type* (ϕ, M), *then* $-A$ *generates a strongly continuous semigroup* $\{T(t)\}$ *having the following additional properties:*

(a) T(t) *can be continued analytically into the sector* $\Delta_\phi = \{t; t \neq 0, |\arg t| < \phi\}$;

(b) AT(t), dT(t)/dt *are bounded operators for each* $t \in \Delta_\phi$ *and*

$$\frac{dT(t)x}{dt} = -AT(t)x \qquad (x \in X); \tag{2.2}$$

(c) *For any* $0 < \varepsilon < \phi$ *there exists a constant* $C = C(\varepsilon)$ *such that*

$$\|T(t)\| \leq C, \quad \|AT(t)\| \leq \frac{C}{|t|} \qquad \text{if } t \in \Delta_{\phi-\varepsilon}; \tag{2.3}$$

(d) *For any* $x \in X$, $0 < \varepsilon < \phi$, $T(t)x \to x$ *if* $t \to 0$, $t \in \Delta_{\phi-\varepsilon}$.

DEFINITION. A strongly continuous semigroup having the additional properties (a)–(c) is called an *analytic semigroup.*

Thus, if A is of type (ϕ, M), then e^{-tA} is an analytic semigroup.

Proof. Let Γ be a smooth curve lying in the resolvent set of $-A$ and consisting of three curves: a segment $\{re^{i\theta_1}, 1 \le r < \infty\}$ with $\frac{1}{2}\pi < \theta_1 < \frac{1}{2}\pi + \phi$, a segment $\{re^{i\theta_2}; 1 \le r < \infty\}$ with $\frac{3}{2}\pi - \phi < \theta_2 < \frac{3}{2}\pi$ and the curve $\{(r, \theta), r = 1, \theta_2 - 2\pi \le \theta \le \theta_1\}$ connecting $e^{i\theta_2}$ to $e^{i\theta_1}$. We orient Γ such that for $\lambda = re^{i\theta_1}$, $d\lambda = ie^{i\theta_1}\, dr$. We first consider the case where $-(\phi - \varepsilon) \le \arg t \le \phi - \varepsilon$. We then restrict θ_1, θ_2 by

$$\tfrac{1}{2}\pi + \phi - \varepsilon < \theta_1 < \tfrac{1}{2}\pi + \phi, \qquad \tfrac{3}{2}\pi - \phi < \theta_2 < \tfrac{3}{2}\pi - \phi + \varepsilon. \tag{2.4}$$

Consider the integral

$$e^{-tA} \equiv \frac{1}{2\pi i} \int_{\Gamma} e^{\lambda t}(\lambda I + A)^{-1}\, d\lambda. \tag{2.5}$$

The integrand is a continuous function from $\lambda \in \Gamma$ into $B(X)$. Hence the integral over $\Gamma_n = \Gamma \cap \{\lambda; |\lambda| < n\}$ is well defined. The integral over Γ is then defined as an improper integral. It is easily verified that the integral exists (in fact it converges absolutely) and is a bounded operator.

We shall prove the semigroup property. Let Γ' be a curve obtained from Γ by translating each point λ of Γ to the right by a fixed small positive distance. We claim:

$$e^{-sA} = \frac{1}{2\pi i} \int_{\Gamma'} e^{\lambda' s}(\lambda' I + A)^{-1}\, d\lambda' \qquad (|\arg s| \le \phi - \varepsilon). \tag{2.6}$$

Indeed, denote the right-hand side of (2.6) by P and let f be any bounded linear functional in X. Consider $f(Px)$ for any $x \in X$. Clearly

$$f(Px) = \frac{1}{2\pi i} \int_{\Gamma'} e^{\lambda' s} f((\lambda' I + A)^{-1} x)\, d\lambda'$$

$$= \frac{1}{2\pi i} \int_{\Gamma} e^{\lambda s} f((\lambda I + A)^{-1} x)\, d\lambda$$

by Cauchy's theorem. Since the right-hand side is equal to $f(Qx)$, where Q is the right-hand side of (2.5) with $t = s$, (2.6) follows.

From (2.5), (2.6), and Fubini's theorem, we get

$$e^{-tA}e^{-sA} = \frac{1}{(2\pi i)^2} \int_{\Gamma} \int_{\Gamma'} e^{\lambda t + \lambda' s}(\lambda I + A)^{-1}(\lambda' I + A)^{-1}\, d\lambda\, d\lambda'.$$

Since $\lambda \neq \lambda'$, we can use the resolvent equation (1.19) and get

$$e^{-tA}e^{-sA} = \frac{1}{(2\pi i)^2} \int_\Gamma \int_{\Gamma'} e^{\lambda t + \lambda' s} \frac{1}{\lambda' - \lambda} [(\lambda I + A)^{-1} - (\lambda' I + A)^{-1}] \, d\lambda \, d\lambda'.$$

Since Γ lies to the left of Γ',

$$\int_\Gamma e^{\lambda t} \frac{d\lambda}{\lambda' - \lambda} = 0, \qquad \int_{\Gamma'} e^{\lambda' s} \frac{d\lambda'}{\lambda' - \lambda} = 2\pi i e^{\lambda s}.$$

Using Fubini's theorem, we find that

$$e^{-tA}e^{-sA} = \frac{1}{2\pi i} \int_\Gamma e^{\lambda(t+s)}(\lambda I + A)^{-1} \, d\lambda = e^{-(t+s)A}.$$

It is clear that $e^{-tA}x$ is continuous in t for $|\arg t| \leq \phi - \varepsilon$. Thus, to prove that e^{-tA} $(t > 0)$ is a strongly continuous semigroup it remains to show that

$$e^{-tA}x \to x \qquad \text{if } t > 0, \, t \to 0 \qquad (x \in X). \tag{2.7}$$

It will be convenient first to prove (a)–(c).

It is clear that

$$\left\| \left(\frac{\lambda}{|t|} I + A \right)^{-1} \right\| \leq \frac{C}{|\lambda/t|} \leq \frac{C|t|}{|\lambda|} \qquad \text{if } \quad t \neq 0, \quad \lambda \in \Gamma, \tag{2.8}$$

where C is a generic constant depending on ε. We substitute $\lambda = \lambda'/|t|$ in (2.5) and denote by Γ' the new contour (that is, $\Gamma' = |t|\Gamma$). Using Cauchy's theorem, we find that the contour Γ' can be modified into the contour Γ. Thus,

$$e^{-tA} = \frac{1}{2\pi i} \int_\Gamma e^{\lambda' \gamma} \left(\frac{\lambda'}{|t|} I + A \right)^{-1} \frac{d\lambda'}{|t|} \qquad (\gamma = \arg t). \tag{2.9}$$

Using (2.8), we find that

$$\|e^{-tA}\| \leq C \int_\Gamma |e^{\lambda' \gamma}| \frac{|d\lambda'|}{|\lambda'|} \leq C.$$

Since A is closed, we have

$$A e^{-tA} = \frac{1}{2\pi i} \int_\Gamma e^{\lambda' \gamma} A \left(\frac{\lambda'}{|t|} I + A \right)^{-1} \frac{d\lambda'}{|t|},$$

provided the latter integral is convergent. Writing

$$A = \left(A + \frac{\lambda'}{|t|} I\right) - \frac{\lambda'}{|t|} I$$

for the first A in the integrand, we easily derive

$$\|Ae^{-tA}\| \leq \frac{C}{|t|}. \tag{2.10}$$

This completes the proof of (c).

Next, from (2.5) it easily follows that the complex t-derivative of e^{-tA} exists and equals

$$\frac{d}{dt} e^{-tA} = \frac{1}{2\pi i} \int_\Gamma \lambda e^{\lambda t} (\lambda I + A)^{-1} \, d\lambda.$$

Since $\lambda(\lambda I + A)^{-1} = I - A(\lambda I + A)^{-1}$ and

$$\int_\Gamma e^{\lambda t} \, d\lambda = 0,$$

we obtain, after using the fact that A is closed,

$$\frac{d}{dt} e^{-tA}x = -A \frac{1}{2\pi i} \int_\Gamma e^{\lambda t} (\lambda I + A)^{-1} x \, d\lambda = -Ae^{-tA}x$$

for any $x \in X$. Similarly, if $x \in D_A$,

$$\frac{d}{dt} e^{-tA}x = -\frac{1}{2\pi i} \int_\Gamma e^{\lambda t} (\lambda I + A)^{-1} Ax \, d\lambda = -e^{-tA}Ax.$$

We have thus completed the proof of (a)–(c) for t with $|\arg t| \leq \phi - \varepsilon$. We proceed to prove (d) (which, of course, implies (2.7)). If $x \in D_A$, then, by Cauchy's theorem and Cauchy's formula, if $t \in \Delta_{\phi-\varepsilon}$, $t \to 0$,

$$e^{-tA}x - x = \frac{1}{2\pi i} \int_\Gamma e^{\lambda t} [(\lambda I + A)^{-1} x - \lambda^{-1} x] \, d\lambda$$

$$= -\frac{1}{2\pi i} \int_\Gamma e^{\lambda t} (\lambda I + A)^{-1} Ax \, \frac{d\lambda}{\lambda}$$

$$\to -\frac{1}{2\pi i} \int_\Gamma (\lambda I + A)^{-1} Ax \, \frac{d\lambda}{\lambda} = 0.$$

Since $\|e^{-tA}\| \leq C$ if $|\arg t| \leq \phi - \varepsilon$, (d) follows for any $x \in X$.

So far we have dealt with the case $|\arg t| \leq \phi - \varepsilon$. The definition of e^{-tA} is independent of θ_1, θ_2 (restricted by (2.4)), as follows by using Cauchy's theorem. Hence, if $|\arg t| \leq \phi - \varepsilon_0$ for some $0 < \varepsilon_0 < \varepsilon$ and if we define e^{-tA} as before with the analogous restrictions on θ_1, θ_2, then the two definitions coincide for $|\arg t| \leq \phi - \varepsilon$. This completes the proof that e^{-tA} is a strongly continuous semigroup satisfying the properties (a)–(d).

It remains to prove that e^{-tA} is generated by $-A$. We have, for $x \in D_A$,

$$\frac{e^{-tA}x - x}{t} = -\frac{1}{t}\left(\int_0^t e^{-\tau A}\,d\tau\right)Ax \to -Ax.$$

Hence the infinitesimal generator B of e^{-tA} satisfies: $D_B \supset D_A$ and $B = -A$ on D_A. The proof that $D_B = D_A$ is the same as in the corresponding case of Theorem 1.2 (in the proof of sufficiency).

PROBLEMS.   (1) If the condition (2.1) is replaced by the stronger condition:

$$\|R(\lambda; A)\| \leq \frac{M}{|\lambda| + \omega} \qquad (\lambda \in S_\phi) \tag{2.11}$$

for some $\omega > 0$, then instead of (2.3) one can derive the stronger inequalities:

$$\|T(t)\| \leq Ce^{-\delta\,\mathrm{Re}\,(t)}, \qquad \|AT(t)\| \leq \frac{C}{|t|}e^{-\delta\,\mathrm{Re}\,(t)}, \tag{2.12}$$

where δ is some positive number.

(2) Under the conditions of Theorem 2.1, prove that if A^{-1} is bounded then for any $\mu > 0$ and integer $k \geq 1$,

$$(\mu I + A)^{-k} = \frac{1}{2\pi i}\int_\Gamma (\mu - \lambda)^{-k}(\lambda I + A)^{-1}\,d\lambda.$$

(3) Let A be any closed operator. Verify that

$$R(\lambda; A) = \frac{1}{\lambda} + \frac{A}{\lambda^2} + \cdots + \frac{A^{m+1}}{\lambda^{m+2}}R(\lambda; A) \qquad (m \geq 0).$$

THEOREM 2.2.   Let A be of type (ϕ, M). For any integer $m \geq 1$, $e^{-tA}x$ belongs to the domain of A^m for any $x \in X$, $t > 0$, and

$$\|A^m e^{-tA}\| \leq Ct^{-m} \qquad (t > 0) \tag{2.13}$$

where C is a constant depending only on A, m.

Proof. We can write

$$e^{-tA} = e^{-(t/m)A}e^{-(t/m)A} \cdots e^{-(t/m)A}.$$

From this it easily follows that, for any $x \in X$, $e^{-tA}x$ belongs to the domain of A^m and

$$\|A^m e^{-tA}x\| \leq \|Ae^{-(t/m)A}\|^m \|x\| \leq \left(\frac{C}{t}\right)^m \|x\|,$$

—that is, (2.13) holds.

We conclude this section with a theorem that is a converse to Theorem 2.1.

THEOREM 2.3.   Let $\{T(t)\}$ be a strongly continuous semigroup having the infinitesimal generator A, and assume that AT(t) is a bounded operator for all $t > 0$, and that

$$\|T(t)\| \leq C, \qquad \|AT(t)\| \leq \frac{C}{t}. \tag{2.14}$$

Then, for any $\delta > 0$, $-(A - \delta I)$ is of type (ϕ, M).

Proof. By the results of Section 1, (1.25) must hold with $\omega = 0$. If we prove that also

$$\limsup_{\lambda \to \infty} |\lambda| \|R(1 + i\lambda; A)\| < \infty \qquad (\lambda \text{ real}), \tag{2.15}$$

then the assertion of the theorem will follow.

Writing

$$\frac{dT(t)}{dt} = AT(t) = T(t - s)AT(s)$$

and noting that $AT(s)$ is a bounded operator, we see that $d^2T(t)/dt^2$ exists and is equal to

$$\frac{d}{dt}(T(t - s)AT(s)) = \frac{dT(t - s)}{dt}AT(s) = AT(t - s)AT(s) = \left(AT\left(\frac{t}{2}\right)\right)^2.$$

Similarly one proves by induction that $d^n T(t)/dt^n$ exists and

$$\frac{d^n T(t)}{dt^n} = \left(AT\left(\frac{t}{n}\right) \right)^n. \tag{2.16}$$

In fact, assume the assertion to hold for n. Then from $dT(t)/dt = T(t - s)AT(s)$ it follows that $d^{n+1} T(t)/dt^{n+1}$ exists and is equal to

$$\frac{d^n T(t - s)}{dt^n} AT(s) = \left(AT\left(\frac{t - s}{n}\right) \right)^n AT(s).$$

Now take $s = t/(n + 1)$. Next, by (2.14),

$$t\|AT(t)\| \le C < \infty.$$

Hence, if we apply (2.16) we get

$$\left(\frac{t}{n}\right)^n \left\| \frac{d^n T(t)}{dt^n} \right\| \le C^n.$$

It follows that, for fixed $t > 0$, the series

$$W_t(z) \equiv T(t) + \sum_{n=1}^{\infty} \frac{1}{n!} (z - t)^n T^{(n)}(t)$$

is absolutely convergent for all complex z satisfying $|z - t| < t/eC$. From Taylor's formula we also deduce that $T(s)$ for $|s - t| < t/eC$, coincides with $W_t(s)$. This implies that the functions $W_t(z)$ define a unique analytic continuation $T(z)$ of $T(t)$ into the sector $|z - t| < t/eC, 0 < t < \infty$. Further, if $|z - t| \le t/2eC$,

$$\|T(z)\| \le c \qquad (c \text{ constant}). \tag{2.17}$$

We now recall the relation

$$R(1 + i\lambda; A) = \int_0^{\infty} e^{-(1 + i\lambda)t} T(t) \, dt = \int_0^{\infty} e^{-i\lambda t} e^{-t} T(t) \, dt.$$

Consider first the case where $\lambda > 0$. The path $t \ge 0$ may be deformed into the path $te^{-i\phi}, t \ge 0$ for some ϕ with $0 < \tan \phi < 1/2eC$. Indeed, this follows by using Cauchy's theorem and the fact that for $t = \rho e^{-i\psi} (\rho > 0, 0 \le \psi \le \phi)$ the integrand is bounded by const. $e^{-\alpha \rho}$ (α positive constant).

We get

$$R(1 + i\lambda; A) = \int_0^\infty \exp\{-i\lambda t e^{-i\phi}\} \exp\{-t e^{-i\phi}\} T(t e^{-i\phi}) \, dt.$$

Using (2.17), the assertion (2.15) follows.

For $\lambda < 0$ the proof is similar; ϕ is now replaced by $-\phi$.

PROBLEMS. (4) If A generates a strongly continuous semigroup, then for any integer $m > 0$, the domain of A^m is dense in X.

(5) Show that if in Theorem 2.3 one replaces (2.14) by (2.12), then $-A$ is of type (ϕ, M).

3 | FUNDAMENTAL SOLUTIONS AND THE CAUCHY PROBLEMS

In this section we begin to consider evolution equations

$$\frac{du}{dt} + A(t)u = 0 \qquad (0 < t \le t_0) \tag{3.1}$$

in a Banach space X with $A(t)$ being, in general, a nonconstant operator. We shall need the following assumptions:

(B_1) The domain D_A of $A(t)$ ($t \in [0, t_0]$) is dense in X and is independent of t, and $A(t)$ is a closed operator.

(B_2) For each $t \in [0, t_0]$ the resolvent $R(\lambda; A(t))$ of $A(t)$ exists for all λ with Re $\lambda \le 0$ and

$$\|R(\lambda; A(t))\| \le \frac{C}{|\lambda| + 1} \qquad (\text{Re } \lambda \le 0). \tag{3.2}$$

(B_3) For any t, s, τ in $[0, t_0]$,

$$\|[A(t) - A(\tau)]A^{-1}(s)\| \le C|t - \tau|^\alpha \qquad (0 < \alpha < 1). \tag{3.3}$$

The constants C, α are independent of t, s, τ.

Note that (B_2) implies that, for each $\sigma \in [0, t_0]$, $-A(\sigma)$ generates an analytic semigroup $\{T(t)\}$.

Introducing the notation

$$e^{-tA(\sigma)} = \exp\{-tA(\sigma)\} = T(t),$$

we conclude (by Section 2) that $\{e^{-tA(\sigma)}\}$ is a strongly continuous semigroup, $A(\sigma)e^{-tA(\sigma)}$ is a bounded operator for $t > 0$, and

$$\|e^{-tA(\sigma)}\| \leq C, \tag{3.4}$$

$$\|A(\sigma)e^{-tA(\sigma)}\| \leq \frac{C}{t} \tag{3.5}$$

for all $t > 0$. The constant C is independent of t, σ.

DEFINITION. An operator-valued function $U(t, \tau)$ (with values in $B(X)$) defined and strongly continuous in t, τ for $0 \leq \tau \leq t \leq t_0$ is called a *fundamental solution* of (3.1) if (i) the derivative $\partial U(t, \tau)/\partial t$ exists in the strong topology and belongs to $B(X)$ for $0 \leq \tau < t \leq t_0$, and it is also strongly continuous in t $(\tau < t \leq t_0)$; (ii) if the range of $U(t, \tau)$ is in D_A, and (iii) if

$$\frac{\partial U(t, \tau)}{\partial t} + A(t)U(t, \tau) = 0 \qquad (\tau < t \leq t_0), \tag{3.6}$$

$$U(\tau, \tau) = I. \tag{3.7}$$

THEOREM 3.1. Let the assumptions (B_1)–(B_3) hold. Then there exists a unique fundamental solution.

Consider the problem of finding a solution u satisfying

$$\begin{cases} \dfrac{du}{dt} + A(t)u = f(t) & (0 < t \leq t_0), \\[2mm] \quad u(0) = u_0, \end{cases} \tag{3.8}$$

where $f(t)$ is a given function with values in X and where u_0 is any element of X. We call such a problem a *Cauchy problem.* The solutions $u(t)$ are always required to be continuous for $0 \leq t \leq t_0$ and continuously differentiable for $0 < t \leq t_0$.

The function $f(t)$ is said to be *uniformly Hölder continuous* (exponent β) in $[0, t_0]$ if $\|f(t) - f(\tau)\| \leq C|t - \tau|^\beta$ for all t, τ in $[0, t_0]$; here C, β are positive constants and $\beta \leq 1$.

THEOREM 3.2. Let the assumptions (B_1)–(B_3) hold. Then for any $u_0 \in X$ and for any $f(t)$ uniformly Hölder continuous (exponent β) in $[0, t_0]$ there exists a unique solution $u(t)$ of the Cauchy problem (3.8). Furthermore, the solution is given by

$$u(t) = U(t, 0)u_0 + \int_0^t U(t, s)f(s)\, ds. \tag{3.9}$$

Note that the integrand in the last integral is a continuous function (with values in X).

The construction of a fundamental solution is given in Sections 4 and 5, and uniqueness is proved in Section 6. Theorem 3.2 is proved in Section 7.

PROBLEM.   (1) The following condition implies (B_3): For any t, τ in $[0, t_0]$,

$$\|[A(t) - A(\tau)]A^{-1}(\tau)\| \leq C|t - \tau|^\alpha \qquad (0 < \alpha < 1),$$

where C, α are constants independent of t, τ.

4 | CONSTRUCTION OF FUNDAMENTAL SOLUTIONS

We shall follow the parametrix method (cf. Section 1.4). For any $x \in X$ the function $U(t, \tau)x - e^{-(t-\tau)A(\tau)}x$ must satisfy the equation

$$\frac{dv}{dt} + A(t)v = -[A(t) - A(\tau)]e^{-(t-\tau)A(\tau)}x. \tag{4.1}$$

Hence, if (3.9) holds with the role of $t = 0$ given to $t = \tau$, and if the expression on the right in (4.1) is uniformly Hölder continuous, then

$$U(t, \tau)x = e^{-(t-\tau)A(\tau)}x + \int_\tau^t U(t, s)[A(\tau) - A(s)]e^{-(s-\tau)A(\tau)}x \, ds. \tag{4.2}$$

The solution of (4.2) can be given formally by a series

$$U(t, \tau) = \sum_{k=0}^\infty U_k(t, \tau), \tag{4.3}$$

where $U_0(t, \tau) = e^{-(t-\tau)A(\tau)}$ and

$$U_{k+1}(t, \tau) = \int_\tau^t U_k(t, s)[A(\tau) - A(s)]e^{-(s-\tau)A(\tau)} \, ds. \tag{4.4}$$

Set

$$\phi_1(t, \tau) = [A(\tau) - A(t)]e^{-(t-\tau)A(\tau)}, \tag{4.5}$$

$$\phi_{k+1}(t, \tau) = \int_\tau^t \phi_k(t, s)\phi_1(s, \tau) \, ds \qquad (k = 1, 2, \ldots). \tag{4.6}$$

Assuming for the moment that the integrals in (4.4), (4.5), (4.6) make sense and that Fubini's theorem can be applied, one easily shows by induction that

$$U_k(t, \tau) = \int_\tau^t e^{-(t-s)A(s)} \phi_k(s, \tau) \, ds \qquad (k = 1, 2, \ldots). \tag{4.7}$$

Setting $\Phi(t, \tau) = \sum_{k=1}^\infty \phi_k(t, \tau)$, we conclude that

$$U(t, \tau) = e^{-(t-\tau)A(\tau)} + \int_\tau^t e^{-(t-s)A(s)} \Phi(s, \tau) \, ds. \tag{4.8}$$

Note, finally, that formally

$$\Phi(t, \tau) = \phi_1(t, \tau) + \int_\tau^t \Phi(t, s) \phi_1(s, \tau) \, ds. \tag{4.9}$$

Our program is first to study the equation (4.9), and then prove that $U(t, \tau)$, as given by (4.8), is a fundamental solution. Before studying equation (4.9), however, we shall derive some properties of $e^{-\tau A(t)}$.

LEMMA 4.1. *The following inequalities hold for all* τ, t, s, ξ, η *in* $[0, t_0]$:

$$\|e^{-\tau A(t)} - e^{-\tau A(s)}\| \le C|t - s|^\alpha, \tag{4.10}$$

$$\|A(\xi)[e^{-\tau A(t)} - e^{-\tau A(s)}]\| \le \frac{C}{\tau} |t - s|^\alpha, \tag{4.11}$$

$$\|A(\xi)[e^{-\tau A(t)} - e^{-\tau A(s)}]A^{-1}(\eta)\| \le C|t - s|^\alpha. \tag{4.12}$$

Proof. We denote by C a generic constant. For any $v \in X$ the function $\phi(\xi) = -e^{-(\tau-\xi)A(t)}e^{-\xi A(s)}v$ is differentiable and $\phi'(\xi) = e^{-(\tau-\xi)A(t)}[A(s) - A(t)]e^{-\xi A(s)}v$. If $v \in D_A$ then we can write

$$\phi'(\xi) = e^{-(\tau-\xi)A(t)}\{[A(s) - A(t)]A^{-1}(s)\}e^{-\xi A(s)}(A(s)v),$$

from which it follows that $\phi'(\xi)$ is a continuous function. The relation

$$\phi(\tau) - \phi(0) = \int_0^\tau \phi'(\xi) \, d\xi,$$

which is then valid, takes the form

$$[e^{-\tau A(t)} - e^{\tau A(s)}]v = \int_0^\tau e^{-(\tau - \xi)A(t)}[A(s) - A(t)]e^{-\xi A(s)} v \, d\xi$$

$$= \left\{ \int_0^\tau e^{-(\tau - \xi)A(t)}[A(s) - A(t)]A^{-1}(s)e^{-\xi A(s)} \, d\xi \right\} A(s)v. \qquad (4.13)$$

Here we have taken $A(s)v$ outside the integral on the right. To justify this, note that the integrand is a continuous function, say $\Gamma(\xi)$, with values in $B(X)$. Now verify that

$$\int_0^\tau \Gamma(\xi)x \, d\xi = \left(\int_0^\tau \Gamma(\xi) \, d\xi \right)x.$$

Using (4.13) we can write, for any $x \in X$,

$$e^{-\tau A(t)}x - e^{-\tau A(s)}x$$

$$= [e^{-(\tau/2)A(t)} - e^{-(\tau/2)A(s)}]e^{-(\tau/2)A(s)}x + e^{-(\tau/2)A(t)}[e^{-(\tau/2)A(t)} - e^{-(\tau/2)A(s)}]x$$

$$= \left\{ \int_0^{\tau/2} e^{-[(\tau/2) - \xi]A(t)}[A(s) - A(t)]A^{-1}(s)e^{-\xi A(s)} \, d\xi \right\} A(s)e^{-(\tau/2)A(s)}x$$

$$+ A(t)e^{-(\tau/2)A(t)} \int_0^{\tau/2} e^{-[(\tau/2) - \xi]A(t)}[A(s) - A(t)]A^{-1}(s)e^{-\xi A(s)}x \, d\xi$$

$$+ e^{-(\tau/2)A(t)}[A(t) - A(s)]A^{-1}(s)e^{-(\tau/2)A(s)}x - e^{-\tau A(t)}[A(t) - A(s)]A^{-1}(s)x.$$

$$(4.14)$$

Using (3.3), (3.4), (3.5), we get

$$\|e^{-\tau A(t)} - e^{-\tau A(s)}\| \le \int_0^{\tau/2} \frac{C}{\tau} |t - s|^\alpha \, d\xi + \frac{C}{\tau} \int_0^{\tau/2} |t - s|^\alpha \, d\xi + C|t - s|^\alpha$$

$$\le C|t - s|^\alpha.$$

This proves (4.10)

To prove (4.11) we shall make use of the inequality

$$\|A(\zeta)A^{-1}(\eta)\| \le C, \qquad (4.15)$$

which follows from (3.3). If we apply $A(\xi)$ to the second and third terms on the right-hand side of (4.14) and use (3.3), (3.4), (3.5), (4.15), and (2.13) (with $m = 2$), then we get the bound $C\|x\| \, |t - s|^\alpha/\tau$ for each of the two terms. Thus it remains to estimate

$$I = A(\xi)[e^{-(\tau/2)A(t)} - e^{-(\tau/2)A(s)}]e^{-(\tau/2)A(s)}x.$$

In view of (4.15), we may take $\xi = s$. It is then easily seen that

$$I = [e^{-(\tau/2)A(t)} - e^{-(\tau/2)A(s)}]A(s)e^{-(\tau/2)A(s)}x + [A(s) - A(t)]e^{-(\tau/2)A(s)}e^{-(\tau/2)A(s)}x$$

$$+ e^{-(\tau/2)A(t)}[A(t) - A(s)]e^{-(\tau/2)A(s)}x$$

$$\equiv I_1 + I_2 + I_3.$$

Using (3.3), (3.4), (3.5), and (4.15), we find that

$$\|I_i\| \leq \frac{C}{\tau}|t - s|^{\alpha}\|x\| \qquad \text{for } i = 2, 3.$$

To estimate I_1, we note, using (4.13), that

$$I_1 = \left\{\int_0^{\tau/2} e^{-[(\tau/2)-\xi]A(t)}[A(s) - A(t)]A^{-1}(s)e^{-\xi A(s)}\,d\xi\right\}A^2(s)e^{-(\tau/2)A(s)}x.$$

Using (3.3), (3.4), (3.5), and (2.13) (with $m = 2$), we find that $\|I_1\| \leq C\|x\|\,|t - s|^{\alpha}/\tau$. We have thus proved that

$$\|I\| \leq \frac{C}{\tau}|t - s|^{\alpha}\|x\|,$$

and the proof of (4.11) is thereby completed.

To prove (4.12), we write (using (4.13))

$$e^{-\tau A(t)}x - e^{-\tau A(s)}x = [e^{-(\tau/2)A(t)} - e^{-(\tau/2)A(s)}]e^{-(\tau/2)A(s)}x$$

$$+ e^{-(\tau/2)A(t)}\left\{\int_0^{\tau/2} e^{-[(\tau/2)-\xi]A(t)}\right.$$

$$\left. \times [A(s) - A(t)]A^{-1}(s)e^{-\xi A(s)}\,d\xi\right\}A(s)x$$

$$\equiv J_1 + J_2,$$

where $x = A^{-1}(\eta)y$, $y \in X$. Using (3.3), (3.4), (3.5), and (4.15), it follows that

$$\|A(\xi)J_2\| \leq C\|y\|\,|t - s|^{\alpha}.$$

As for $A(\xi)J_1$, we can treat it in the same way that we have treated I above. (4.12) thus follows.

From the relation

$$e^{-tA(\tau)}v - e^{-sA(\tau)}v = -\int_s^t A(\tau)e^{-\xi A(\tau)}v\,d\xi \qquad (v \in D_A), \tag{4.16}$$

(3.5), and (4.15), we get

LEMMA 4.2.   For all t, s, τ, ξ *in* $[0, t_0]$,

$$\|[e^{-tA(\tau)} - e^{-sA(\tau)}]A^{-1}(\xi)\| \leq C|t - s|, \tag{4.17}$$

$$\|A(\xi)[e^{-tA(\tau)} - e^{-sA(\tau)}]A^{-1}(\tau)\| \leq \frac{C|t - s|}{\min(t, s)}. \tag{4.18}$$

In proving (4.18) we also make use of Lemma 1.2.

LEMMA 4.3.   The function $A(t)e^{-\tau A(s)}$ *is uniformly continuous in the uniform topology (that is, in the norm of* $B(X)$*) in the variables* (t, τ, s) *where* $0 \leq t \leq t_0, \varepsilon \leq \tau \leq t_0, 0 \leq s \leq t_0$, *and* ε *is any positive number.*

Proof.   If $0 \leq t + \Delta t \leq t_0, \varepsilon \leq \tau + \Delta\tau \leq t_0, 0 \leq s + \Delta s \leq t_0$, then

$$
\begin{aligned}
A(t &+ \Delta t)e^{-(t + \Delta\tau)A(s + \Delta s)} - A(t)e^{-\tau A(s)} \\
&= [A(t + \Delta t) - A(t)]A^{-1}(s + \Delta s)A(s + \Delta s)e^{-(t + \Delta\tau)A(s + \Delta s)} \\
&\quad + A(t)[e^{-[(\tau/2) + \Delta\tau]A(s + \Delta s)} - e^{-(\tau/2)A(s + \Delta s)}]A^{-1}(s + \Delta s) \\
&\quad \times A(s + \Delta s)e^{-(\tau/2)A(s + \Delta s)} \\
&\quad + A(t)[e^{-\tau A(s + \Delta s)} - e^{-\tau A(s)}].
\end{aligned}
$$

We estimate the first term on the right by using (3.3), (3.5); the second term by using (4.18), (3.5), and the last term by using (4.11). We thus find that the norm of the left side is bounded by

$$\frac{C}{\varepsilon}|\Delta t|^\alpha + \frac{C}{\varepsilon}|\Delta\tau| + \frac{C}{\varepsilon}|\Delta s|^\alpha.$$

This completes the proof.

COROLLARY.   The operator-valued functions $[A(\tau) - A(t)]e^{-(t-\tau)A(\tau)}$, $[A(\tau) - A(t)]e^{-(t-\tau)A(t)}$, $e^{-(t-\tau)A(\tau)}$, $e^{-(t-\tau)A(t)}$ *are uniformly continuous in the uniform topology in the variables* (t, τ) *where* $0 \leq \tau \leq t - \varepsilon, \varepsilon \leq t \leq t_0$, *for any* $\varepsilon > 0$.

Indeed, the continuity of the first two functions is obvious. The continuity of the last two functions follows from Lemma 4.3 and the boundedness of $A^{-1}(s)$.

We finally prove:

LEMMA 4.4. *For each* $x \in X$, *the function* $A(t)e^{-\tau A(s)}A^{-1}(\xi)x$ *is continuous in* (t, s, τ, ξ), *where each variable varies in* $[0, t_0]$.

Proof. Write

$$A(t + \Delta t)e^{-(\tau + \Delta \tau)A(s + \Delta s)}A^{-1}(\xi + \Delta \xi) - A(t)e^{-\tau A(s)}A^{-1}(\xi)$$

$$= [A(t + \Delta t) - A(t)]A^{-1}(s + \Delta s)e^{-(\tau + \Delta \tau)A(s + \Delta s)}A(s + \Delta s)A^{-1}(\xi + \Delta \xi)$$

$$+ A(t)e^{-(\tau + \Delta \tau)A(s + \Delta s)}A^{-1}(\xi)[A(\xi) - A(\xi + \Delta \xi)]A^{-1}(\xi + \Delta \xi)$$

$$+ A(t)[e^{-(\tau + \Delta \tau)A(s + \Delta s)} - e^{-(\tau + \Delta \tau)A(s)}]A^{-1}(\xi)$$

$$+ A(t)A^{-1}(s)[e^{-(\tau + \Delta \tau)A(s)} - e^{-\tau A(s)}]A(s)A^{-1}(\xi)$$

$$\equiv J_1 + J_2 + J_3 + J_4.$$

By (3.3), (4,15), $\|J_1\| \leq C|\Delta t|^\alpha$, $\|J_2\| \leq C|\Delta \xi|^\alpha$. By (4.12), $\|J_3\| \leq C|\Delta s|^\alpha$. Finally, using (4.17) we see that if $x \in D_A$, then

$$\lim_{\Delta \tau \to 0} \|e^{-(\tau + \Delta \tau)A(s)}x - e^{-\tau A(s)}x\| = 0. \tag{4.19}$$

Since

$$\|e^{-(\tau + \Delta \tau)A(s)} - e^{-\tau A(s)}\| \leq C,$$

it follows that (4.19) holds for all $x \in X$. From this and (4.15) we conclude that $\|J_4 x\| \to 0$ if $\Delta \tau \to 0$. This completes the proof.

COROLLARY. *The functions* $e^{-(t - \tau)A(\tau)}$, $e^{-(t - \tau)A(t)}$, $[A(\tau) - A(t)]$ $e^{-(t - \tau)A(\tau)}A^{-1}(\tau)$ *are strongly continuous in* (t, τ), *where* $0 \leq \tau \leq t \leq t_0$.

We shall now study the equation (4.9). Recalling (4.5) and using (3.3), (3.5), we get

$$\|\phi_1(t, \tau)\| \leq \frac{C}{|t - \tau|^{1-\alpha}}. \tag{4.20}$$

From the corollary to Lemma 4.3 we also have that $\phi_1(t, \tau)$ is uniformly continuous in (t, τ) in the uniform topology, provided $t - \tau \geq \varepsilon$.

We can now use the classical arguments of successive approximations and show that the Volterra integral equation (4.9) has a unique solution $\Phi(t, \tau)$ $(0 \leq \tau < t \leq t_0)$ that is uniformly continuous in the uniform topology in t, τ provided $0 \leq \tau \leq t - \varepsilon$, $\varepsilon \leq t \leq t_0$ for any $\varepsilon > 0$. Furthermore,

$$\|\Phi(t, \tau)\| \leq \frac{C}{|t - \tau|^{1-\alpha}}. \tag{4.21}$$

An alternate way of constructing $\Phi(t, \tau)$ is to note that formally

$$\Phi(t, \tau) = \sum_{k=1}^{\infty} \phi_k(t, \tau) \tag{4.22}$$

is a solution of (4.9), where the ϕ_k are defined by (4.5), (4.6). Now one verifies by induction that $\phi_k(t, \tau)$ is continuous in (t, τ) in the uniform topology for $t - \tau \geq \varepsilon > 0$ and

$$\|\phi_k(t, \tau)\| \leq C^k \frac{|t - \tau|^{k\alpha - 1}}{\Gamma(k\alpha)}, \tag{4.23}$$

where $\Gamma(m)$ is the gamma function. It then follows that the integral defining $\phi_{k+1}(t, \tau)$ makes sense as an improper integral with integrand which is continuous in proper subintervals. In proving the continuity of $\phi_k(t, \tau)$, we make use of the following elementary lemma, the proof of which is left to the reader:

LEMMA 4.5.   *Let* $\phi(t, \tau), \psi(t, \tau)$ *be functions with values in* B(X) *defined for* $0 \leq \tau < t \leq t_0$ *and (uniformly) continuous in* (t, τ) *for* $t - \tau \geq \varepsilon$ *and for any* $\varepsilon > 0$. *Assume that*

$$\int_{\tau}^{\tau + \delta} \|\phi(t, s)\psi(s, \tau)\| \, ds + \int_{t - \delta}^{t} \|\phi(t, s)\psi(s, \tau)\| \, ds \to 0$$

as $\delta \to 0$ *uniformly with respect to* (t, τ), $t - \tau \geq \varepsilon$, *for any* $\varepsilon > 0$. *Then the function* $\int_{\tau}^{t} \phi(t, s)\psi(s, \tau) \, ds$ *is (uniformly) continuous in* (t, τ) *for* $t - \tau \geq \varepsilon$, *and for any* $\varepsilon > 0$.

Using the inequalities (4.23), one can justify the relation

$$\int_{\tau}^{t} \Phi(t, s)\phi_1(s, \tau) \, ds = \sum_{k=1}^{\infty} \int_{\tau}^{t} \phi_k(t, s)\phi_1(s, \tau) \, ds.$$

Hence Φ is a solution of (4.9). (4.21) is clearly valid. We also have

$$\Phi(t, \tau) = \phi_1(t, \tau) + \int_{\tau}^{t} \phi_1(t, s)\Phi(s, \tau) \, ds. \tag{4.24}$$

In proving (4.24) we use Fubini's theorem. Note that all the integrands in question are continuous functions (with values in $B(X)$) in proper subintervals and, therefore, the proof of the classical Fubini's theorem applies. By going to the limit we obtain the assertion of Fubini's theorem also for the improper integrals under consideration.

LEMMA 4.6. *For any* $0 < \eta \le \alpha,\, 0 \le \tau < t \le t + \Delta t \le t_0$,

$$\|\Phi(t + \Delta t, \tau) - \Phi(t, \tau)\| \le C(\Delta t)^{\alpha - \eta}|t - \tau|^{\eta - 1} \qquad (4.25)$$

where C is a constant depending on η.

Proof. From (4.20),

$$\|\phi_1(t + \Delta t, \tau) - \phi_1(t, \tau)\| \le C|t - \tau|^{\alpha - 1}. \qquad (4.26)$$

Writing

$$\phi_1(t + \Delta t, \tau) - \phi_1(t, \tau) = [A(t) - A(t + \Delta t)]e^{-(t + \Delta t - \tau)A(\tau)} \\ + [A(\tau) - A(t)][e^{-(t + \Delta t - \tau)A(\tau)} - e^{-(t - \tau)A(\tau)}],$$

we note that the first term on the right is bounded in norm by

$$C(\Delta t)^{\alpha}(t - \tau)^{-1}.$$

The second term is bounded in norm by

$$C(t - \tau)^{\alpha - 1}.$$

We can write this term in the form

$$-[A(\tau) - A(t)]A^{-1}(\tau)\left\{\int_0^{\Delta t} e^{-\xi A(\tau)}\, d\xi\right\}A^2(\tau)e^{-(t - \tau)A(\tau)}$$

and thus, using Theorem 2.2, obtain another bound, namely,

$$C\,\Delta t \cdot (t - \tau)^{\alpha - 2}.$$

From the last two bounds we obtain the bound

$$\{C(t - \tau)^{\alpha - 1}\}^{1 - \alpha}\{C\,\Delta t \cdot (t - \tau)^{\alpha - 2}\}^{\alpha} = C(\Delta t)^{\alpha}(t - \tau)^{-1}.$$

Hence

$$\|\phi_1(t + \Delta t, \tau) - \phi_1(t, \tau)\| \le C(\Delta t)^{\alpha}(t - \tau)^{-1}.$$

From this inequality and (4.26) we conclude that

$$\|\phi_1(t + \Delta t, \tau) - \phi_1(t, \tau)\| \le C(\Delta t)^{\alpha - \eta}(t - \tau)^{\eta - 1}. \qquad (4.27)$$

Using (4.24), we can now write

$$\Phi(t + \Delta t, \tau) - \Phi(t, \tau) = [\phi_1(t + \Delta t, \tau) - \phi_1(t, \tau)] + \int_t^{t + \Delta t} \phi_1(t + \Delta t, s)\Phi(s, \tau) \, d\tau$$

$$+ \int_\tau^t [(\phi_1(t + \Delta t, s) - \phi_1(t, s)]\Phi(s, \tau) \, ds. \qquad (4.28)$$

We estimate the norm of the first term on the right by using (4.27), the norm of the second term by using (4.20), (4.21), and the norm of the last term by using (4.27), (4.21). After simple calculations (4.25) follows.

PROBLEMS. (1) Under the assumptions (B_1)–(B_3), prove that

$$\|e^{-\tau A(t)} - e^{-\tau A(s)}\| \leq C|t - s|^\alpha e^{-\delta\tau} \qquad (C > 0, \delta > 0)$$

for all t, s in $[0, t_0]$ and for all τ in $[0, \infty)$.
(*Hint:* Use Problem 2.1 and the proof of (4.10).]
(2) Under the assumptions (B_1)–(B_3), prove that

$$\|A^{-1}(t) - A^{-1}(\tau)\| \leq C|t - \tau|^\alpha.$$

[*Hint:* Use (1.24).]

5 | CONSTRUCTION OF FUNDAMENTAL SOLUTIONS (Continued)

We turn to the equation (4.8). Recalling that $\Phi(t, \tau)$ is (uniformly) continuous in (t, τ) provided $t - \tau \geq \varepsilon > 0$, and using (4.21), one can verify without difficulty that

$$\int_\tau^t e^{-(t-s)A(s)}\Phi(s, \tau) \, ds$$

is (uniformly) continuous (in the norm of $B(X)$) in $(t, \tau), 0 \leq \tau \leq t \leq t_0$.
Since, for any $x \in X$, $e^{-(t-\tau)A(\tau)}x$ is continuous in $(t, \tau), 0 \leq \tau \leq t \leq t_0$, we conclude that $U(t, \tau)x$ is also continuous in (t, τ), for $0 \leq \tau \leq t \leq t_0$. It is also obvious that $U(\tau, \tau) = I$.
Next, for any $x \in X$,

$$\int_\tau^t U(t, s)\phi_1(s, \tau)x \, ds = \int_\tau^t e^{-(t-s)A(s)}\phi_1(s, \tau)x \, ds$$

$$+ \int_\tau^t \left[\int_s^t e^{-(t-\xi)A(\xi)}\Phi(\xi, s) \, d\xi \right] \phi_1(s, \tau)x \, ds.$$

Using Fubini's theorem and (4.9), we get

$$\int_\tau^t U(t,s)[A(\tau) - A(s)]e^{-(s-\tau)A(\tau)}x\,ds = \int_\tau^t e^{-(t-s)A(s)}\Phi(s,\tau)x\,ds.$$

By (4.8), the last term is equal to $U(t,\tau)x - e^{-(t-\tau)A(\tau)}x$. We have thus derived the relation (4.2). This relation will actually not be needed in proving Theorem 3.1.

We proceed now to prove that for any $x \in X$, $\partial U(t,\tau)x/\partial t$ exists and is continuous if $t > \tau$, that $U(t,\tau)x \in D_A$, and that (3.6) holds.

Consider, for $t - \tau \geq \varepsilon > 0, 0 < \rho < \varepsilon$, the expression

$$U_\rho(t,\tau)x = e^{-(t-\tau)A(\tau)}x + \int_\tau^{t-\rho} e^{-(t-s)A(s)}\Phi(s,\tau)x\,ds. \tag{5.1}$$

$U_\rho(t,\tau)x$ is differentiable in t and

$$\frac{\partial U_\rho(t,\tau)x}{\partial t} = -A(\tau)e^{-(t-\tau)A(\tau)}x + e^{-\rho A(t-\rho)}\Phi(t-\rho,\tau)x$$

$$- \int_\tau^{t-\rho} A(s)e^{-(t-s)A(s)}\Phi(s,\tau)x\,ds. \tag{5.2}$$

PROBLEMS. (1) Prove formula (5.2).

(2) Use (5.2) to show that $\partial U_\rho(t,\tau)x/\partial t$ is a uniformly continuous function in (t,τ) for $0 \leq \tau \leq t - \varepsilon, \varepsilon \leq t \leq t_0\,(\varepsilon > 0)$.

If we use (4.24) and Lemma 1.2, we find from (5.2) that

$$\frac{\partial U_\rho(t,\tau)x}{\partial t} = -A(t)e^{-(t-\tau)A(\tau)}x - A(t)\int_\tau^{t-\rho} e^{-(t-s)A(s)}\Phi(s,\tau)x\,ds$$

$$+ \int_{t-\rho}^t \phi_1(t,s)\Phi(s,\tau)x\,ds + e^{-\rho A(t-\rho)}[\Phi(t-\rho,\tau) - \Phi(t,\tau)]x$$

$$+ [e^{-\rho A(t-\rho)} - I]\Phi(t,\tau)x$$

$$\equiv I_1 + I_2 + I_3 + I_4 + I_5.$$

Using (4.20), (4.21), and (4.25) we find that

$$\|I_3\| \leq \frac{C\rho^\alpha}{\varepsilon^{1-\alpha}}, \qquad \|I_4\| \leq \frac{C\rho^{\alpha-\eta}}{\varepsilon^{1-\eta}} \qquad (0 < \eta \leq \alpha),$$

where $t - \tau \geq \varepsilon$, and where C is a generic constant (depending only on η). Since

$e^{-\rho A(t-\rho)} \to I$ strongly, and since $\Phi(t, \tau)x$ is uniformly continuous in (t, τ) for $t - \tau \geq \varepsilon$, one can easily show that

$$\|I_5\| \to 0 \qquad \text{as } \rho \to 0,$$

uniformly with respect to (t, τ), provided $t - \tau \geq \varepsilon$. We shall prove the following:

LEMMA 5.1.   *If* $\rho \to 0$ *then* $A(t) \int_\tau^{t-\rho} e^{-(t-s)A(s)}\Phi(s, \tau)x \, ds$ *converges to a limit uniformly with respect to* (t, τ), *provided* $t - \tau \geq \varepsilon > 0$.

Suppose the lemma is true. Since $A(t)$ is a closed operator, it then follows that $\int_\tau^t e^{-(t-s)A(s)}\Phi(s, \tau)x \, ds$ is in D_A and, as $\rho \to 0$,

$$A(t) \int_\tau^{t-\rho} e^{-(t-s)A(s)}\Phi(s, \tau)x \, ds \to A(t) \int_\tau^t e^{-(t-s)A(s)}\Phi(s, \tau)x \, ds.$$

Together with (4.8) and the previous results we conclude that $U(t, \tau)x \in D_A$ and, as $\rho \to 0$,

$$\frac{\partial U_\rho(t, \tau)x}{\partial t} \to -A(t)U(t, \tau)x \tag{5.3}$$

uniformly in (t, τ), provided $t - \tau \geq \varepsilon > 0$. Since also, as $\rho \to 0$, $U_\rho(t, \tau)x \to U(t, \tau)x$ uniformly in (t, τ), we conclude (by a standard argument) that $\partial U(t, \tau)x/\partial t$ exists and is continuous in (t, τ) $(t - \tau > 0)$, and

$$\frac{\partial U(t, \tau)x}{\partial t} = \lim_{\rho \to 0} \frac{\partial U_\rho(t, \tau)x}{\partial t} = -A(t)U(t, \tau)x.$$

Proof of Lemma 5.1   Write

$$A(t) \int_\tau^{t-\rho} e^{-(t-s)A(s)}\Phi(s, \tau)x \, ds$$

$$= A(t) \int_\tau^{t-\rho} e^{-(t-s)A(s)}[\Phi(s, \tau) - \Phi(t, \tau)]x \, ds + A(t) \int_\tau^{t-\rho} e^{-(t-s)A(s)}\Phi(t, \tau)x \, ds$$

$$\equiv K_1(\rho) + K_2(\rho).$$

PROBLEM.   (3) Use Lemma 4.6 to show that if $\rho, \rho' \to 0$, then $\|K_1(\rho) - K_1(\rho')\| \to 0$ uniformly with respect to (t, τ) $(t - \tau \geq \varepsilon > 0)$.

We shall consider $K_2(\rho)$. Since $\Phi(t, \tau)x$ is uniformly continuous in (t, τ) for $t - \tau \geq \varepsilon > 0$, it suffices to consider the expression

$$H_\rho y \equiv A(t) \int_\tau^{t-\rho} e^{-(t-s)A(s)}y \, ds, \tag{5.4}$$

where $y \in X$, provided one shows that $\|H_\rho\| \leq C$. It is easily seen that if $y \in D_A$, then

$$\|H_\rho y - H_{\rho'} y\| \leq \text{const. } \rho \qquad (\rho' < \rho),$$

where the constant is independent of (t, τ) $(t - \tau \geq \varepsilon > 0)$. Hence if we prove that $\|H_\rho\| \leq C$ $(t - \tau \geq \varepsilon > 0)$ with C independent of t, τ, ρ, then the assertion of the lemma follows.

Note that $H_\rho \equiv H_\rho(t, \tau)$ is a bounded operator if $\rho > 0$. The operator $H(t, \tau)$ defined by (5.4) with $\rho = 0$ is certainly defined for all $y \in D_A$. Using a variant of (4.13), we get (for $y \in D_A$)

$$H(t, \tau)y = A(t) \int_\tau^t e^{-(t-s)A(t)} y \, ds$$

$$+ A(t) \int_\tau^t \left[\int_s^t e^{-(t-\xi)A(t)} [A(t) - A(s)] e^{-(\xi-s)A(s)} y \, d\xi \right] ds.$$

PROBLEM.   (4) Using the last relation, verify that

$$H(t, \tau)y = [I - e^{-(t-\tau)A(t)}] \left[I - \int_\tau^t \phi_1(t, s) \, ds \right] y$$

$$+ A(t) \int_\tau^t e^{-(t-\xi)A(t)} \left[\int_\xi^t \phi_1(t, s) \, ds + \int_\tau^\xi [\phi_1(t, s) - \phi_1(\xi, s)] \, ds \right] d\xi \cdot y$$

$$+ A(t) \int_\tau^t e^{-(t-\xi)A(t)} [A(t) - A(\xi)] A^{-1}(\xi) H(\xi, \tau) y \, d\xi. \tag{5.5}$$

Using (4.20), (4.27), we get from (5.5)

$$\|H(t, \tau)y\| \leq C\|y\| + C\|y\| \int_\tau^t \frac{1}{t-\xi} \left[\int_\xi^t (t - s)^{\alpha-1} \, ds \right.$$

$$\left. + \int_\tau^\xi (t - \xi)^{\alpha-\eta}(\xi - s)^{\eta-1} \, ds \right] d\xi + C \int_\tau^t \frac{1}{t-\xi} (t - \xi)^\alpha \|H(\xi, \tau)y\| \, d\xi$$

$$(0 < \eta < \alpha).$$

Hence,

$$\|H(t, \tau)y\| \leq C\|y\| + C \int_\tau^t (t - \xi)^{\alpha-1} \|H(\xi, \tau)y\| \, d\xi.$$

This leads to

$$\|H(t, \tau)y\| \leq C\|y\|, \tag{5.6}$$

for $y \in D_A$. Note that the constant C occurring in the last inequality is independent of t, τ. Therefore,

$$\|H_\rho(t, \tau)y\| \le \|H(t, \tau)y - H(t, t - \rho)y\| \le C\|y\|$$

for all $y \in D_A$, where C is indendent of ρ. Hence, $\|H_\rho(t, \tau)\| \le C$ if $t - \tau \ge \varepsilon > 0$. This completes the proof of the lemma.

The inequality (5.6) is equivalent to the following:

COROLLARY.   For all t, τ with $0 \le \tau < t \le t_0$,

$$\left\| A(t) \int_\tau^t e^{-(t-s)A(s)} \, ds \right\| \le C. \tag{5.7}$$

6 | UNIQUENESS OF FUNDAMENTAL SOLUTIONS

To complete the proof of Theorem 3.1 it remains to prove uniqueness. This follows from the following lemma:

LEMMA 6.1.   Let the assumptions (B_1)–(B_3) hold. Then for any $u_0 \in X$ and $\tau \in (0, t_0)$ there exists a unique solution of

$$\frac{du}{dt} + A(t)u = 0 \quad (\tau < t \le t_0), \qquad u(\tau) = u_0, \tag{6.1}$$

and it is given by $u(t) = U(t, \tau)u_0$, where $U(t, \tau)$ is any fundamental solution.

Indeed, if there is another fundamental solution, say $\hat{U}(t, \tau)$, then the solution $u(t)$ is given also by $u(t) = \hat{U}(t, \tau)u_0$. It follows that $\hat{U}(t, \tau) = U(t, \tau)$.

The following lemma will be needed in the proof of Lemma 6.1.

LEMMA 6.2.   For any $x \in X$, the function $A(t)U(t, \tau)A^{-1}(\tau)x$ is uniformly continuous in t, τ for $0 \le \tau < t \le t_0$.

Proof.   Set $W(t, \tau) = A(t)U(t, \tau)A^{-1}(\tau)$. The function

$$\psi(s) = e^{-(t-s)A(t)}U(s, \tau)A^{-1}(\tau)x \qquad (\tau < s < t)$$

is continuously differentiable and

$$\psi'(s) = e^{-(t-s)A(t)}[A(t) - A(s)]U(s, \tau)A^{-1}(\tau)x.$$

Integrating this relation and applying $A(t)$ to the resulting equation, we get

$$W(t, \tau)x = A(t)e^{-(t-\tau)A(t)}A^{-1}(\tau)x$$

$$+ \int_{\tau}^{t} A(t)e^{-(t-s)A(t)}[A(t) - A(s)]A^{-1}(s)W(s, \tau)x \, ds. \qquad (6.2)$$

By Lemma 4.4, the first term on the right is uniformly continuous in t, τ for $t > \tau$. Since $W(t, \tau)x$ is obviously uniformly continuous in (t, τ) for $t - \tau \geq \varepsilon > 0$, it remains to show that the second term on the right-hand side of (6.2) converges uniformly to a limit, as $t - \tau \to 0$. Writing this term in the form

$$\int_{\tau}^{t} \Psi(t, s)W(s, \tau)x \, ds, \qquad (6.3)$$

we have

$$\|\Psi(t, s)\| \leq C(t - s)^{\alpha - 1}, \qquad (6.4)$$

where C is a generic constant. Hence,

$$\|W(t, \tau)x\| \leq C\|x\| + C\int_{\tau}^{t}(t - s)^{\alpha - 1}\|W(s, \tau)x\| \, ds.$$

This leads to the bound

$$\|W(t, \tau)x\| \leq C\|x\|. \qquad (6.5)$$

From (6.4), (6.5) it follows that the integral in (6.3) converges to 0 (uniformly in t, τ) if $t - \tau \to 0$. This completes the proof of the lemma.

Incidentally we have also proved (cf. (6.5)):

COROLLARY.   *For all* t, τ *with* $0 \leq \tau < t \leq t_0$,

$$\|A(t)U(t, \tau)A^{-1}(\tau)\| \leq C. \qquad (6.6)$$

Proof of Lemma 6.1.   It is clear that $U(t, s)u_0$ is a solution of (6.1). Thus it remains to prove uniqueness.

We introduce the bounded operators $A_n(t) = A(t)(I + (1/n)A(t))^{-1}$, $n = 1, 2, \ldots$. One easily verifies that

$$A_n(t) + \lambda I = \frac{n + \lambda}{n} \left[\frac{n\lambda}{n + \lambda} I + A(t) \right] \left[I + \frac{1}{n} A(t) \right]^{-1}$$

and that if Re $\lambda \geq 0$,

$$[A_n(t) + \lambda I]^{-1} = \frac{1}{n + \lambda} I + \frac{n^2}{(n + \lambda)^2} \left[A(t) + \frac{\lambda n}{n + \lambda} I \right]^{-1}.$$

It follows that

$$\|[A_n(t) + \lambda I]^{-1}\| \leq \frac{1}{|n + \lambda|} + \frac{n}{|n + \lambda|} \frac{C}{|\lambda| + 1} \leq \frac{C}{|\lambda| + 1}, \tag{6.7}$$

where C is a generic constant independent of n.

Writing

$$[A_n(t) - A_n(\tau)]A_n^{-1}(s)$$

$$= [A(t) - A(\tau)]\left[I + \frac{1}{n} A(t) \right]^{-1} \left[I + \frac{1}{n} A(s) \right] A^{-1}(s)$$

$$- A(\tau) \left[I + \frac{1}{n} A(\tau) \right]^{-1} \frac{A(t) - A(\tau)}{n} \left[I + \frac{1}{n} A(t) \right]^{-1} \left[I + \frac{1}{n} A(s) \right] A^{-1}(s),$$

we find that

$$\|[A_n(t) - A_n(\tau)]A_n^{-1}(s)\| \leq C |t - \tau|^\alpha. \tag{6.8}$$

From (6.7), (6.8) and the results of Sections 4, 5, it follows that the fundamental solution $U_n(t, \tau)$ corresponding to $A_n(t)$ satisfies

$$\|U_n(t, \tau)\| \leq C, \tag{6.9}$$

where C is independent of n.

Let $v(t)$ be a continuously differentiable function for $\tau \leq t < t_0$ satisfying (6.1), and let v_n be the solution of

$$\begin{cases} \dfrac{dw}{dt} + A_n(t)w = 0 & \text{for } \tau < t < t_0, \\ w(\tau) = u_0. \end{cases} \tag{6.10}$$

The function $w_n(t) = v(t) - v_n(t)$ then satisfies

$$\begin{cases} \dfrac{dw_n}{dt} + A_n(t)w_n = [A_n(t) - A(t)]v(t), \\[2mm] w_n(\tau) = 0. \end{cases} \qquad (6.11)$$

We now note that if $B(t)$ is a bounded operator that varies continuously with t (in the uniform topology), then there exists a unique solution

$$\begin{cases} \dfrac{dz}{dt} + B(t)z = g(t) \qquad (\tau < t < t_0), \\[2mm] z(0) = 0 \end{cases} \qquad (6.12)$$

for any continuous function $g(t)$ (with values in X), and it is given by

$$z(t) = \int_0^t W(t, s)g(s)\, ds, \qquad (6.13)$$

where W is the fundamental solution. Indeed, $\partial W(t, s)/\partial t$ is continuous for $t \geq s$ since it is equal to $B(t)W(t, s)$. It therefore follows that $z(t)$, given by (6.13), is a solution of (6.11). To prove uniqueness, suppose $g(t) = 0$. Then any solution $z(t)$ satisfies

$$\|z(t)\| \leq \int_\tau^t \|B(s)z(s)\|\, ds \leq C \int_\tau^t \|z(s)\|\, s, \qquad z(\tau) = 0.$$

It follows that $z(t) \equiv 0$.

Applying the previous remarks to the system (6.11), we have

$$w_n(t) = \int_\tau^t U_n(t, s)[A_n(s) - A(s)]v(s)\, ds. \qquad (6.14)$$

We next claim that for any $x \in X$, if $n \to \infty$,

$$[A_n(s) - A(s)]A^{-1}(s)x \to 0 \qquad \text{uniformly with respect to } s. \qquad (6.15)$$

In fact,

$$\|[A_n(s) - A(s)]A^{-1}(s)\| = \left\|\left[I + \frac{1}{n}A(s)\right]^{-1} - I\right\| \leq C + 1,$$

whereas for every $x \in D_A$

$$\|[A_n(s) - A(s)]A^{-1}(s)x\| = \left\| \frac{1}{n}\left[I + \frac{1}{n}A(s)\right]^{-1}A(s)x \right\| \leq \frac{C}{n}\|A(0)x\| \to 0$$

uniformly with respect to s.

Since the function $A(s)v(s)$ is continuous for $\tau \leq s < t_0$, it follows from (6.15) that, if $n \to \infty$,

$$[A_n(s) - A(s)]v(s) \to 0 \qquad \text{uniformly with respect to } s, \tau \leq s \leq t.$$

From this and (6.9), (6.14) we conclude that $w_n(t) \to 0$ if $n \to \infty$—that is,

$$v(t) = \lim_{n \to \infty} v_n(t).$$

Since $v_n(t)$ is defined uniquely as the solution of (6.10), also $v(t)$ is unique.

Now let $v(t)$ be any solution of (6.1) (not necessarily continuously differentiable near $t = \tau$). For any $\tau < s < t_0$, $v(t)$ is a continuously differentiable solution for $s \leq t < t_0$. Writing

$$U(t, s)v(s) = U(t, s)A^{-1}(s)[A(s)v(s)],$$

we see, using Lemma 6.2, that $U(t, s)v(s)$ is also a continuously differentiable solution for $s \leq t < t_0$ and $U(s, s)v(s) = v(s)$. By the previous result on uniqueness we conclude that $v(t) = U(t, s)v(s)$. Taking $s \to \tau$, it follows that $v(t) = U(t, \tau)u_0$. This completes the proof.

PROBLEM.   (1) Prove: if $0 \leq s \leq \tau \leq t \leq t_0$, then

$$U(t, s) = U(t, \tau)U(\tau, s). \tag{6.16}$$

7 | SOLUTION OF THE CAUCHY PROBLEM

In this section we prove Theorem 3.2. In view of Lemma 6.1 it remains to prove that the function

$$w(t) = \int_0^t U(t, s)f(s) \, ds \tag{7.1}$$

satisfies the equation

$$\frac{\partial w}{\partial t} + A(t)w = f(t) \qquad (0 < t \leq t_0), \tag{7.2}$$

$$w(0) = 0. \tag{7.3}$$

Since, by (4.8), $\|U(t, \tau)\| \leq C$, (7.3) is obvious.

We shall need the following lemma:

LEMMA 7.1.   The following inequalities hold for $0 \leq \tau < t \leq t_0$:

$$\|U(t, \tau) - e^{-(t-\tau)A(t)}\| \leq C\,|t - \tau|^{\alpha}, \tag{7.4}$$

$$\|A(t)[U(t, \tau) - e^{-(t-\tau)A(t)}]\| \leq C\,|t - \tau|^{\alpha-1}. \tag{7.5}$$

From (7.5) and (3.5) we obtain:

COROLLARY.   The following inequality holds for $0 \leq \tau < t \leq t_0$:

$$\|A(t)U(t, \tau)\| \leq \frac{C}{|t - \tau|}. \tag{7.6}$$

Proof of Lemma 7.1.   From (4.8),

$$U(t, \tau) - e^{-(t-\tau)A(t)} = [e^{-(t-\tau)A(\tau)} - e^{-(t-\tau)A(t)}] + \int_{\tau}^{t} e^{-(t-s)A(s)}\Phi(s, \tau)\,ds$$

$$\equiv I(t, \tau) + J(t, \tau). \tag{7.7}$$

Using (4.10) we find that $\|I(t, \tau)\| \leq C(t - \tau)^{\alpha}$, where C is a generic constant, and using (4.21) we find that $\|J(t, \tau)\| \leq C(t - \tau)^{\alpha}$. Thus (7.4) follows.

To prove (7.5), we write

$$A(t)[U(t, \tau) - e^{-(t-\tau)A(t)}] = A(t)I(t, \tau) + A(t)J(t, \tau), \tag{7.8}$$

$$A(t)J(t, \tau) = A(t)\int_{\tau}^{t} e^{-(t-s)A(s)}[\Phi(s, \tau) - \Phi(t, \tau)]\,ds$$

$$+ A(t)\left[\int_{\tau}^{t} e^{-(t-s)A(s)}\,ds\right]\Phi(t, \tau). \tag{7.9}$$

Using (4.25) and (5.7), (4.21), we get

$$\|A(t)J(t, \tau)\| \le C \int_{\tau}^{t} |t - s|^{-1} |t - s|^{\alpha - \eta} |s - \tau|^{\eta - 1} \, ds + C |t - \tau|^{\alpha - 1}$$

for any $0 < \eta < \alpha$. Hence $\|A(t)J(t, \tau)\| \le C |t - \tau|^{\alpha - 1}$. Since, by (4.11), $\|A(t)I(t, \tau)\| \le C |t - \tau|^{\alpha - 1}$, (7.5) follows.

We shall need also the following lemma:

LEMMA 7.2.   For any $x \in X$, $0 \le t \le t_0$

$$\frac{U(t + \Delta t, t) - I}{\Delta t} A^{-1}(t)x \to -x \qquad \text{as } \Delta t \to 0. \tag{7.10}$$

Proof.   Consider the function $\psi(s) = e^{-(t-s)A(t)} U(s, \tau)y$, for $y \in X$. It is continuously differentiable for $\tau < s < t$ and

$$\psi'(s) = e^{-(t-s)A(t)} [A(t) - A(s)] U(s, \tau)y.$$

Integrating this relation, we get

$$U(t, \tau)y = e^{-(t-\tau)A(t)}y + \int_{\tau}^{t} e^{-(t-s)A(t)} [A(t) - A(s)] U(s, \tau)y \, ds. \tag{7.11}$$

We can now write, with $y = A^{-1}(t)x$,

$$\frac{U(t + \Delta t, t) - I}{\Delta t} A^{-1}(t)x = \frac{e^{-\Delta t A(t + \Delta t)} - I}{\Delta t} A^{-1}(t)x + \frac{1}{\Delta t} \int_{t}^{t + \Delta t} e^{(t + \Delta t - s)A(t + \Delta t)}$$

$$\times [A(t + \Delta t) - A(s)]A^{-1}(s)[A(s)U(s, t)A^{-1}(t)x] \, ds$$

$$\equiv A(t, \Delta t) + B(t, \Delta t).$$

Using (3.3) and (6.6), we obtain

$$\|B(t, \Delta t)\| \le \frac{C}{|\Delta t|} \left| \int_{t}^{t + \Delta t} |t + \Delta t - s|^{\alpha} \, ds \right| \le C |\Delta t|^{\alpha}.$$

Since

$$A(t, \Delta t) = - \left[\frac{1}{\Delta t} \int_{0}^{\Delta t} e^{-\sigma A(t + \Delta t)} \, d\sigma \right] A(t + \Delta t)A^{-1}(t)x \to -x,$$

(7.10) follows.

Consider now the function $w(t)$ given in (7.1), and assume that

$$\|f(t) - f(s)\| \le C|t - s|^{\beta}. \tag{7.12}$$

Employing (6.16), we can formally write

$$\frac{w(t + \Delta t) - w(t)}{\Delta t} = \frac{1}{\Delta t} \int_{t}^{t + \Delta t} U(t + \Delta t, s) f(s) \, ds$$

$$+ \frac{1}{\Delta t} [U(t + \Delta t, t) - I] \int_{0}^{t} U(t, s) f(s) \, ds$$

$$= \frac{1}{\Delta t} \int_{t}^{t + \Delta t} U(t + \Delta t, s) f(s) \, ds + \frac{1}{\Delta t} [U(t + \Delta t, t) - I] A^{-1}(t)$$

$$\times \left\{ A(t) \int_{0}^{t} U(t, s)[f(s) - f(t)] \, ds \right.$$

$$+ \left[A(t) \int_{0}^{t} [U(t, s) - e^{-(t-s)A(t)}] \, ds \right] f(t) + [I - e^{-tA(t)}] f(t) \right\}. \tag{7.13}$$

Denote by $G(t)$ the expression in braces. Recalling (7.6), (7.12), and (7.5), we see that each of the terms that make up $G(t)$ is well defined. Consequently, (7.13) is valid. Taking $\Delta t \to 0$ and using Lemma 7.2, we conclude that $dw(t)/dt$ exists and

$$\frac{dw(t)}{dt} = f(t) - \left\{ A(t) \int_{0}^{t} U(t, s)[f(s) - f(t)] \, ds \right.$$

$$+ \left[A(t) \int_{0}^{t} [U(t, s) - e^{-(t-s)A(t)}] \, ds \right] f(t) + [I - e^{-tA(t)}] f(t) \right\}$$

$$= f(t) - A(t) \int_{0}^{t} U(t, s) f(s) \, ds$$

$$= f(t) - A(t) w(t), \tag{7.14}$$

—that is, (7.2) holds. Note our tacit assumption that $\int_{0}^{t} U(t, s) f(s) \, ds$ is in D_A. This, however, follows by Lemma 1.2 if we observe (employing a variant of the second equation in (7.14)) that $\int_{0}^{\sigma} A(t) U(t, s) f(s) \, ds$ is convergent as $\sigma \nearrow t$.

We shall need the following fact:

LEMMA 7.3.   If $g(s, t)$ *is a uniformly continuous function (with values in* X)
for $0 \leq s \leq t_0$, $0 \leq t \leq t_0$, $|t - s| > \varepsilon$ *(for any* $\varepsilon > 0$) *and if*

$$\int_{|s-t| < \varepsilon} \|g(s, t)\| \, ds \to 0 \qquad as \; \varepsilon \to 0,$$

uniformly with respect to t, *then* $\int_0^t g(s, t) \, ds$ *is a uniformly continuous function*
for $0 \leq t \leq t_0$.

The proof is left to the reader.

From the lemma it easily follows that each of the terms in the first expression for dw/dt in (7.14) is a continuous function of t. Hence, dw/dt is also continuous in t. This completes the proof of Theorem 3.2.

8 | DIFFERENTIABILITY OF SOLUTIONS

We shall assume that $A(t)$ is k-smooth in the following sense:
(C_k) For any $x \in X$ the function $A(t)A^{-1}(0)x$ has continuous derivatives

$$d^j[A(t)A^{-1}(0)x]/dt^j \equiv A^{(j)}(t)A^{-1}(0)x$$

for all $1 \leq j \leq k$. The operators $A^{(j)}(t)A^{-1}(0)$ are uniformly bounded for
$0 \leq t \leq t_0$, and

$$\|A^{(k)}(t)A^{-1}(0) - A^{(k)}(s)A^{-1}(0)\| \leq C|t - s|^{\alpha}. \tag{8.1}$$

LEMMA 8.1.   Let (C_k) *hold, and assume that* $A(0)A^{-1}(t)$ *is a bounded*
operator for each t, $0 \leq t \leq t_0$. *Then for any* $x \in X$ *the function* $A(0)A^{-1}(t)x$ *has*
continuous derivatives $d^j[A(0)A^{-1}(t)x]/dt^j \equiv A(0)(A^{-1}(t))^{(j)}x$ *for all* $1 \leq j \leq k$.
Furthermore, the operators $A(0)(A^{-1}(t))^{(j)}$ *are uniformly bounded for* $0 \leq t \leq t_0$.

Proof.   Set $B(t) = A(t)A^{-1}(0)$. Then $B^{-1}(t) = A(0)A^{-1}(t)$. We have

$$B(t + h)y - B(t)y = hB(t, h)y,$$

where $\|B(t, h)x\| \leq c$, c independent of h. Multiplying both sides of this equation
by $B^{-1}(t)$ on the left and taking $y = B^{-1}(t + h)x$, we get

$$B^{-1}(t + h)x - B^{-1}(t)x = -hB^{-1}(t)B(t, h)B^{-1}(t + h)x. \tag{8.2}$$

Hence

$$\|B^{-1}(t + h)x - B^{-1}(t)x\| \le C|h| \, \|B^{-1}(t + h)x\|.$$

The continuity of $B^{-1}(t)x$ now readily follows. From (8.2) we now conclude that $B^{-1}(t)x$ is differentiable and

$$\frac{d}{dt} B^{-1}(t)x = -B^{-1}(t)B'(t)B^{-1}(t)x.$$

From this relation it is clear that all the first k-derivatives of $B^{-1}(t)x$ exist and are continuous, and $d^j B^{-1}(t)/dt^j$ are uniformly bounded operators (for $1 \le j \le k$). Since, finally, $B^{-1}(t) = A(0)A^{-1}(t)$, the proof of the lemma is complete.

THEOREM 8.1. *Let the assumptions* (B_1)–(B_3) *and* (C_k) *hold, and assume that* f(t) *has* k *continuous derivatives for* $0 \le t \le t_0$ *and that* $f^{(k)}(t)$ *is uniformly Hölder continuous (exponent* β) *in* $[0, t_0]$. *Then the solution* u(t) *of* (3.8) *has* k + 1 *continuous derivatives in any interval* $\varepsilon < t \le t_0$ ($\varepsilon > 0$).

Proof.   For any function $v(t)$, set

$$v_h(t) = \frac{v(t + h) - v(t)}{h}.$$

Then

$$\frac{du_h(t)}{dt} + A(t)u_h(t) = f_h(t) - \frac{A(t + h) - A(t)}{h} u(t + h).$$

Applying Theorem 3.2, we find that for any $\tau \in (0, t)$

$$u_h(t) = U(t, \tau)u_h(\tau) + \int_\tau^t U(t, s)\left[f_h(s) - \frac{A(s + h) - A(s)}{h} u(s + h)\right] ds. \qquad (8.3)$$

We claim that, as $h \to 0$,

$$f_h(s) \to f(s) \qquad \text{uniformly with respect to } s. \qquad (8.4)$$

In fact, for any bounded linear functional γ, the mean value theorem gives

$$\gamma[f_h(s) - f(s)] = \gamma[f'(s + \tilde{h}) - f'(s)] \qquad \text{for some } 0 < \tilde{h} < h;$$

here we assume, for simplicity, that γ is a real-valued functional. Hence, by the uniform continuity of $f'(t)$,

$$|\gamma[f_h(s) - f(s)]| \le \|\gamma\| \, \delta(h),$$

where $\delta(h) \to 0$ if $h \to 0$ and $\delta(h)$ is independent of s. It follows that

$$\|f_h(s) - f(s)\| \le \delta(h),$$

which is the assertion (8.4).

Next we write

$$\frac{A(s + h) - A(s)}{h} u(s + h)$$

$$= \frac{A(s + h)A^{-1}(0) - A(s)A^{-1}(0)}{h} A(0)A^{-1}(s + h)A(s + h)u(s + h).$$

Using the assumption (C_k), Lemma 8.1, and the uniform continuity of $A(t)u(t)$ for $\varepsilon \le t \le t_0$ ($\varepsilon > 0$), we conclude that

$$\frac{A(s + h) - A(s)}{h} u(s + h) \to [A'(s)A^{-1}(0)]A(0)u(s),$$

uniformly with respect to s, $\tau \le s \le t$.

Combining this with (8.4), and taking $h \to 0$ in (8.3), we deduce the relation

$$u'(t) = U(t, \tau)u'(\tau) + \int_\tau^t U(t, s)[f'(s) - A'(s)u(s)] \, ds, \qquad (8.5)$$

where $A'(s)u(s)$ stands for $[A'(s)A^{-1}(0)]A(0)u(s)$.

We now need the following lemma:

LEMMA 8.2.   Let $g(s)$ be a continuous function for $\tau \le s \le t_0$. Then the function $\int_\tau^t U(t, s)g(s) \, ds$ is uniformly Hölder continuous in $t \in [\tau, t_0]$ with any exponent $\gamma < 1$.

Proof.   We claim that if $0 < h < 1$, $|t - s| \ge h$, then

$$\|U(t + h, s) - U(t, s)\| \le \frac{Ch^\gamma}{|t - s|^\gamma}. \qquad (8.6)$$

Indeed, for any $x \in X$ and for any bounded linear functional p, the mean value theorem gives, for some $0 < \tilde{h} < h$,

$$|p(U(t + h, s)x) - p(U(t, s)x)| \leq h \left| p\left(\frac{\partial}{\partial t} U(t + \tilde{h}, s)x\right) \right|$$

$$\leq \|p\| \|x\| h \left\| \frac{\partial}{\partial t} U(t + \tilde{h}, s) \right\|$$

$$\leq C\|p\| \|x\| h |t - s|^{-1} \leq C\|p\| \|x\| h^\gamma |t - s|^{-\gamma},$$

where (7.6) has been used. (Here we assume, for simplicity, that p is a real valued functional.) (8.6) immediately follows.

We can now write (for $0 < h < t - \tau, h < 1$)

$$\int_\tau^{t+h} U(t + h, s)g(s)\, ds - \int_\tau^t U(t, s)g(s)\, ds$$

$$= \int_{t-h}^{t+h} U(t + h, s)g(s)\, ds + \int_{t-h}^t U(t, s)g(s)\, ds$$

$$+ \int_\tau^{t-h} [U(t + h, s) - U(t, s)]g(s)\, ds. \qquad (8.7)$$

The first two integrals on the right are obviously bounded by Ch. In view of (8.6), the third integral on the left is bounded by Ch^γ. If $t - \tau < h$, then clearly each term on the left side of (8.7) is bounded by Ch.

Consider now the integral in (8.5). Since $A'(s)u(s)$ and $f'(s)$ are continuous, Lemma 8.2 shows that this integral is uniformly Hölder continuous with any exponent $\gamma < 1$. The same is therefore true of $u'(t)$ in $[\tau', t_0]$ for any $\tau' > \tau$. Since τ is an arbitrary point in $(0, t_0)$, $u'(t)$ is uniformly Hölder continuous (exponent γ) in $[\varepsilon, t_0]$, for any $\varepsilon > 0$.

By the results of Section 7 it now follows from (8.5) that $u'(t)$ is continuously differentiable in $[\tau', t_0]$ for any $\tau' > \tau$, and

$$\frac{d^2u}{dt^2} + A(t)\frac{du}{dt} = f'(t) - A'(t)u(t). \qquad (8.8)$$

Since τ is an arbitrary point in $(0, t_0)$, $u''(t)$ is continuous in every interval $[\varepsilon, t_0]$, $\varepsilon > 0$.

Writing $A'(t)u(t) = [A'(t)A^{-1}(0)][A(0)A^{-1}(t)]A(t)u(t)$ and using Lemma 8.1, it follows that $A'(t)u(t)$ is continuously differentiable. We can now proceed as before and write

$$u''(t) = U(t, \tau)u''(\tau) + \int_\tau^t U(t, s)[f''(s) - A''(s)u(s) - 2A'(s)u'(s)]\, ds,$$

where $A''(s)u(s)$ stands for

$$[A''(s)A^{-1}(0)]A(0)u(s) = [A''(s)A^{-1}(0)][A(0)A^{-1}(s)]A(s)u(s).$$

Using Lemma 8.2, we find that $u''(t)$ is uniformly Hölder continuous in every interval $[\varepsilon, t_0]$ ($\varepsilon > 0$). Hence, $u'''(t)$ exists and is continuous in every interval $[\varepsilon, t_0]$.

We leave it for the reader to prove, by induction, that $u^{(j+1)}(t)$ exists, for all $1 \leq j \leq k$, and that

$$u^{(j+1)}(t) + A(t)u^{(j)}(t) = f^{(j)}(t) - \sum_{i=0}^{j-1} \binom{j}{i} A^{(j-i)}(t)u^{(i)}(t). \tag{8.9}$$

9 | THE INITIAL-BOUNDARY VALUE PROBLEM FOR PARABOLIC EQUATIONS

Let Ω be a bounded domain in R^n. For any $0 < T < \infty$ denote by Q_T the cylinder $\{(x, t); x \in \Omega, 0 < t < T\}$ and by S_T the lateral boundary

$$\{(x, t); x \in \partial\Omega, 0 < t \leq T\},$$

where $\partial\Omega$ is the boundary of Ω. We also write $\Omega_s = \{(x, s); x \in \Omega\}$.

We consider differential operators

$$Lu \equiv \frac{\partial u}{\partial t} + A(x, t, D)u \equiv \frac{\partial u}{\partial t} + \sum_{|\alpha| \leq 2m} a_\alpha(x, t)D^\alpha u \tag{9.1}$$

with coefficients defined in $\bar{Q}_T$.

DEFINITION. L is said to be *parabolic* at a point (x^0, t^0) if $A(x, t^0, D)$ is strongly elliptic at x^0. L is *parabolic* on a set A if L is parabolic at each point of A. If the coefficients a_α are bounded in $\bar{Q}_T$ and if

$$(-1)^m \text{Re}\left\{\sum_{|\alpha|=2m} a_\alpha(x, t)\xi^\alpha\right\} \geq c|\xi|^{2m} \qquad (c > 0)$$

for all $(x, t) \in \bar{Q}_T$ and for all real ξ, then L is said to be *uniformly parabolic* in $\bar{Q}_T$.

We consider the following system:

$$Lu = f \qquad \text{in } Q_T \cup \Omega_T, \tag{9.2}$$

$$\frac{\partial^j u}{\partial v^j} = \phi_j \qquad \text{on } S_T \qquad (0 \le j \le m - 1) \tag{9.3}$$

$$u(x, 0) = \psi(x) \qquad \text{on } \Omega_0, \tag{9.4}$$

where v is the outward normal. The conditions (9.3) are called *boundary conditions* and the condition (9.4) is called an *initial condition*. Given f, ϕ_j, ψ, the problem of solving (9.2)–(9.4) is called an *initial-boundary value problem*. More general initial-boundary value problems can be obtained by replacing in the conditions (9.3) the operators $\partial^j / \partial v^j$ by more general differential operators.

For simplicity we assume that $\phi_j \equiv 0$ $(0 \le j \le m - 1)$, $\psi \equiv 0$. Thus, we consider the equation (9.2) and the conditions

$$\frac{\partial^j u}{\partial v^j} = 0 \qquad \text{on } S_T \qquad (0 \le j \le m - 1), \tag{9.5}$$

$$u(x, 0) = 0 \qquad \text{on } \Omega_0. \tag{9.6}$$

DEFINITION. A function $u(x, t)$, continuous in $\bar{Q}_T$, is called a *classical solution* of (9.2), (9.5), (9.6) if (i) the derivatives $\partial u(x, t) / \partial t$, $D^\alpha u(x, t)$ $(0 \le |\alpha| \le 2m)$ exist and are continuous in $Q_T \cup \Omega_T$; (ii) the derivatives $D^\alpha u(x, t)$ $(0 \le |\alpha| \le m - 1)$ are continuous in $Q_T \cup \Omega_T \cup S_T$; and (iii) the equations (9.2), (9.5), (9.6) hold.

Consider now the evolution equation

$$\frac{du}{dt} + A(t)u = f(t) \tag{9.7}$$

in the Hilbert space $X = L^2(\Omega)$, where, for each t, $f(t)$ is the function $f(x, t)$ belonging to $L^2(\Omega)$, and $A(t)$ is the operator with domain $D_A = H^{2m}(\Omega) \cap H_0^m(\Omega)$ given by $A(t)v(x) = A(x, t, D)v(x)$. $u(t)$ is a function with values in X. Thus, for each t, it is a function $u(x, t)$ in $L^2(\Omega)$.

(9.7) is an abstract form of (9.2). The condition that $u(t) \in H_0^m(\Omega)$ implies that u satisfies (9.5) in a generalized sense. Finally, the condition

$$u(0) = 0 \tag{9.8}$$

is a generalized form of (9.6).

DEFINITION. If $u(t)$ is a solution of (9.7), (9.8), then we say that $u(x, t)$ is a *generalized solution* of (9.2), (9.5), (9.6).

Recall that the solutions of (9.7), (9.8) are assumed to be continuous for $0 \leq t \leq T$ and continuously differentiable for $0 < t \leq T$.

We shall need the following assumption:

(**D$_0$**) The coefficients $a_\alpha(x, t)$ are continuous in $\bar{Q}_T$, and

$$|a_\alpha(x, t) - a_\alpha(x, t')| \leq C |t - t'|^\beta \qquad (0 < \beta \leq 1)$$

for all $x \in \bar{\Omega}$, $t \in [0, T]$, $t' \in [0, T]$, where C, β are constants.

THEOREM 9.1. *Assume that* L *is uniformly parabolic in* $\bar{Q}_T$, *that* (D$_0$) *holds, and that* $\partial\Omega$ *is of class* C^{2m}. *If* f(t) *is uniformly Hölder continuous in* [0, T], *then there exists a unique solution of* (9.7), (9.8).

Proof. For each fixed t, $A(t) + kI$ is of type (ϕ, M) for some constant $k > 0$—that is, the assumptions (i), (ii) of Section 2 are satisfied for $A(t) + kI$. In fact, this follows from Sections 1.18, 1.19. From the results of these sections it also follows that (3.2) holds for $A(t) + kI$, with constants k, C independent of t. Set

$$v = e^{-kt}u.$$

Then (9.7), (9.8) take the equivalent form

$$\frac{dv}{dt} + (A(t) + kI)v = e^{-kt}f(t), \tag{9.9}$$

$$v(0) = 0. \tag{9.10}$$

We now claim that (3.3) holds for $A(t) + kI$. Indeed, since Theorem I.18.2 gives

$$\|(A(t) + kI)^{-1}v\|_{2m}^\Omega \leq C \|v\|_0^\Omega,$$

(3.3), with $\alpha = \beta$, is a consequence of (D$_0$).

We can thus apply Theorem 3.2 to deduce the assertion of the theorem.

We introduce the following assumption:

(**D$_h$**) The derivatives $\partial^j a_\alpha(x, t)/\partial t^j$ $(0 \leq |\alpha| \leq 2m, 0 \leq j \leq h)$ exist and are continuous in $\bar{Q}_T$, and

$$\left| \frac{\partial^h}{\partial t^h} a_\alpha(x, t) - \frac{\partial^h}{\partial t^h} a_\alpha(x, t') \right| \leq C |t - t'|^\beta \qquad (0 < \beta \leq 1)$$

for all $x \in \bar{\Omega}$, $t \in [0, T]$, $t' \in [0, T]$, where C, β are constants.

THEOREM 9.2.   Assume that L *is uniformly parabolic in* $\bar{Q}_T$, *that* (D_h) *holds, and that* $\partial\Omega$ *is of class* C^{2m}. *If* f(t) *has* h *continuous derivatives in* [0, T] *and if* $f^{(h)}(t)$ *is uniformly Hölder continuous, then the solution* u(t) *of* (9.7), (9.8) *has* (h + 1) *continuous derivatives in each interval* $(\varepsilon, T]$, $\varepsilon > 0$.

Proof.   It is easily verified that the condition (C_h) of Section 8 holds for $A(t) + kI$. Now apply Theorem 8.1.

In the next section we shall prove the smoothness of the generalized solution $u(x, t)$ with respect to (x, t).

10 | SMOOTHNESS OF THE SOLUTIONS OF THE INITIAL-BOUNDARY VALUE PROBLEM

Let $-\infty \leq a < b \leq \infty$ and let X be a Banach space. By $C^m((a, b); X)$ we denote the space of all functions $u(t)$ from the interval $a < t < b$ into X that have m continuous derivatives in (a, b). We introduce the norm

$$\|u\|_{H^m((a, b); X)} = \left\{ \sum_{j=0}^{m} \int_a^b \|u^{(j)}(t)\|^2 \, dt \right\}^{1/2} \tag{10.1}$$

and denote by $H^m((a, b); X)$ the completion with respect to the norm (10.1) of the subset of $C^m((a, b); X)$ whose elements have a finite norm. Strong derivatives, with values in X, are defined in the obvious way.

LEMMA 10.1.   If $u \in H^j((a, b); H^j(\Omega))$, *then* $u \in H^j(Q)$, *where* $Q = \Omega \times (a, b)$; *more precisely, there exists a function* $U(x, t)$ *in* $H^j(Q)$ *such that for almost all* t *it is equal to* u(t), *as an element of* $H^j(\Omega)$.

Proof.   Set $Y = H^j((\jmath, b); H^j(\Omega))$. Let $\{e_m\}$ be an orthonormal basis in the Hilbert space $H^j(\Omega)$. Then we have, for almost all $t \in (a, b)$,

$$u(t) = \sum_{m=1}^{\infty} u_m(t) e_m \qquad (u_m(t) = (u(t), e_m)_j^{\Omega})$$

and the series is convergent in the norm of $H^j(\Omega)$. For the strong derivative $D_t^i u(t)$ $(1 \leq i \leq j)$ of $u(t)$ we also have, for almost all t,

$$D_t^i u(t) = \sum_{m=1}^{\infty} w_{m, i}(t) e_m \qquad (w_{m, i}(t) = (D_t^i u(t), e_m)_j^{\Omega}).$$

If u is in $C^j((a, b); H^j(\Omega))$, then clearly $w_{m,i}(t) = D_t^i u_m(t)$. For general u in Y this relation still holds, with $D_t^i u_m(t)$ being strong derivatives. Indeed, if $W_\nu \in C^j((a, b); H^j(\Omega))$, and

$$\| W_\nu - u \|_Y \to 0 \qquad \text{as } \nu \to \infty,$$

then for any $\phi(t) \in C_0^\infty(a, b)$,

$$\int w_{m,i}(t)\overline{\phi(t)} \, dt = \lim_{\nu \to \infty} (D_t^i W_\nu, e_m \phi)_Y = (-1)^i \lim_{\nu \to \infty} (W_\nu, e_m D_t^i \phi)_Y$$

$$= (-1)^i \int u_m(t) D_t^i \overline{\phi(t)} \, dt,$$

which is the assertion.

Thus

$$D_t^i u(t) = \sum_{m=1}^\infty [D_t^i u_m(t)] e_m \qquad \text{in } H^j(\Omega), \qquad \text{for } 0 \le i \le j. \qquad (10.2)$$

We conclude that

$$\infty > \| u \|_Y^2 = \sum_{i=0}^j \int_a^b (\| D_t^i u(t) \|_{H^j(\Omega)})^2 \, dt = \sum_{m=1}^\infty \sum_{i=0}^j \int_a^b |D_t^i u_m(t)|^2 \, dt. \qquad (10.3)$$

Let $M < N < \infty$ and introduce the function

$$U_{MN}(x, t) = \sum_{m=M}^N u_m(t) e_m(x).$$

Clearly $U_{MN}(x, t)$ belongs to $H^j(Q)$ and

$$\| U_{MN}(x, t) \|_j^Q \le \| U_{MN}(x, t) \|_Y \to 0 \qquad \text{if } M \to \infty.$$

Hence there exists a function $U(x, t)$ in $H^j(Q)$ such that

$$\| U_{1N}(x, t) - U(x, t) \|_j^Q \to 0 \qquad \text{if } N \to \infty.$$

By Fubini's theorem and Fatou's lemma, for almost all $t \in (a, b)$

$$\liminf_{N \to \infty} \int_\Omega |U_{1N}(x, t) - U(x, t)|^2 \, dx = 0.$$

Since for almost all $t \in (a, b)$, we also have

$$\int_\Omega |U_{1N}(x, t) - u(x, t)|^2 \, dx \to 0 \qquad \text{as } N \to \infty,$$

it follows that, for almost all t, $U(x, t) = u(x, t)$ almost everywhere.

We now make the following assumptions:

(A) L is uniformly parabolic in $\bar{Q}_T$ and all its coefficients belong to $C^\infty(\bar{Q}_T)$. The boundary $\partial\Omega$ is of class C^∞.

(B) For every $t \in [0, T]$ the null space of $A(t)$ is the zero function.

In the final differentiability theorem proved below the assumption (B) is not needed.

THEOREM 10.1.   Assume that (A), (B) hold and let f be a fixed function in $H^p(\Omega)$. Then the mapping $t \to A^{-1}(t)f$ from the interval $[0, T]$ into $H^{2m+p}(\Omega)$ has continuous derivatives of any order.

Proof.   For any function $w(x, t)$, introduce the notation

$$w^h(x, t) = \frac{w(x, t + h) - w(x, t)}{h} \qquad \text{for } t \in [0, T], t + h \in [0, T].$$

Let $u(x, t) = (A^{-1}(t)f)(x)$. One easily sees that $u^h \in H^{2m}(\Omega) \cap H_0^m(\Omega)$ and

$$Au^h = F_h \qquad \text{in } \Omega,$$

where $F_h(x, t) = -\sum a_\alpha^h(x, t)D^\alpha u(x, t + h)$. From Problem 3, Section I.19, it follows that $\|u^h\|_{2m+p}$ is bounded independently of h. Hence

$$\|u(\cdot, t + h) - u(\cdot, t)\|_{2m+p} \le C|h|, \qquad (10.4)$$

where $u(\cdot, t)$ is the function whose value at each point x is $u(x, t)$. (10.4) shows that the mapping $t \to A^{-1}(t)f$ is a continuous map from $[0, T]$ into $H^{2m+p}(\Omega)$.

Differentiating the relation $A(t)u(t) = f$ formally with respect to t, we find that $v = \partial u/\partial t$, if existing, must satisfy

$$A(t)v = \hat{F} \qquad \text{in } \Omega, \qquad (10.5)$$

where $\hat{F}(x, t) = -\sum [\partial a_\alpha(x, t)/\partial t]D^\alpha u(x, t)$. Observing that $\hat{F}(\cdot, t) \in H^p(\Omega)$ for each $t \in [0, T]$, we conclude that there exists a unique solution $v(\cdot, t)$ of (10.5).

Using (10.4), it follows that

$$\|F_h(\cdot, t) - \hat{F}(\cdot, t)\|_p \to 0 \qquad \text{if } h \to 0.$$

By Problem 3, Section I.19, it then follows that

$$\|u^h(\cdot, t) - v(\cdot, t)\|_{2m+p} \to 0 \qquad \text{if } h \to 0.$$

This shows that the map $t \to A^{-1}(t)f$, from $[0, T]$ into $H^{2m+p}(\Omega)$, is differentiable in t.

One can now proceed (taking finite differences) as before to show that $v(\cdot, t)$ varies continuously in t, as a function with values in $H^{2m+p}(\Omega)$. Next one shows that $\partial v(\cdot, t)/\partial t$ exists—that is, the mapping $t \to A^{-1}(t)f$ is twice continuously differentiable.

Proceeding similarly step by step, the proof of the theorem is completed. From the previous proof one obtains:

COROLLARY.   *The inequalities*

$$\|D_t^i A^{-1}(t)f\|_{2m+p} \le C_j \|f\|_p \qquad (C_j \text{ constant})$$

hold for $0 \le t \le T$, $j = 1, 2, \ldots$.

PROBLEM.   (1) Let the assumptions of Theorem 10.1 hold and let k be any integer ≥ 0. Then the mapping $f(t) \to A^{-1}(t)f(t)$ is a continuous linear map from $H^k((0, T); H^p(\Omega))$ into $H^k((0, T); H^{2m+p}(\Omega))$.

THEOREM 10.2.   *Let the assumption* (A) *hold and let* $f(x, t) \in C^\infty(\bar{Q}_T)$. *Then the unique solution of* (9.7), (9.8) *belongs to* $C^\infty(\bar{\Omega} \times (0, T])$.

Proof.   Without loss of generality we may assume that (B) holds, for otherwise we perform a transformation $v = e^{-kt}u$ with k such that $A(t) + kI$ has a zero null space. (9.7) gives, for any $\varepsilon > 0$,

$$u(t) = A^{-1}(t)(f(t) - u'(t)) \qquad (\varepsilon \le t \le T). \tag{10.6}$$

It is clear that, for any integer $k \ge 0$, $f \in H^k([0, T]; H^0(\Omega))$. By Theorem 9.2 also $u' \in H^k([\varepsilon, T]; H^0(\Omega))$. Applying the result of Problem 1 to the right-hand side of (10.6), we conclude that $u \in H^k([\varepsilon, T], H^{2m}(\Omega))$.

Next, $f \in H^{k-1}([0, T], H^{2m}(\Omega))$ and $u' \in H^{k-1}([\varepsilon, T]; H^{2m}(\Omega))$. It follows (by Problem 1 and (10.6)) that $u \in H^{k-1}([\varepsilon, T]; H^{4m}(\Omega))$. Proceeding step by step, we conclude $u \in H^{k-i}([\varepsilon, T]; H^{2m+2mi}(\Omega))$. Since k and i are arbitrary, it follows that $u \in H^j([\varepsilon, T]; H^j(\Omega))$ for any integer $j \ge 0$. By Lemma 10.1, $u \in H^j(\Omega \times [\varepsilon, T])$. The assertion of the theorem now follows by Problem 1, Section I.11.

A review of the proof of Theorem 10.2 gives:

COROLLARY. Let L *be uniformly parabolic in* $\bar{Q}_T$ *and assume that the coefficients* $a_\alpha(x, t)$ *belong to* $C^k(\bar{Q}_T)$ *and* $f \in H^{k-i+2m(i+1)}(\bar{Q}_T)$, *where* $i = 1 + [(p + n/2 + 1)/2m]$, $k = 2m + p + i + [(n + 1)/2]$ *and* p *is a nonnegative integer. Assume also that* $\partial\Omega$ *is of class* $C^{2m(i+1)}$. *Then* $u \in C^{2m+p}(\bar{\Omega} \times [\varepsilon, T])$ *for any* $\varepsilon > 0$.

We next consider the differentiability at $t = 0$.

THEOREM 10.3. Let the assumption (A) *hold and let* $f(x, t) \in C^\infty(\bar{\Omega} \times [-1, T])$ *and* $f(x, t) \equiv 0$ *for* $-1 \le t \le 0$. *Then the unique solution of* (9.7), (9.8) *belongs to* $C^\infty(\bar{\Omega} \times [0, T])$.

Proof. We shall prove that for any $p \ge 0$, $u \in C^{2m+p}(\bar{Q}_T)$. Extend the coefficients of L into $\bar{\Omega} \times [-1, 0]$, so that L remains uniformly parabolic and its coefficients belong to $C^k(\bar{\Omega} \times [-1, T])$ with k as in the corollary to Theorem 10.2. Let w be the unique solution of

$$\frac{dw}{dt} + A(t)w = f(t) \qquad (-1 < t < T),$$

$$w(-1) = 0.$$

By uniqueness we find that $w(t) \equiv 0$ for $-1 < t < 0$. Hence $w(0) = 0$. But then, again, by uniqueness, $w(t) = u(t)$ for $0 < t \le T$. Applying the corollary of Theorem 10.2 to w, the assertion that $u \in C^{2m+p}(\bar{Q}_T)$ follows.

COROLLARY. Let L *be uniformly parabolic in* $\bar{Q}_T$ *and assume that* $a_\alpha \in C^k(\bar{Q}_T)$, $f \in H^{k-i+2m(i+1)}(\bar{\Omega} \times [-1, T])$, *where* $i = 1 + [(p + n/2 + 1)/2m$ *If* $k = 2m + p + i + [(n + 1)/2]$, *and that* $f(x, t) \equiv 0$ *if* $-1 < t < 0$.], $\partial\Omega \in C^{2m(i+1)}$, *then* $u \in C^{2m+p}(\bar{Q}_T)$.

The reader may verify that if $f \in C^{k-i+2m(i+1)}(\bar{Q}_T)$, then f satisfies the conditions of the corollary if

$$\frac{\partial^j f(x, 0)}{\partial t^j} = 0 \qquad \text{for } x \in \Omega, 0 \le j \le j_0, \tag{10.7}$$

where $j_0 = k - i + 2m(i + 1)$.

It is actually sufficient to assume that

$$\frac{\partial^j f(x, 0)}{\partial t^j} = 0 \qquad \text{for } x \in \Omega_0, 0 \le j \le j_0 \tag{10.8}$$

(with a different j_0), where Ω_0 is some neighborhood of $\partial\Omega$. Indeed, one can then decompose f into a sum $f_1 + f_2$ such that f_2 satisfies (10.7) and f_1 vanishes in a neighborhood of S_T.

Write $U(t, \tau)$ in the form $K(x, t; \xi, \tau)$ and consider the function

$$u_1(x, t) = -\int_0^t \int_\Omega K(x, t; \xi, \tau) f_1(\xi, \tau) \, d\xi \, d\tau.$$

It can be shown (Friedman [5]) that u_1 satisfies (10.8). Let w be such that $w = u_1$ on S_T and $\partial w/\partial t + A(t)w$ satisfies (10.7). Let u_2 be the solution of

$$\frac{\partial u_2}{\partial t} + A(t)u_2 = \left(\frac{\partial w}{\partial t} + A(t)w\right) + f_2$$

with zero initial and boundary data. Applying the corollary of Theorem 10.3 to u_2, and noting that $u = u_1 - w + u_2$, it follows that $u \in C^{2m+p}(\bar{Q}_T)$.

We finally note that if there exists a solution u of (9.2), (9.5), (9.6) that is smooth in $\bar{Q}_T$, then $f(x, 0) = 0$ for $x \in \partial\Omega$.

Using the results of Section I.19, it is clear that Theorems 9.1, 9.2, 10.1, 10.2, 10.3 extend (without any change in the proofs) to the case where the boundary conditions (9.5) are replaced by the more general conditions

$$B_j(x, D)u = 0 \quad \text{on } S_T \quad (1 \le j \le m); \tag{10.9}$$

here $(A(t), \{B_j\}, \Omega)$ is a regular elliptic boundary value problem satisfying the conditions of Theorem I.19.4 for each θ in the interval $\pi/2 \le \theta \le 3\pi/2$, and the B_j are independent of t.

The differentiability theorems of this section are not the best possible ones. Thus, the smoothness assumptions and the assumption (10.8) can be significantly weakened. In particular, to prove that $u(x,t)$ is continuous for $x \in \bar{\Omega}, 0 \le t \le T$, it suffices to assume, instead of (10.8), that $f(x, 0) = 0$ on $\partial\Omega$. But the proofs of these stronger results depend on tools that go beyond the scope of this book, such as singular integrals and interpolation theory.

PROBLEMS. (2) Consider the *heat equation*

$$\frac{\partial u}{\partial t} - \sum_{i=1}^n \frac{\partial^2 u}{\partial x_i^2} = 0. \tag{10.10}$$

Prove that (i) the function

$$\Gamma(x, t; \xi, \tau) = \frac{1}{(2\sqrt{\pi})^n} (t - \tau)^{-n/2} \exp\left\{\frac{|x - \xi|^2}{4(t - \tau)}\right\} \quad (\tau < t) \tag{10.11}$$

is a solution of (10.10) (for fixed (ξ, τ)), and (ii) for any continuous bounded function $f(x)$ in R^n, the integral

$$u(x, t) = \int_{R^n} \Gamma(x, t; \xi, 0) f(\xi) \, d\xi \qquad (t > 0) \qquad (10.12)$$

satisfies

$$u(x, t) \to f(x) \qquad \text{if } t \to 0. \qquad (10.13)$$

(3) For $n = 1$, consider the function

$$G(x, t; \xi, \tau) = \sum_{m=-\infty}^{\infty} \Gamma(x + 4ma, t; \xi, \tau) - \sum_{m=-\infty}^{\infty} \Gamma(2a - x + 4ma, t; \xi, \tau)$$

in the rectangle $-a < x < a, 0 < t < T$. Prove that the function $U(t, \tau) = G(\cdot, t; \cdot, \tau)$ is a fundamental solution of (10.10) under the boundary condition: $u(\pm a, t) = 0$.

(4) Extend the construction of the previous problem to $n = 2$.

(5) Prove that the operator $U(t)f$ defined by the right-hand side of (10.12) satisfies the semigroup property.

11 | A DIFFERENTIABILITY THEOREM IN HILBERT SPACE

It is sometimes more convenient to write the evolution equation in the form

$$Lu \equiv \frac{1}{i} \frac{du}{dt} - A(t)u = f(t). \qquad (11.1)$$

In Sections 3–8 we have proved theorems of existence, uniqueness, and differentiability. In the present section we deal only with the question of differentiability of solutions (even when there may not generally exist solutions of the Cauchy problem). We consider the equation (11.1) in a Hilbert space E. We shall make use of the tool of Fourier transforms. Recall that the Fourier transform of a function $u(t)$ with values in E is defined by

$$\hat{u}(\lambda) = \frac{1}{\sqrt{2\pi}} \int_{-\infty}^{\infty} e^{-i\lambda t} u(t) \, dt.$$

We then have the Plancherel theorem:

$$\int_{-\infty}^{\infty} \|\hat{u}(\lambda)\|^2 \, d\lambda = \int_{-\infty}^{\infty} \|u(t)\|^2 \, dt. \tag{11.2}$$

To be more precise, we first define the Fourier transform $\hat{u}(\lambda)$ for a function $u(t)$ in $C((-\infty, \infty); E)$ with compact support. For such a function we can prove (11.2) as follows:

Let $\{e_m\}$ be an orthonormal basis in E, and write

$$u(t) = \sum_{m=1}^{\infty} u_m(t)e_m, \quad \text{where } u_m(t) = (u(t), e_m).$$

By the classical Plancherel theorem, each function

$$\hat{u}_m(\lambda) = \frac{1}{\sqrt{2\pi}} \int_{-\infty}^{\infty} e^{-i\lambda t} u_m(t) \, dt$$

satisfies:

$$\int_{-\infty}^{\infty} |\hat{u}_m(\lambda)|^2 \, d\lambda = \int_{-\infty}^{\infty} |u_m(t)|^2 \, dt \quad (m = 1, 2, \ldots).$$

Since the space $Y \equiv L^2((-\infty, \infty); E) \equiv H^0((-\infty, \infty); E)$ is complete, and since, by the last sequence of equalities,

$$\left\| \sum_{m=M}^{N} \hat{u}_m(\lambda)e_m \right\|_Y^2 = \left\| \sum_{m=M}^{N} u_m(t)e_m \right\|_Y^2 \to 0 \quad \text{if } N > M \to \infty,$$

it follows that $\sum_{m=1}^{\infty} \hat{u}_m(\lambda)e_m$ converges, in Y, to some element $v = v(\lambda)$, and

$$\int_{-\infty}^{\infty} \|v(\lambda)\|^2 \, d\lambda = \int_{-\infty}^{\infty} \|u(t)\|^2 \, dt.$$

Since $(v(\lambda), e_m) = (\hat{u}(\lambda), e_m)$ for all m, $v(\lambda) = \hat{u}(\lambda)$, so that (11.2) holds.

Now let $u(t)$ be any element of Y. Take any sequence $\{u^j(t)\}$ from $C((-\infty, \infty); E)$ of functions with compact support such that $\|u^j - u\|_Y \to 0$. Then the sequence $\{\hat{u}^j(t)\}$ lies in Y and

$$\|\hat{u}^j - \hat{u}^i\|_Y = \|u^j - u^i\|_Y \to 0 \quad \text{if } j > i \to \infty.$$

It follows that there exists an element $\hat{u}$ in Y such that $\hat{u}^j \to \hat{u}$ in Y; also (11.2) holds. We define the Fourier transform of $u(t)$ to be this function $\hat{u}(\lambda)$. (Of course, it is defined just for almost all λ.)

By the method that we proved (11.2) one proves that

$$u(t) = \frac{1}{\sqrt{2\pi}} \int_{-\infty}^{\infty} e^{i\lambda t} \hat{u}(\lambda) \, d\lambda.$$

Let $-\infty < a < b < \infty$. In the space $C^n([a, b]; X)$ we introduce the norm

$$|u|_n = \max_{0 \le j \le n} \max_{a \le t \le b} \left\| \frac{d^j u(t)}{dt^j} \right\|.$$

We shall now state a necessary condition for differentiability of solutions of (11.1) in a Banach space X.

THEOREM 11.1.   Let $A(t) \equiv A$ *be a closed linear operator in* X *with a domain* D_A. *Let* $a < \alpha < \beta < b$. *Assume that if* $u(t) \in C^1([a, b]; X)$ *and* $f = Lu \in C^k([a, b]; X)$ *for some integer* $k \ge 1$, *then* $u \in C^{k+1}([\alpha, \beta]; X)$. *Then*

$$\|(A - \lambda I)x\| \ge c|\lambda| \, \|x\| \qquad \text{if } \lambda \text{ real}, \, |\lambda| \ge N, \, x \in D_A, \tag{11.3}$$

where c, N *are some positive constants.*

Proof.   Consider the set S of elements (Lu, u) for all $u \in C^1([a, b]; X)$ with $u(t) \in D_A$ such that $Lu \in C^k([a, b]; X)$. Using the fact that A is closed, one easily verifies that S is a closed subset of the product space $Y = C^k([a, b]; X) \times C^1([a, b]; X)$. Since S is also linear, it is a Banach space with the norm induced by Y. Consider the mapping W from S into $C^{k+1}([\alpha, \beta]; X)$ defined by $W(Lu, u) = u$. By the assumptions of the theorem, this map is well defined on the whole space S. Furthermore, it is a closed map. We now use the closed-graph theorem, which states the following:

A closed linear map from the whole Banach space Z into a Banach space V is a continuous map.

We conclude that W is continuous—that is,

$$|u|'_{k+1} \le K(|Lu|_k + |u|_1) \qquad \text{if } (Lu, u) \in S, \tag{11.4}$$

where K is a positive constant and $|u|'_{k+1}$ is the norm of u in $C^{k+1}([\alpha, \beta]; X)$.

Applying (11.4) to $u(t) = xe^{i\lambda t}$, where $x \in D_A$ and λ is real, (11.3) easily follows.

THEOREM 11.2.   Assume that $A(t)$ $(a < t < b)$ *is a closed operator in a Hilbert space* E *with a domain* D_A *independent of* t *and dense in* E. *Assume further that the resolvent of* $A(t)$ *exists for all real* λ, $|\lambda| \ge N(t)$, *and that*

$$\|R(\lambda; A(t))\| \le \frac{C(t)}{|\lambda|} \qquad \text{if } \lambda \text{ real}, \, |\lambda| \ge N(t), \tag{11.5}$$

where $C(t)$, $N(t)$ *are constants. Assume, finally, that for each* $s \in (a, b)$, $A^{-1}(s)$ *and* $A(t)A^{-1}(s)$ *are bounded operators, and, for any* $x \in X$,

$$A(t)A^{-1}(s)x \quad has \text{ m } continuous \text{ } derivatives \tag{11.6}$$

for $a < t < b$, *where* m *is an integer* ≥ 1. *If* u *is a continuously differentiable solution of* (11.1) *in* (a, b) *and if* $f \in H^m((a, b); E)$, *then* $u \in H^{m+1}((\alpha, \beta); E)$ *for any* $a < \alpha < \beta < b$.

It is left to the reader to verify that the assumptions of the theorem imply that the strong t-derivatives of $A(t)A^{-1}(s)$ of orders $\leq m$ are uniformly bounded operators (for fixed s and for t is any closed subset of (a, b)); for $m = 1$, cf. Problem 3.

Proof. We introduce the notation

$$H^m(a, b) = H^m((a, b); E), \qquad H^m = H^m(-\infty, \infty)$$

and denote by H_0^m the subspace of H^m consisting of all functions with compact support.

In what follows we shall consider solutions $u(t)$ of (11.1) that are not necessarily in $C^1([a', b']; E)$ but merely in $H^1(a', b')$. Thus the derivative du/dt is taken as a "strong" derivative.

LEMMA 11.1. *If* $A(t) \equiv A$, $f \in H_0^m$ $(m \geq 0)$, $u \in H_0^1$ *and* (11.1) *holds for* $-\infty < t < \infty$, *then* $u \in H_0^{m+1}$ *and*

$$\|u\|_{m+1} \leq C(\|f\|_m + \|u\|_0), \tag{11.7}$$

where $\|u\|_m$ *is defined by the right-hand side of* (10.1) *with* $a = -\infty$, $b = \infty$.

Proof.   Taking the Fourier transform of (11.1) we get $(\lambda I - A)\hat{u}(\lambda) = \hat{f}(\lambda)$. Hence

$$\sqrt{2\pi}\, u(t) = \int_{-N}^{N} e^{i\lambda t}\hat{u}(\lambda)\, d\lambda + \int_{-\infty}^{-N} e^{i\lambda t}R(\lambda; A)\hat{f}(\lambda)\, d\lambda + \int_{N}^{\infty} e^{i\lambda t}R(\lambda; A)\hat{f}(\lambda)\, d\lambda$$

$$\equiv u_1 + u_2 + u_3.$$

By Schwarz's inequality and Plancherel's theorem,

$$\|u_1\|_{m+1}^2 \leq C \int_{-N}^{N} \|\hat{u}(\lambda)\|^2\, d\lambda \leq C \|u\|_0^2,$$

where C is a generic constant. Next, it can easily be verified that

$$u_2^{(j)}(t) = \int_{-\infty}^{-N} e^{i\lambda t}(i\lambda)^j R(\lambda; A)\hat{f}(\lambda)\, d\lambda \qquad (0 \le j \le m+1),$$

where the derivative on the left is a strong derivative. Hence, by Plancherel's theorem and (11.5),

$$\|u_2\|_{m+1}^2 \le C \sum_{j=1}^{m+1} \int_{-\infty}^{-N} |\lambda^{j-1}\hat{f}(\lambda)|^2\, d\lambda \le C\|f\|_m^2.$$

Since, finally, u_3 can be handled similarly to u_2, (11.7) follows. From (11.1), (11.7) we get

$$\|Au\|_m \le C(\|f\|_m + \|u\|_0). \tag{11.8}$$

LEMMA 11.2. Let the assumptions of Theorem 11.2 hold for $(a, b) = (-\infty, \infty)$, *let the derivatives in* (11.6) *be uniformly bounded in* t, *and let* $\|B(t)\| < \delta$, *where* $B(t) = [A(t) - A(s)]A^{-1}(s)$. *If* u *is a solution of* (11.1) *in* $(-\infty, \infty)$, *if* $f \in H_0^m$ $(m \ge 0)$, $u \in H_0^1$ *and* $A(s)u \in H_0^m$, *and if, finally,* δ *is sufficiently small* (*depending only on* A(s), m), *then* $u \in H_0^{m+1}$ *and*

$$\|u\|_{m+1} \le C(\|f\|_m + \|u\|_0). \tag{11.9}$$

Proof. u satisfies

$$\frac{1}{i}\frac{du}{dt} - A(s)u = B(t)A(s)u(t) + f(t), \tag{11.10}$$

from which it follows that $u \in H_0^{m+1}$. Applying (11.8) with $m = 0$ and taking $\delta < 1/2C$ (C as in (11.8)), we get $\|A(s)u\|_0 \le C(\|f\|_0 + \|u\|_0)$. Next we apply (11.8) with $m = 1$ and use the last inequality. We find that $\|A(s)u\|_1 \le C(\|f\|_1 + \|u\|_0)$ provided δ is sufficiently small. Proceeding step by step, one finally gets

$$\|A(s)u\|_m \le C(\|f\|_m + \|u\|_0). \tag{11.11}$$

(11.9) follows from (11.10), (11.11).

For any function $v(t)$, set $v_h(t) = [v(t+h) - v(t)]/h$.

LEMMA 11.3. Let $u \in H_0^m$ $(m \ge 0)$. *Then* $u \in H^{m+1}$ *if and only if* $\|u_h\|_m \le M$ *for all* h *with* $|h| \le 1$, *and, in that case,* $\|u\|_{m+1} \le CM$ *and* $\|u_h\|_m \le C\|u\|_{m+1}$.

If $u(t)$ is a complex-valued function, then the assertion is well known (see Section I.15). In the general case we write

$$u(t) = \sum_{n=1}^{\infty} u_n(t)e_n,$$

where $\{e_n\}$ is an orthonormal basis in E, and use the fact that for each function $u_n(t)e_n$ the assertion of the lemma is true. The details are left to the reader.

LEMMA 11.4. Lemma 11.2 remains true even if the assumption that $A(s)u \in H^m$ is dropped.

Proof. Taking finite differences in (11.1), we get

$$\frac{1}{i}\frac{du_h}{dt} - A(t)u_h = [A_h(t)A^{-1}(s)]A(s)u(t+h) + f_h(t) \equiv g(t; h).$$

Writing

$$A(s)u(t) = [A(s)A^{-1}(t)]A(t)u(t) \tag{11.12}$$

and noting that $A(t)u(t) \in H^0$ and $A(s)A^{-1}(t)x$ is continuously differentiable for any $x \in X$ (by Lemma 8.1), we conclude that $A(s)u(t) \in H^0$. Clearly also $A(s)u_h(t) \in H^0$. We can therefore apply Lemma 11.2 to u_h with $m = 0$. We find (using Lemma 11.3) that $\|u_h\|_1 \leq C$. Hence, by Lemma 11.3, $u \in H^2$. From (11.1) it follows that $A(t)u(t) \in H^1$. Applying Lemma 8.1, we then deduce from (11.12) that $A(s)u(t) \in H^1$. We can therefore apply Lemma 11.2 with $m = 1$ and thus conclude that $u \in H^3$. Proceeding in this manner step by step, the proof is completed.

From the proof of Lemma 11.4 we deduce:

COROLLARY. Let (γ, γ') be an interval containing the support of u. If the assumptions of Lemma 11.4 hold with $(-\infty, \infty)$ replaced by (γ, γ'), then the assertion of Lemma 11.4 is still valid.

Let $\zeta(t)$ be a C^∞ function satisfying: $\zeta(t) = 1$ if $|t - s| < \varepsilon$, $\zeta(t) = 0$ if $|t - s| > 2\varepsilon$, where ε is sufficiently small. The function $v = \zeta u$ satisfies

$$\frac{1}{i}\frac{dv}{dt} - A(t)v = \zeta f + i\zeta' u. \tag{11.13}$$

On the support of v the operator $B(t)$ introduced in Lemma 11.2 satisfies $\|B(t)\| < \delta$, where $\delta \to 0$ if $\varepsilon \to 0$. Hence, by the corollary to Lemma 11.4, with $m = 1$, we find that $u \in H^2(s - \varepsilon, s + \varepsilon)$. Since this is true for any s, $u \in H^2(\alpha, \beta)$ for any $a < \alpha < \beta < b$.

Next we consider again the equation (11.13) and apply the corollary to Lemma 11.4 with $m = 2$. In this way we find that $u \in H^3(\alpha, \beta)$. Proceeding step by step, the proof of Theorem 11.2 is completed.

Note, in view of Theorem 11.1, that the sufficient condition (11.5) cannot (essentially) be improved.

PROBLEMS. (1) Show that if $u \in H^{m+1}(a, b)$, then $u(t)$ can be identified with a function in $C^m(a, b)$.

(2) Consider (11.1) with $A(t) \equiv A$ a closed operator in a Banach space X, and assume the following: There exist numbers $0 < \alpha < a$ and integers $k \geq 2$, $s \geq -1$ such that if $u(t) \in C^1[-a, a]$ and $f = Lu \in C^{k+s}[-a, a]$, then $u \in C^k[-\alpha, \alpha]$. Prove that for any $x \in D_A$,

$$\|(A - \lambda I)x\| \geq c \, |\lambda|^{-s} \exp \{-(a - \alpha)\,|\text{Im}\,(\lambda)|\} \, \|x\|$$

for all complex numbers λ satisfying

$$|\text{Im}\,(\lambda)| \leq \frac{k - 1}{a - \alpha} \log |\text{Re}\,(\lambda)| - C, \qquad |\lambda| \geq N,$$

where C, c, N are certain positive constants.

(3) Show that the assumption (11.6) for $m = 1$ and all $x \in X$ imply that $\|[A(t) - A(\sigma)]A^{-1}(s)\| \leq C\,|t - \sigma|$ for some constant C. [*Hint:* Use the principle of uniform boundedness.]

12 | A UNIQUENESS THEOREM IN HILBERT SPACE

Consider an ordinary differential inequality

$$\left\| \frac{1}{i} \frac{du}{dt} - A(t)u \right\| \leq \eta \, \|A(t)u\| + \kappa \, \|u\| \qquad (a \leq t \leq b), \tag{12.1}$$

where u, du/dt belong to the Hilbert space E and η, κ are positive constants. u is assumed to be continuous and du/dt is assumed to be piecewise continuous.

We shall make the following assumptions:

(i) $A(t)$ is a closed linear operator with domain D_A independent of t. The resolvent $R(\lambda; A(t))$ exists for $\text{Im } \lambda = \sigma_n, a \le t \le b$, where $\{\sigma_n\}$ is a sequence of negative numbers that converges to $-\infty$, and

$$\|R(\lambda; A(t))\| \le \frac{C}{1 + |\lambda|} \qquad (\text{Im } \{\lambda\} = \sigma_n; n = 1, 2, \ldots; a \le t \le b). \qquad (12.2)$$

(ii) $A^{-1}(t)$ exists for $a \le t \le b$, and, for $a \le t, \tau \le b$,

$$\|[A(t) - A(\tau)]A^{-1}(t)\| \le \delta(t - \tau),$$

where $\delta(s) \to 0$ if $s \to 0$.

THEOREM 12.1. *Assume that* (i), (ii) *hold and that* η *is a sufficiently small constant (depending only on* C). *If* u *satisfies* (12.1) *and if* u(a) = 0, *then* u(t) $\equiv$ 0 *for* a $\le$ t $\le$ b.

Proof. Set $u(t) = 0$, $A(t) = A(a)$ if $t < a$ and $u(t) = 0$, $A(t) = A(b)$ if $t > b$. Clearly u is a solution of (12.1) for $-\infty < t < b$. Given $d > 0$, let $\zeta(t)$ be a C^∞ function satisfying: $\zeta(t) = 0$ if $t \le 0$, $\zeta(t) = 0$ if $t \ge 5d$ and $\zeta(t) = 1$ if $d \le t \le 4d$. Set $\zeta_c(t) = \zeta(t - c)$, $v(t) = e^{\sigma t}\zeta_c(t)u(t)$, where σ is any real number and $c + 5d < b$. Then $v(t)$ is continuous and dv/dt is piecewise continuous for $-\infty < t < \infty$. Clearly

$$\frac{1}{i}\frac{dv}{dt} - B_\sigma(t)v = g(t), \qquad (12.3)$$

where

$$g(t) = \frac{1}{i} e^{\sigma t}\zeta_c' u + e^{\sigma t}\zeta_c f, \qquad f = \frac{1}{i}\frac{du}{dt} - A(t)u \qquad (12.4)$$

and $B_\sigma(t) = A(t) - i\sigma I$. From the relation $A(t)B_\sigma^{-1}(t) = I + i\sigma B_\sigma^{-1}(t)$ for $\sigma = \sigma_m$ and (i) it follows that

$$\|A(t)x\| \le (C + 1)\|B_\sigma(t)x\| \qquad \text{for all } \sigma = \sigma_m, x \in D_A. \qquad (12.5)$$

We now write (12.3) in the form

$$\frac{1}{i}\frac{dv}{dt} - B_\sigma(c)v = g(t) + \{[A(t) - A(c)]A^{-1}(t)\}A(t)v \equiv h(t).$$

Choose $\sigma = \sigma_m$ and take the Fourier transform of both sides. Using the assumption

$$\|R(\lambda; B_\sigma(c))\| = \|R(\lambda + i\sigma; A(c))\| \leq C/(1 + |\sigma| + |\lambda|)$$

for $\sigma = \sigma_m$, λ real, we get

$$(1 + |\sigma|) \|\hat{v}(\lambda)\| + |\lambda| \|\hat{v}(\lambda)\| \leq C \|\hat{h}(\lambda)\|.$$

Using Plancherel's theorem, it follows that

$$(1 + |\sigma|)^2 \int \|v(t)\|^2 \, dt + \int \|v'(t)\|^2 \, dt \leq C \int \|h\|^2 \, dt. \tag{12.6}$$

By (ii), $\|h(t)\| \leq \|g(t)\| + \delta(5d) \|A(t)v\|$. By (12.3), (12.5),

$$\|v'\| \geq \|B_\sigma(t)v\| - \|g\| \geq \|A(t)v\| (C + 1)^{-1} - \|g\| \qquad \text{for } \sigma = \sigma_m.$$

Substituting these inequalities into (12.6) and then using (12.1), (12.4), we get

$$(1 + |\sigma|)^2 \int \|v(t)\|^2 \, dt + \int \|A(t)v\|^2 \, dt$$

$$\leq C' \int \|e^{\sigma t} \zeta'_c u\|^2 \, dt + C'[\eta^2 + \delta^2(5\delta)] \int \|A(t)v\|^2 \, dt + C'\kappa^2 \int \|v(t)\|^2 \, dt$$

for $\sigma = \sigma_m$, where C' is a constant depending only on C. Hence, if $2C'\eta^2 < 1$ and if $d < d_0$, where d_0 is sufficiently small (depending on C and δ), and, finally, if m is sufficiently large such that $(1 + |\sigma_m|)^2 > C'\kappa^2 + 1$, then

$$\int \|v(t)\|^2 \, dt + \int \|A(t)v(t)\|^2 \, dt \leq C'' \int \|e^{\sigma t} \zeta'_c u\|^2 \, dt, \tag{12.7}$$

where C'' is a constant depending only on C. (12.7) implies that

$$\int_{c+2d}^{c+3d} \|u(t)\|^2 \, dt \leq C_0 \left\{ e^{-6\sigma d} \int_c^{c+d} \|u(t)\|^2 \, dt + e^{2\sigma d} \int_{c+4d}^{c+5d} \|u(t)\|^2 \, dt \right\} \tag{12.8}$$

for all $\sigma = \sigma_m$ with m sufficiently large, where C_0 is a constant depending only on C, ζ. Taking, in particular, c such that $c + d = a$, we obtain

$$\int_{c+2d}^{c+3d} \|u(t)\|^2 \, dt \leq C \, e^{2\sigma d} \int_{a+4d}^{a+5d} \|u(t)\|^2 \, dt \to 0 \qquad \text{if } \sigma = \sigma_m \to -\infty.$$

Thus $u(t) = 0$ if $a + d \leq t \leq a + 2d$. Varying d in the interval $0 < d < d_0$, we find that $u(t) = 0$ if $a \leq t \leq a + 2d_0$. We can now proceed step by step to show that $u(t) = 0$ for $a + 2d_0 \leq t \leq a + 4d_0$, and so on, and the proof is completed.

Theorem 12.1 can be applied to parabolic inequalities:

$$\int_{\Omega} \left| \frac{\partial u}{\partial t} + A(x, t, D)u \right|^2 dx$$

$$\leq \eta \int_{\Omega} \sum_{|\alpha|=2m} |D^{\alpha}u(x, t)|^2 \, dx + \kappa \int_{\Omega} \sum_{|\alpha| \leq 2m-1} |D^{\alpha}u(x, t)|^2 \, dx \qquad (12.9)$$

for $a \leq t \leq b$, where $u(x, t)$ are subject to the conditions

$$\frac{\partial^j u}{\partial v^j} = 0 \qquad \text{on } \partial\Omega \times [a, b] \qquad (0 \leq j \leq m - 1) \qquad (12.10)$$

(or to more general elliptic boundary conditions independent of t). We have:

COROLLARY. If the coefficients of A(x, t, D) *are continuous in* $\bar{\Omega} \times [a, b]$ *and if* u *satisfies* (12.9) *for* $a \leq t \leq b$ *and* (12.10), *where* η *is sufficiently small and* κ *is any constant, then* $u(x, t) \equiv 0$ *in* $\bar{\Omega} \times [a, b]$ *whenever* $u(x, a) \equiv 0$ *in* $\bar{\Omega}$.

Note that in view of Theorems I.8.1, I.18.1, the inequality (12.9) is equivalent to the inequality (12.1) with η replaced by $\eta/2$ (and with a different κ). Note also that in contrast to the uniqueness assertion of Theorem 3.2, we here assume that the coefficients of $A(x, t, D)$ are merely continuous (instead of uniformly continuous in x and uniformly Hölder continuous in t).

PROBLEMS. (1) Let u satisfy $\Delta u - u_t \geq 0$ ($\Delta = $ Laplacian) in a cylinder $Q_T = \Omega \times (0, T]$, and u continuous in $\bar{Q}_T$. Prove the *maximum principle*: For any $\sigma \in (0, T]$,

$$\max_{\bar{Q}_{\sigma}} u \leq \max_{\Omega_0 \cup S_{\sigma}} u$$

where $Q_{\sigma} = \Omega \times (0, \sigma]$, $\Omega_0 = \{(x, 0); x \in \Omega\}$ and $S_{\sigma} = \partial\Omega \times (0, \sigma]$. [*Hint:* If $u(x^0, \sigma) > \max_{\Omega_0 \cup S_{\sigma}} u$, then $v = u + \varepsilon|x - x^0|^2$ takes its maximum in $\bar{Q}_{\sigma}$ at an interior point, for ε sufficiently small.]

(2) Let v be a solution of the heat equation in $R^n \times (0, T]$, continuous in $R^n \times [0, T]$. Assume that $|v(x, t)| \leq Ce^{\beta|x|^2}$ for some $C > 0$, $\beta > 0$. Prove: If $v(x, 0) \equiv 0$, then $v(x, t) \equiv 0$. [*Hint:* Consider

$$u = v + \varepsilon w, \quad w = \exp\{2\beta(|x|^2 + 1)e^{\alpha t}\} \qquad \text{for } 0 \leq t \leq 1/\alpha,$$

α large, on a sequence of domains $|x| = R_m$, $R_m \to \infty$.]

(3) Extend the results of Problems 1, 2 to the case where Δ is replaced by $\sum a_{ij}(x, t) \, \partial^2/\partial x_i \, \partial x_j + \sum b_i \, \partial/\partial x_i$.

13 | CONVERGENCE OF SOLUTIONS AS $t \to \infty$

Consider the equation

$$\frac{du}{dt} + A(t)u = f(t) \qquad (0 < t < \infty). \tag{13.1}$$

We shall study the behavior of the solutions as $t \to \infty$. We need the following assumptions:

(i) The assumptions (B_1)–(B_3) of Section 3 hold for all $t_0 < \infty$, with C independent of t_0, and

$$\sup_{0 < t, s < \infty} \|A(t)A^{-1}(s)\| < \infty.$$

(ii) There exists a closed operator $A(\infty)$ with domain D_A and with bounded inverse such that

$$\|[A(t) - A(\infty)]A^{-1}(0)\| \to 0 \qquad \text{if } t \to \infty. \tag{13.2}$$

(iii) $f(t)$ is uniformly Hölder continuous in $0 \le t < \infty$—that is, for $0 \le t$, $s < \infty$.

$$\|f(t) - f(s)\| \le C|t - s|^\beta \qquad (C > 0, 0 < \beta \le 1). \tag{13.3}$$

Furthermore, there is an element $f(\infty)$ in X such that

$$\|f(t) - f(\infty)\| \to 0 \qquad \text{if } t \to \infty. \tag{13.4}$$

THEOREM 13.1. Let the assumptions (i)–(iii) *hold and let* u(t) *be a solution of* (13.1). *Then there exists an element* u(∞) *in* X *such that*

$$A(\infty)u(\infty) = f(\infty), \tag{13.5}$$

and

$$\|u(t) - u(\infty)\| \to 0, \qquad \left\|\frac{du(t)}{dt}\right\| \to 0 \qquad if\, t \to \infty. \tag{13.6}$$

Proof. By Problem 1, Section 2, for all $0 \leq t, \tau < \infty$,

$$\|e^{-tA(\tau)}\| \leq Ce^{-\delta t}, \tag{13.7}$$

$$\|A(\tau)e^{-tA(\tau)}\| \leq \frac{C}{t}\, e^{-\delta t}, \tag{13.8}$$

where (as follows from the considerations of Section 2) C, δ are positive constants independent of t, τ. Set

$$\sup_{\substack{t > \tau \geq \mu \\ 0 < s < \infty}} \|[A(t) - A(\tau)]A^{-1}(s)\| = \eta(\mu). \tag{13.9}$$

By our assumptions, $\eta(\mu) \to 0$ if $\mu \to \infty$. From (13.9) and (3.3) it follows that

$$\|[A(t) - A(\tau)]A^{-1}(s)\| \leq C\sqrt{\eta(\mu)}\,|t - s|^{\alpha/2} \qquad \text{if } \mu \leq \tau < t < \infty, 0 < s < \infty. \tag{13.10}$$

We shall prove the following:

LEMMA 13.1. For any $0 < \theta < \delta$, if $t > \tau \geq \mu \geq 0$ then

$$\|U(t, \tau) - e^{-(t-\tau)A(t)}\| \leq C\sqrt{\eta(\mu)}(t - \tau)^{\alpha/2}e^{-\theta(t-\tau)}, \tag{13.11}$$

$$\|A(t)[U(t, \tau) - e^{-(t-\tau)A(t)}]\| \leq C\sqrt{\eta(\mu)}(t - \tau)^{-1+\alpha/2}e^{-\theta(t-\tau)}, \tag{13.12}$$

where C is a constant independent of μ.

Proof. Using (13.8), (13.9), we find that

$$\|\phi_1(t, \tau)\| \leq K\sqrt{\eta(\mu)}\,|t - \tau|^{-1+\alpha/2}\,e^{-\delta(t-\tau)} \qquad (K \text{ constant}) \tag{13.13}$$

and, by induction, that

$$\|\phi_k(t, \tau)\| \leq (K\sqrt{\eta(\mu)})^k\,|t - \tau|^{k\alpha/2-1}\,\frac{(\Gamma(\alpha/2))^k}{\Gamma(k\alpha/2)}\, e^{-\delta(t-\tau)}. \tag{13.14}$$

Hence, for any $0 < \theta < \delta$,

$$\|\Phi(t, \tau)\| \leq \frac{C\sqrt{\eta(\mu)}}{|t - \tau|^{1-\alpha/2}}\, e^{-\theta(t-\tau)} \qquad (t > \tau \geq \mu), \tag{13.15}$$

where C is a generic constant independent of μ.

Reviewing the proof of Lemma 4.6 and using (13.10), (13.7), (13.8), (13.13), instead of (3.3), (3.4), (3.5), and (4.20), respectively, one finds that

$$\|\Phi(t + \Delta t, \tau) - \Phi(t, \tau)\| \leq C\sqrt{\eta(\mu)}(\Delta t)^{\alpha/2 - \delta}|t - \tau|^{\delta - 1}e^{-\theta(t - \tau)}. \tag{13.16}$$

We now proceed similarly to the proof of Lemma 7.1. From (7.7), (13.7), (13.15), and the inequality

$$\|e^{-\tau A(t)} - e^{-\tau A(s)}\| \leq C\sqrt{\eta(\mu)}|t - s|^{\alpha/2}e^{-\theta\tau} \qquad (t > s \geq \mu) \tag{13.17}$$

(which replaces (4.10)) we find that (13.11) holds.

(13.12) follows from (7.8), (7.9), making use of (13.8), (13.15), and (13.16).

COROLLARY.   *For any* $0 < \theta < \delta$, *if* $t > \tau \geq \mu \geq 0$ *then*

$$\|U(t, \tau)\| \leq Ce^{-\theta(t - \tau)}, \tag{13.18}$$

$$\|A(t)U(t, \tau)\| \leq C(t - \tau)^{-1}e^{-\theta(t - \tau)}. \tag{13.19}$$

We return to the proof of Theorem 13.1. We shall prove that

$$\frac{du}{dt} \to 0 \qquad \text{if } t \to \infty. \tag{13.20}$$

We can write:

$$u(t) = U(t, \tau)u(\tau) + \int_\tau^t U(t, s)f(s)\, ds \equiv U(t, \tau)u(\tau) + w(t). \tag{13.21}$$

For the function w we have (cf. (7.14)):

$$\frac{dw}{dt} = f(t) - \left\{ \int_\tau^t A(t)U(t, s)[f(s) - f(t)]\, ds \right.$$

$$\left. + \left[\int_\tau^t A(t)[U(t, s) - e^{-(t - s)A(t)}]\, ds \right] f(t) + [I - e^{-(t - \tau)A(t)}]f(t) \right\}. \tag{13.22}$$

By assumption, (13.3) holds and

$$\|f(s) - f(t)\| \leq \delta(\mu) \qquad \text{if } t, s \geq \mu,$$

where $\delta(\mu) \to 0$ if $\mu \to \infty$. Hence,

$$\|f(s) - f(t)\| \leq C\delta(\mu)|t - s|^{\beta/2} \qquad \text{if } s, t \geq \mu. \tag{13.23}$$

Using (13.23), (13.19) we get (if $\tau \geq \mu$)

$$\left\| \int_\tau^t A(t)U(t, s)[f(s) - f(t)] \, ds \right\| \leq C\sqrt{\delta(\mu)} \int_\tau^t \frac{e^{-\theta(t-s)}}{(t-s)^{1-\beta/2}} \, ds$$

$$= C\sqrt{\delta(\mu)}. \qquad (13.24)$$

Using (13.12) and the fact that $\sup_{t>0} \|f(t)\| < \infty$, we get

$$\left\| \left[\int_\tau^t A(t)[U(t, s) - e^{-(t-s)A(t)}] \, ds \right] f(t) \right\| \leq C\sqrt{\eta(\mu)}. \qquad (13.25)$$

Using (13.24), (13.25), (13.7), we conclude from (13.22) that $\|dw/dt\|$ can be made arbitrarily small if t is sufficiently large. Noting that the same is true of the derivative of $U(t, \tau)u(\tau)$ and recalling (13.21), (13.20) follows.

From (13.20), (13.1) it follows that

$$-A(t)u(t) + f(t) \to 0 \qquad \text{if } t \to \infty,$$

Since $f(t) \to f(\infty)$ by assumption, we deduce that

$$A(\infty)u(t) = [A(\infty)A^{-1}(t)]A(t)u(t) \to f(\infty).$$

Hence

$$u(t) = A^{-1}(\infty)A(\infty)u(t) \to A^{-1}(\infty)f(\infty).$$

Setting $A^{-1}(\infty)f(\infty) = u(\infty)$, we see that (13.5), (13.6) hold.

Note that the proof of Theorem 13.1 gives quantitative bounds on the rate of convergence of $u(t)$ and $du(t)/dt$ as $t \to \infty$.

Let Lu be a uniformly parabolic operator in Q_∞, given by (9.1), and assume that the condition (D_0) holds in $\bar{Q}_\infty$ and that $\|R(\lambda; A(t))\| \leq C/(|\lambda| + 1)$ if $\text{Re } \lambda \leq 0$ for $0 \leq t < \infty$, where $A(t)$ is considered as an operator in $L^p(\Omega)$ $(1 < p < \infty)$ with domain $H^{2m,p}(\Omega; \{\partial^{j-1}/\partial v^{j-1}\})$. Assume finally that

$$\sup_x |a_\alpha(x, t) - a_\alpha(x, \infty)| \to 0 \qquad \text{as } t \to \infty.$$

Then Theorem 13.1 can be applied. Thus, if

$$\int_\Omega |f(x, t) - f(x, s)|^p \, dx \leq C|t - s|^{p\beta} \qquad (0 < t, s < \infty),$$

$$\int_\Omega |f(x, t) - f(x, \infty)|^p \, dx \to 0 \qquad \text{as } t \to \infty,$$

then

$$\int_\Omega |u(x, t) - u(x, \infty)|^p \, dx \to 0 \qquad \text{as } t \to \infty$$

where $u(x, \infty)$ is the solution of $A(\infty)v = f(\infty)$ ($A(\infty), f(\infty)$ being defined in the obvious way). Since $\|A(\infty)u(\cdot, t)\| \le \text{const.} < \infty$ for $0 < t < \infty$, we get (using Theorem 1.9.1.)

$$|D^\gamma u(x, t)| \le C < \infty \qquad \text{in } Q_\infty, \qquad \text{if } 0 \le |\gamma| \le 2m - 1,$$

provided $p > n$.

PROBLEMS. (1) Let L, given by (9.1), be a uniformly parabolic operator in Q_∞ for which

$$\int_\Omega \overline{\phi(x)} A(x, t, D)\phi(x)\, dx \ge c \int_\Omega |\phi(x)|^2\, dx \qquad (c > 0)$$

for any $\phi \in H^{2m}(\Omega) \cap H_0^m(\Omega)$. Let u be a solution of $Lu = f$ in Q_∞ with $u(x, t) \in H^{2m}(\Omega) \cap H_0^m(\Omega)$ for every t. Prove: if

$$\int_\Omega |f(x, t)|^2\, dx \to 0 \qquad \text{as } t \to \infty,$$

then

$$\int_\Omega |u(x, t)|^2\, dx \to 0 \qquad \text{as } t \to \infty.$$

(2) If one only assumes that the assumption (i) holds, then for all $0 \le \tau < t < \infty$,

$$\|U(t, \tau)\| \le C e^{b(t - \tau)}, \tag{13.26}$$

$$\|A(t)U(t, \tau)\| \le \frac{C e^{b(t - \tau)}}{t - \tau}, \tag{13.27}$$

where C, b are positive constants.

(3) Let u be a solution of $du/dt + Au = \psi(t)$ $(0 < t < = \infty)$, where A is independent of t and satisfies the conditions (B_1), (B_2) of Section 3. Assume that

$$\int_0^\infty \|\psi(t) - \psi_0(t)\|\, dt < \infty,$$

where $\psi_0(t)$ is a continuous periodic function on $(-\infty, \infty)$, with period α. Prove that there exists a continuous periodic function $\psi_0(t)$ on $(-\infty, \infty)$, with period α, such that

$$\|u(t) - \psi_0(t)\| \to 0 \text{ if } t \to \infty.$$

(4) Let A be a self-adjoint operator in a Hilbert space H, whose spectrum consists only of eigenvalues λ_n, and assume that $|\lambda_n| \geq c > 0$ for all n. Consider the equation

$$\frac{du}{dt} + i\,Au = e^{i\alpha t}f \qquad (f \in H, \alpha \text{ real}).$$

Prove that for any $u_0 \in H$ this equation has a unique solution $u(t)$ satisfying $u(0) = u_0$, and

$$\left\| \frac{1}{t} \int_0^t u(s)\,ds - e^{i\alpha t}g \right\| \to 0 \qquad \text{if } t \to \infty, \qquad \text{for some } g \in H.$$

Furthermore, if $\alpha \neq \lambda_n$ for all n, then $g = 0$.

[*Hint:* $e^{iAt}f = \sum e^{i\lambda_n t}(f, \psi_n)\psi_n$, where $\{\psi_n\}$ is an orthonormal sequence of eigenvectors of A corresponding to $\{\lambda_n\}$.]

14 | FRACTIONAL POWERS OF OPERATORS

Let A be a closed operator in a Banach space X, with a domain D_A dense in X, and assume that the resolvent of A exists for Re $\{\lambda\} \leq 0$ and

$$\|R(\lambda; A)\| \leq \frac{C}{|\lambda| + 1} \qquad \text{if Re } \{\lambda\} \leq 0. \tag{14.1}$$

Then (by Problem 1 of Section 2 and the proof of Theorem 2.2)

$$\|e^{-tA}\| \leq Ce^{-\delta t}, \tag{14.2}$$

$$\|A^m e^{-tA}\| \leq \frac{C}{t^m} e^{-\delta t} \qquad (m = 1, 2, \ldots; C = C(m)). \tag{14.3}$$

For any $\alpha > 0$ we define the fractional power $A^{-\alpha}$ by

$$A^{-\alpha} = \frac{1}{\Gamma(\alpha)} \int_0^\infty e^{-sA} s^{\alpha-1}\,ds. \tag{14.4}$$

By the results of Section 1, A^{-1} as defined by (14.4) is, in fact, the inverse of A.

PROBLEM.   (1) Prove that $A^{-\alpha}$ is a bounded linear operator and

$$A^{-\alpha}A^{-\beta} = A^{-(\alpha+\beta)}. \qquad (14.5)$$

If $A^{-\alpha}v = 0$, then, by (14.5),

$$A^{-[\alpha]-1}v = A^{-[\alpha]-1+\alpha}A^{-\alpha}v = 0.$$

Since A^{-1} is one-to-one, we easily conclude that $v = 0$. Hence $A^{-\alpha}$ is one-to-one. We can therefore define

$$A^{\alpha} = (A^{-\alpha})^{-1}. \qquad (14.6)$$

PROBLEMS.   (2) A^{α} is a closed operator and its domain $D(A^{\alpha})$ is dense in X and is contained in the domain $D(A^{\beta})$ of A^{β} if $\alpha \geq \beta$.
(3) For any real numbers α, β, $A^{\alpha}A^{\beta}v = A^{\beta}A^{\alpha}v = A^{\alpha+\beta}v$ if $v \in D(A^{\gamma})$, where $\gamma = \max(\alpha, \beta, \alpha + \beta)$.

THEOREM 14.1.   Let $\alpha < \beta < \gamma$. Then

$$\|A^{\beta}v\| \leq C\|A^{\gamma}v\|^{(\beta-\alpha)/(\gamma-\alpha)}\|A^{\alpha}v\|^{(\gamma-\beta)/(\gamma-\alpha)} \qquad (v \in D(A^{\gamma})), \qquad (14.7)$$

where C is a constant depending only on A, α, β, γ.

Proof.   Suppose first $\alpha = -1$, $\gamma = 0$. For any $\varepsilon > 0$

$$A^{\beta}v = \int_0^{\varepsilon} e^{-sA}v\,\frac{s^{-\beta-1}}{\Gamma(-\beta)}\,ds + \int_{\varepsilon}^{\infty} Ae^{-sA}A^{-1}v\,\frac{s^{-\beta-1}}{\Gamma(-\beta)}\,ds.$$

Using (14.2), (14.3), we get

$$\|A^{\beta}v\| \leq \left(\int_0^{\varepsilon} Ce^{-\delta s}\,\frac{s^{-\beta-1}}{\Gamma(-\beta)}\,ds\right)\|v\| + \left(\int_{\varepsilon}^{\infty} Ce^{-\delta s}\,\frac{s^{-\beta-2}}{\Gamma(-\beta)}\,ds\right)\|A^{-1}v\|$$

$$\leq \frac{C}{(-\beta)\Gamma(-\beta)}\,\varepsilon^{-\beta}\|v\| + \frac{C}{\Gamma(-\beta)(-\beta-1)}\,\varepsilon^{-\beta-1}\|A^{-1}v\|.$$

Taking the minimum with respect to ε, we get

$$\|A^{\beta}v\| \leq C\|A^{-1}v\|^{-\beta}\|v\|^{1+\beta}, \qquad (14.8)$$

—that is, (14.7) is valid. Taking $v = Aw$ in the last inequality we find that (14.7) holds for $\alpha = 0 < \beta < 1 = \gamma$. If we apply this result to $A^m e^{-tA} x$ and make use of (14.3), we obtain

$$\|A^\beta e^{-tA}\| \leq \frac{C}{t^\beta} e^{-\delta t} \tag{14.9}$$

for any $\beta \geq 0$.

We can now prove (14.7) for $\alpha < \beta < \gamma \leq 0$ as follows:

$$A^\beta v = \int_0^\varepsilon A^{-\gamma} e^{-sA} \frac{s^{-\beta-1}}{\Gamma(-\beta)} A^\gamma v \, ds + \int_\varepsilon^\infty A^{-\alpha} e^{-sA} \frac{s^{-\beta-1}}{\Gamma(-\beta)} A^\alpha v \, ds.$$

Hence, by (14.9),

$$\|A^\beta v\| \leq C\varepsilon^{\gamma-\beta} \|A^\gamma v\| + C\varepsilon^{\alpha-\beta} \|A^\alpha v\|,$$

and (14.7) follows by minimizing with respect to ε. The proof of the general case ($\alpha < \beta < \gamma$) now follows by applying the last result with α, β, γ replaced by $\alpha - \gamma, \beta - \gamma, 0$ and with v replaced by $A^\gamma v$.

PROBLEMS.   In the following five problems it is assumed that (B_1)–(B_3) of Section 3 hold.

(4) Show that if $0 \leq \gamma < \beta < 1$,

$$\|A^\gamma(t)[A^{-\beta}(t) - A^{-\beta}(\tau)]\| \leq C(\gamma, \beta)|t - \tau|^\alpha. \tag{14.10}$$

(5) Show that if $0 \leq \beta \leq \gamma < 1 + \alpha$, then

$$\|A^\gamma(t) e^{-(t-\tau)A(t)} A^{-\beta}(\tau)\| \leq C(\gamma, \beta)|t - \tau|^{\beta-\gamma}. \tag{14.11}$$

(6) Show that if $0 \leq \beta \leq \gamma < 1 + \alpha$, then

$$\|A^\gamma(t) U(t, \tau) A^{-\beta}(\tau)\| \leq C(\gamma, \beta)|t - \tau|^{\beta-\gamma}. \tag{14.12}$$

[*Hint:* Prove (14.12) first for $0 < \beta \leq 1 = \gamma$, then for $0 < \beta \leq \gamma < 1 + \alpha$, using (6.2) (or (7.11)) in both cases. For $\beta = 0$, $\gamma < 1$ use (4.8) and for $\beta = 0$, $1 \leq \gamma < 1 + \alpha$, write

$$A^\gamma(t) U(t, \tau) = A^\gamma(t) U(t, \tfrac{1}{2}(t + \tau)) A^{-1/2}(\tfrac{1}{2}(t + \tau)) A^{1/2}(\tfrac{1}{2}(t + \tau)) U(\tfrac{1}{2}(t + \tau), \tau).]$$

(7) Show that if $0 \leq \gamma \leq 1, 0 \leq \beta \leq 1$, then for any $0 \leq \tau < t \leq t_0$, $0 \leq \zeta < t_0$,

$$\|A^\gamma(\zeta)[U(t, \tau) - e^{-(t-\tau)A(t)}] A^{-\beta}(\tau)\| \leq C(\beta, \gamma)|t - \tau|^{\alpha+\beta-\gamma}. \tag{14.13}$$

[*Hint:* If $\beta = 0$, use (7.4), (7.5), and Theorem 14.1. If $\beta > 0$, use (6.2).]

(8) If $0 \leq \gamma < \beta \leq 1$, then for any $0 \leq \tau, t \leq t_0$,

$$\|A^{\gamma}(t)A^{-\beta}(\tau)\| \leq C(\beta, \gamma)[\|A(t)A^{-1}(\tau)\|]^{\gamma}. \tag{14.14}$$

(9) Let A be a closed operator with dense domain and assume that A^{-1} is bounded and that $R(\lambda; A)$ exists if $|\pi - \arg \lambda| \leq \phi_0$ and $\|R(\lambda; A)\| \leq C/(1 + |\lambda|)$. Define, for $0 < \alpha < 1$,

$$A^{-\alpha} = \frac{\sin \pi \alpha}{\pi} \int_0^{\infty} \lambda^{-\alpha}(\lambda I + A)^{-1} \, d\lambda.$$

Let a contour Γ lie in $\rho(A)$ and run from $\infty e^{-i\theta_0}$ to $\infty e^{i\theta_0}$, where $\pi - \phi_0 < \theta_0 < \pi$. Assume that Γ avoids the negative real axis and 0 and that for $\lambda \in \Gamma$, $|\lambda| \geq 1$, we have: $|\pi - \arg \lambda| \leq \phi_0$. Prove that

$$A^{-\alpha} = -\frac{1}{2\pi i} \int_{\Gamma} \lambda^{-\alpha}(\lambda I - A)^{-1} \, d\lambda.$$

(10) Show that the definition of $A^{-\alpha}$ as given in Problem 8 coincides with the definition (14.4), when $\phi_0 \geq \pi/2$ (A is then of type (ϕ, M)).

We conclude this section with a few results that will be needed in Section 16.

LEMMA 14.1.   Assume that (B_1)–(B_3) hold. If $0 \leq \gamma \leq 1$, $0 \leq \beta \leq \delta < 1 + \alpha$, $0 < \delta - \gamma \leq 1$, then for any $0 \leq \tau < t < t + \Delta t \leq t_0$, $0 \leq \zeta \leq t_0$,

$$\|A^{\gamma}(\zeta)[U(t + \Delta t, \tau) - U(t, \tau)]A^{-\beta}(\tau)\| \leq C(\beta, \gamma, \delta)(\Delta t)^{\delta - \gamma}|t - \tau|^{\beta - \delta}. \tag{14.15}$$

Proof.   Write

$A^{\gamma}(\zeta)[U(t + \Delta t, \tau) - U(t, \tau)]A^{-\beta}(\tau)$

$\quad = \{A^{\gamma}(\zeta)[U(t + \Delta t, t) - e^{-\Delta t A(t + \Delta t)}]A^{-\delta}(t)$

$\qquad - A^{\gamma}(\zeta) \int_0^{\Delta t} A^{1-\delta}(t + \Delta t)e^{-sA(t+\Delta t)} \, ds + A^{\gamma}(\zeta) \int_0^{\Delta t} A(t + \Delta t)e^{-sA(t+\Delta t)}$

$\qquad \times [A^{-\delta}(t + \Delta t) - A^{-\delta}(t)] \, ds\}A^{\delta}(t)U(t, \tau)A^{-\beta}(\tau)$

and use (14.9), (14.10), (14.12), (14.13).

LEMMA 14.2. *Assume that* (B_1)–(B_3) *hold. If* $0 \le \beta \le 1, 0 \le \gamma < \alpha$, *then for any* $\varepsilon > 0$ *and for any* $0 \le \zeta \le t_0, 0 \le \tau < t \le t_0$,

$$\|A^\gamma(\zeta)A(t)[U(t, \tau) - e^{-(t-\tau)A(t)}]A^{-\beta}(\tau)\| \le C(\gamma, \beta, \varepsilon)|t - \tau|^{\beta - \gamma - 1 + \alpha - \varepsilon}. \quad (14.16)$$

If $\beta > 0$ *or if* $\beta = 0$ *and* $\zeta = t$, *then* (14.16) *holds with* $\varepsilon = 0$.

Proof. If $\beta > 0$, then this follows easily from (6.2), (14.12). It remains to consider the case $\beta = 0$. We have the identity

$$U(t, \tau) - e^{-(t-\tau)A(t)}$$

$$= \left[U\left(t, \frac{t+\tau}{2}\right) - e^{-[(t-\tau)/2]A(t)} \right] A^{-1}\left(\frac{t+\tau}{2}\right) A\left(\frac{t+\tau}{2}\right) A^{-1}(t) A(t) e^{-[(t-\tau)/2]A(t)}$$

$$+ U\left(t, \frac{t+\tau}{2}\right) A^{-1}\left(\frac{t+\tau}{2}\right) A\left(\frac{t+\tau}{2}\right) \left[U\left(\frac{t+\tau}{2}, \tau\right) - e^{-[(t-\tau)/2]A[(t+\tau)/2]} \right]$$

$$+ U\left(t, \frac{t+\tau}{2}\right) A^{-1}\left(\frac{t+\tau}{2}\right) A\left(\frac{t+\tau}{2}\right) [e^{-[(t-\tau)/2]A[(t+\tau)/2]} - e^{-[(t-\tau)/2]A(t)}].$$

Applying $A^\gamma(\zeta)A(t)$ to both sides and using (14.16) with $\beta = 1$, (14.14) with $\beta = \gamma + \varepsilon$, $t = \zeta$, $\tau = t$, (14.13) with $\gamma = 1$, $\beta = 0$, (14.12) and (4.11), (14.16) follows.

COROLLARY. *If* $0 \le \gamma < \alpha, 0 \le \beta \le 1$, *then for any* $\varepsilon > 0, 0 \le \zeta \le t_0$, $0 \le \tau < t \le t_0$,

$$\|A^\gamma(\zeta)A(t)U(t, \tau)A^{-\beta}(\tau)\| \le C(\gamma, \beta, \varepsilon)|t - \tau|^{\beta - \gamma - 1 - \varepsilon}. \quad (14.17)$$

LEMMA 14.3. *Assume that* (B_1)–(B_3) *hold and let* $0 \le \gamma < \alpha, 0 \le \beta \le 1$. *Then for any* $0 \le \tau < t + \Delta t \le t_0, 0 \le \zeta \le t_0, 0 < \rho < \gamma, \varepsilon > 0$,

$$\|A^\rho(\zeta)[A(t + \Delta t)U(t + \Delta t, \tau) - A(t)U(t, \tau)]A^{-\beta}(\tau)\|$$
$$\le C(\beta, \gamma, \rho, \varepsilon)(\Delta t)^{\gamma - \rho}|t - \tau|^{\beta - \gamma - 1 - \varepsilon}. \quad (14.18)$$

Proof. Using (14.15) with $\delta = 1 + \gamma$, $\zeta = t + \Delta t$, one easily deduces that

$$\|[A(t + \Delta t)U(t + \Delta t, \tau) - A(t)U(t, \tau)]A^{-\beta}(\tau)\|$$
$$\le C(\Delta t)^\gamma |t - \tau|^{\beta - \gamma - 1} + C(\Delta t)^\alpha |t - \tau|^{\beta - 1}$$
$$\le C(\Delta t)^\gamma |t - \tau|^{\beta - \gamma - 1} \quad (14.19)$$

if $\Delta t > |t - \tau|$. By Theorem 14.1, (14.17), (14.19), we get

$$\|A^{\rho}(\zeta)[A(t + \Delta t)U(t + \Delta t, \tau) - A(t)U(t, \tau)]A^{-\beta}(\tau)\|$$
$$\leq C\|A^{\gamma}(\zeta)[A(t + \Delta t)U(t + \Delta t, \tau) - A(t)U(t, \tau)]A^{-\beta}(\tau)\|^{\rho/\gamma}$$
$$\times \|[A(t + \Delta t)U(t + \Delta t, \tau) - A(t)U(t, \tau)]A^{-\beta}(\tau)\|^{1 - \rho/\gamma}$$
$$\leq C|t - \tau|^{(\beta - \gamma - 1)(\rho/\gamma) - \varepsilon}(\Delta t)^{\gamma(1 - \rho/\gamma)}|t - \tau|^{(\beta - \gamma - 1)(1 - \rho/\gamma)}$$

and the assertion follows. If $\Delta t \leq |t - \tau|$, then (14.18) follows easily from (14.17).

LEMMA 14.4. Assume that (B_1)–(B_3) *hold and let* $0 \leq \gamma < 1$. *Then for any* $0 \leq \tau \leq t \leq t + \Delta t \leq t_0$, *and for any continuous function* f(s),

$$\left\| A^{\gamma}(\zeta) \left[\int_t^{t + \Delta t} U(t + \Delta t, s)f(s) \, ds - \int_{\tau}^t U(t, s)f(s) \, ds \right] \right\|$$
$$\leq C(\gamma)(\Delta t)^{1 - \gamma}(|\log (\Delta t)| + 1) \max_{\tau \leq s \leq t + \Delta t} \|f(s)\|. \quad (14.20)$$

Proof. From (14.13) we get

$$\|A^{\gamma}(\zeta)U(t, s)\| \leq C(t - s)^{-\gamma}.$$

Using this inequality and the analogous inequality for $t + \Delta t$, we easily obtain (14.20) in case $t - \tau \leq \Delta t$. If $t - \tau > \Delta t$, we estimate

$$A^{\gamma}(\zeta)\left[\int_{t - \Delta t}^{t + \Delta t} U(t + \Delta t, s)f(s) \, ds - \int_{t - \Delta t}^t U(t, s)f(s) \, ds \right]$$

as before. Finally,

$$\left\| A^{\gamma}(\zeta) \int_{\tau}^{t - \Delta t} [U(t + \Delta t, s) - U(t, s)]f(s) \, ds \right\|$$

is estimated by using (14.15) with $\beta = 0$, $\delta = 1$.

If we modify the proof above by estimating integrals of the form $\int A^{\gamma}Vf \, ds$ by Hölder's inequality

$$\left\| \int A^{\gamma}Vf \, ds \right\| \leq \left(\int \|A^{\gamma}V\|^q \, ds \right)^{1/q} \left(\int \|f\|^p \, ds \right)^{1/p} \quad \left(\frac{1}{p} + \frac{1}{q} = 1 \right),$$

where $(p - 1)/p > \gamma$, and if we apply (14.15) with $\delta > (p - 1)/p$, then we get the following result:

COROLLARY. If $(p - 1)/p > \gamma$, then the left-hand side of (14.20) is bounded by

$$C(\Delta t)^{(p-1)/p-\gamma}\left[\int_{\tau}^{t+\Delta t}\|f(s)\|^{p}\,ds\right]^{1/p} \tag{14.21}$$

We shall, finally, need the following results:

LEMMA 14.5. Assume that (B_1)–(B_3) hold and that $f(s)$ is uniformly Hölder continuous (exponent ρ) in $(0, t_0]$. Let $0 \leq \beta \leq \gamma < \min(\alpha, \rho)$. Then for any $\varepsilon > 0, 0 \leq \zeta \leq t_0, 0 \leq \tau < t < t \leq \Delta t \leq t_0$,

$$\left\|A^{\beta}(\zeta)\left[\frac{\partial}{\partial t}\int_{\tau}^{t+\Delta t}U(t+\Delta t,s)f(s)\,ds - \frac{\partial}{\partial t}\int_{\tau}^{t}U(t,s)f(s)\,ds\right]\right\|$$

$$\leq C(\gamma, \beta, \varepsilon)(t-\tau)^{-\gamma-\varepsilon}(\Delta t)^{\gamma-\beta}\left\{\|f(t)\| + \sup_{\tau \leq s \leq t}\frac{\|f(t)-f(s)\|}{|t-s|^{\rho}}\right.$$

$$\left. + \sup_{t \leq s \leq t \leq \Delta t}\frac{\|f(t+\Delta t)-f(s)\|}{|t+\Delta t-s|^{\rho}}\right\}. \tag{14.22}$$

The proof is quite involved and is given in the following section.

15 | PROOF OF LEMMA 14.5

Denote the expression in braces on the right-hand side of (14.22) by $\Gamma_{\rho}(f)$ and denote by C a generic constant. From (7.14) we have, for $0 < \gamma < \min(\alpha, \rho)$,

$$\left\|A^{\gamma}(\zeta)\frac{\partial}{\partial t}\int_{\tau}^{t}U(t,s)f(s)\,ds\right\|$$

$$\leq \int_{\tau}^{t}\|A^{\gamma}(\zeta)A(t)U(t,s)\|\,|t-s|^{\rho}\Gamma_{\rho}(f)\,ds$$

$$+ \int_{\tau}^{t}\|A^{\gamma}(\zeta)A(t)[U(t,s)-e^{-(t-s)A(t)}]\|\,ds\|f(t)\| + \|A^{\gamma}(\zeta)e^{-(t-\tau)A(t)}\|\,\|f(t)\|.$$

Using (14.17), (14.16), and (14.9), we find that

$$\left\|A^{\gamma}(\zeta)\frac{\partial}{\partial t}\int_{\tau}^{t}U(t,s)f(s)\,ds\right\| \leq C|t-\tau|^{-\gamma}\Gamma_{\rho}(f). \tag{15.1}$$

We shall now use (5.5)—that is,

$$A(t) \int_\tau^t e^{-(t-s)A(s)} \, ds = H(t, \tau)$$

$$= I - \int_\tau^t \phi_1(t, s) \, ds - e^{-(t-\tau)A(t)}$$

$$+ e^{-(t-\tau)A(t)} \int_\tau^t \phi_1(t, s) \, ds + \int_\tau^t A(t) e^{-(t-\zeta)A(t)} \int_\zeta^t \phi_1(t, s) \, ds \, d\zeta$$

$$+ \int_\tau^t A(t) e^{-(t-\zeta)A(t)} \int_\tau^\zeta [\phi_1(t, s) - \phi_1(\zeta, s)] \, ds \, d\zeta$$

$$+ \int_\tau^t A(t) e^{-(t-\zeta)A(t)} [A(t) - A(\zeta)] A^{-1}(\zeta) H(\zeta, \tau) \, d\zeta. \quad (15.2)$$

We shall prove that

$$\|H(t + \Delta t, \tau) - H(t, \tau)\| \leq C \frac{(\Delta t)^\gamma}{|t - \tau|^\gamma} \qquad \text{for any } 0 \leq \gamma < \alpha. \quad (15.3)$$

Writing

$$\int_\tau^{t+\Delta t} \phi_1(t + \Delta t, s) \, ds - \int_\tau^t \phi_1(t, s) \, ds$$

$$= \int_t^{t+\Delta t} \phi_1(t + \Delta t, s) \, ds + \int_\tau^t [\phi_1(t + \Delta t, s) - \phi_1(t, s)] \, ds$$

and using (4.20), (4.27), we get

$$\left\| \int_\tau^{t+\Delta t} \phi_1(t + \Delta t, s) \, ds - \int_\tau^t \phi_1(t, s) \, ds \right\| \leq C(\Delta t)^\alpha + C(\Delta t)^\gamma (t - \tau)^{\alpha - \gamma} \leq C(\Delta t)^\gamma.$$

$$(15.4)$$

Next note, by (4.16), that

$$\|e^{-(t+\Delta t - \tau)A(t)} - e^{-(t-\tau)A(t)}\| \leq C(t - \tau)^{-\gamma}(\Delta t)^\gamma.$$

Using this inequality and (4.10), we find that

$$\|e^{-(t+\Delta t - \tau)A(t+\Delta t)} - e^{-(t-\tau)A(t)}\| \leq C(t - \tau)^{-\alpha}(\Delta t)^\alpha. \quad (15.5)$$

Combining (15.4), (15.5), and using (4.20), we get

$$\left\| e^{-(t+\Delta t-\tau)A(t+\Delta t)} \int_\tau^{t+\Delta t} \phi_1(t+\Delta t, s)\, ds - e^{-(t-\tau)A(t)} \int_\tau^t \phi_1(t, s)\, ds \right\| \le C(\Delta t)^\gamma.$$

(15.6)

Next,

$$J \equiv \left\| \int_\tau^{t+\Delta t} A(t+\Delta t)e^{-(t+\Delta t-\zeta)A(t+\Delta t)} \int_\zeta^{t+\Delta t} \phi_1(t+\Delta t, s)\, ds\, d\zeta \right.$$

$$\left. - \int_\tau^t A(t)e^{-(t-\zeta)A(t)} \int_\zeta^t \phi_1(t, s)\, ds\, d\zeta \right\|$$

$$\le \left\| \int_t^{t+\Delta t} A(t+\Delta t)e^{-(t+\Delta t-\zeta)A(t+\Delta t)} \int_\zeta^{t+\Delta t} \phi_1(t+\Delta t, s)\, ds\, d\zeta \right\|$$

$$+ \left\| \int_\tau^t A(t+\Delta t)e^{-(t+\Delta t-\zeta)A(t+\Delta t)} \left[\int_\zeta^{t+\Delta t} \phi_1(t+\Delta t, s)\, ds - \int_\zeta^t \phi_1(t, s)\, ds \right] d\zeta \right\|$$

$$+ \int_\tau^t \left\| [A(t+\Delta t)e^{-(t+\Delta t-\zeta)A(t+\Delta t)} - A(t)e^{-(t-\zeta)A(t)}] \int_\zeta^t \phi_1(t, s)\, ds \right\| d\zeta$$

$$\equiv J_1 + J_2 + J_3.$$

By (4.20), $J_1 \le C(\Delta t)^\alpha$. By (15.4)

$$J_2 \le C \int_\tau^t \frac{1}{|t+\Delta t - \zeta|} [(\Delta t)^\alpha + (\Delta t)^\gamma |t - \zeta|^{\alpha-\gamma}]\, d\zeta \le C(\Delta t)^\gamma.$$

Finally, the proof of Lemma 4.3 shows that

$$J_3 \le C \int_\tau^t \left\| \int_\zeta^t \phi_1(t, s)\, ds \right\| \frac{(\Delta t)^\alpha}{|t - \zeta|}\, d\zeta \le C(\Delta t)^\alpha,$$

where (4.20) has been used. Combining the estimates of J_1, J_2, J_3, we get

$$J \le C(\Delta t)^\gamma.$$

(15.7)

We have established so far a uniform Hölder condition (exponent γ) for the first five terms on the right-hand side of (15.2). The Hölder condition (exponent α) for the sixth term is derived by slightly different calculations than those used for the fifth term. The details are therefore left to the reader.

Consider finally the last term on the right-hand side of (15.2). We can write

$$L = \left\| \int_\tau^{t+\Delta t} A(t + \Delta t) e^{-(t+\Delta t-\zeta)A(t+\Delta t)} [A(t + \Delta t) - A(\zeta)] A^{-1}(\zeta) H(\zeta, \tau) \, d\zeta \right.$$

$$\left. - \int_\tau^t A(t) e^{-(t-\zeta)A(t)} [A(t) - A(\zeta)] A^{-1}(\zeta) H(\zeta, \tau) \, d\zeta \right\|$$

$$\leq \left\| \int_t^{t+\Delta t} A(t + \Delta t) e^{-(t+\Delta t-\zeta)A(t+\Delta t)} [A(t + \Delta t) - A(\zeta)] A^{-1}(\zeta) H(\zeta, \tau) \, d\zeta \right\|$$

$$+ \left\| \int_\tau^t A(t + \Delta t) e^{-(t+\Delta t-\zeta)A(t+\Delta t)} \{ [A(t + \Delta t) - A(\zeta)] A^{-1}(\zeta) \right.$$

$$\left. - [A(t) - A(\zeta)] A^{-1}(\zeta) \} H(\zeta, \tau) \, d\zeta \right\|$$

$$+ \left\| \int_\tau^t [A(t + \Delta t) e^{-(t+\Delta t-\zeta)A(t+\Delta t)} - A(t) e^{-(t-\zeta)A(t)}] \right.$$

$$\left. \times [A(t) - A(\zeta)] A^{-1}(\zeta) H(\zeta, \tau) \, d\zeta \right\|$$

$$= L_1 + L_2 + L_3.$$

Using (5.6), we obtain

$$L_1 \leq C(\Delta t)^\alpha.$$

The expression in braces in L_2 is bounded by both $C(\Delta t)^\alpha$ and $C|t + \Delta t - \zeta|^\alpha$. Hence it is also bounded by $C(\Delta t)^\gamma |t + \Delta t - \zeta|^{\alpha-\gamma}$. We now easily obtain (using (5.6)) the bound

$$L_2 \leq C(\Delta t)^\gamma.$$

Finally, by (5.6) and the proof of Lemma 4.3,

$$L_3 \leq C \int_\tau^t \frac{(\Delta t)^\alpha}{|t - \zeta|} |t - \zeta|^\alpha \, d\zeta \leq C(\Delta t)^\alpha.$$

We conclude that $L \leq C(\Delta t)^\gamma$. This completes the proof of (15.3).

LEMMA 15.1.   *If ψ is a function (with values in X) satisfying*

$$\|\psi(t) - \psi(\tau)\| \leq C(t - \tau)^\delta \qquad (0 < \delta < \alpha) \tag{15.8}$$

then the function $\Psi(t) \equiv A(t) \int_\tau^t e^{-(t-s)A(s)} \psi(s) \, ds$ *satisfies*

$$\|\Psi(t + \Delta t) - \Psi(t)\| \le C(\Delta t)^\alpha |t - \tau|^{-\alpha} \|\psi(t)\|$$

$$+ C(\Delta t)^\delta \sup_{\tau \le s \le t} \frac{\|\psi(t) - \psi(s)\|}{|t - s|^\delta} + \sup_{t \le s \le t + \Delta t} \frac{\|\psi(t + \Delta t) - \psi(s)\|}{|t + \Delta t - s|^\delta}. \qquad (15.9)$$

Proof. We can write

$$\|\Psi(t + \Delta t) - \Psi(t)\|$$

$$\le \|H(t + \Delta t, \tau) - H(t, \tau)\| \, \|\psi(t)\| + \|H(t + \Delta t, t)\| \, \|\psi(t + \Delta t) - \psi(t)\|$$

$$+ \left\| \int_t^{t + \Delta t} A(t + \Delta t) e^{-(t + \Delta t - s)A(s)} [\psi(s) - \psi(t + \Delta t)] \, ds \right\|$$

$$+ \left\| \int_\tau^t [A(t + \Delta t) e^{-(t + \Delta t - s)A(s)} - A(t) e^{-(t - s)A(s)}] [\psi(s) - \psi(t)] \, ds \right\|.$$

Using (15.3), (5.6), and the proof of Lemma 4.3, we get the bound

$$C(\Delta t)^\gamma |t - \tau|^{-\gamma} \|\psi(t)\| + C\|\psi(t + \Delta t) - \psi(t)\| + C(\Delta t)^\delta \sup_{t \le s \le t + \Delta t} \frac{\|\psi(t + \Delta t) - \psi(s)\|}{|t + \Delta t - s|^\delta}$$

$$+ C(\Delta t)^\alpha \sup_{\tau \le s \le t} \frac{\|\psi(t) - \psi(s)\|}{|t - s|^\delta},$$

from which (15.9) follows.

We want to apply Lemma 15.1 to the function

$$\psi(t) = \int_\tau^t \Phi(t, s) \, ds.$$

Thus we have to verify (15.8). Using (4.21), (4.25), we get

$$\|\psi(t + \Delta t) - \psi(t)\| \le \left\| \int_t^{t + \Delta t} \Phi(t + \Delta t, s) \, ds \right\| + \left\| \int_\tau^t [\Phi(t + \Delta t, s) - \Phi(t, s)] \, ds \right\|$$

$$\le C(\Delta t)^\alpha + C(\Delta t)^\delta$$

for any $0 \le \delta < \alpha$. Thus (15.8) holds for any $\delta < \alpha$.

Applying Lemma 15.1 and noting that $\|\psi(t)\| \le C|t - \tau|^\alpha$, we conclude that

$$\left\| \int_\tau^{t + \Delta t} A(t + \Delta t) e^{-(t + \Delta t - \zeta)A(\zeta)} \int_\tau^\zeta \Phi(\zeta, s) \, ds \, d\zeta \right.$$

$$\left. - \int_\tau^t A(t) e^{-(t - \zeta)A(\zeta)} \int_\tau^\zeta \Phi(\zeta, s) \, ds \, d\zeta \right\| \le C(\Delta t)^\gamma \qquad (0 < \gamma < \alpha). \qquad (15.10)$$

Integrating both sides of (4.8) with respect to τ and using (15.10), (15.3), we find that

$$\left\| A(t+\Delta t) \int_\tau^{t+\Delta t} U(t+\Delta t, s)\, ds - A(t) \int_\tau^t U(t, s)\, ds \right\| \le C(\Delta t)^\gamma |t-\tau|^{-\gamma}. \quad (15.11)$$

We can now proceed as in the proof of Lemma 15.1 and deduce the inequality (for $0 \le \gamma < \rho < \alpha$)

$$\left\| A(t+\Delta t) \int_\tau^{t+\Delta t} U(t+\Delta t, s)f(s)\, ds - A(t)\int_\tau^t U(t, s)f(s)\, ds \right\|$$

$$\le C(\Delta t)^\gamma |t-\tau|^{-\gamma}\|f(t)\| + C(\Delta t)^\rho \Gamma_\rho(f)$$

$$\le C(\Delta t)^\gamma |t-\tau|^{-\gamma}\Gamma_\rho(f). \quad (15.12)$$

Since $\int_\tau^t U(t, s)f(s)\, ds$ is a solution of $dw/dt + A(t)w = f$, we easily find that

$$\left\| \frac{\partial}{\partial t} \int_\tau^{t+\Delta t} U(t+\Delta t, s)f(s)\, ds - \frac{\partial}{\partial t}\int_\tau^t U(t, s)f(s)\, ds \right\|$$

$$\le C(\Delta t)^\gamma |t-\tau|^{-\gamma}\Gamma_\rho(f). \quad (15.13)$$

(14.22) now follows by applying Theorem 14.1 and using (15.1), (15.13).

16 | NONLINEAR EVOLUTION EQUATIONS

We consider the Cauchy problem

$$\frac{du}{dt} + A(t, u)u = f(t, u), \quad (16.1)$$

$$u(0) = u_0, \quad (16.2)$$

where (16.1) is, in general, a nonlinear equation with respect to u.

We shall make the following assumptions:

(F_1) The operator $A_0 = A(0, u_0)$ is a closed operator with a domain D_0 dense in X and

$$\|(\lambda I - A_0)^{-1}\| \le \frac{C}{1+|\lambda|} \quad \text{for all } \lambda \text{ with Re } \lambda \le 0. \quad (16.3)$$

(F_2) A_0^{-1} is a completely continuous operator.

(F$_3$) For some $\alpha \in [0, 1)$ and $R > 0$ and for any $v \in X$ with $\|v\| < R$ the operator $A(t, A_0^{-\alpha}v)$ is well defined on D_0, for all $0 \le t \le t_0$. Furthermore, for any t, τ in $[0, t_0]$ and v, w in X with $\|v\| < R$, $\|w\| < R$,

$$\|[A(t, A_0^{-\alpha}v) - A(\tau, A_0^{-\alpha}w)]A^{-1}(\tau, A_0^{-\alpha}w)\| \le C(R)(|t - \tau|^\sigma + \|v - w\|^\rho) \quad (16.4)$$

where $0 < \sigma \le 1$, $0 < \rho \le 1$.

(F$_4$) For every t, τ in $[0, t_0]$, and v, w in X with $\|v\| < R$, $\|w\| < R$,

$$\|f(t, A_0^{-\alpha}v) - f(\tau, A_0^{-\alpha}w)\| \le C(R)(|t - \tau|^\sigma + \|v - w\|^\rho). \quad (16.5)$$

Note that (F$_1$), (F$_2$) imply that $A_0^{-\gamma}$ is also completely continuous for any $0 < \gamma \le 1$. Indeed, set

$$A_{0,\varepsilon}^{-\gamma} = \frac{1}{\Gamma(\gamma)} \int_\varepsilon^\infty e^{-tA_0} t^{\gamma-1} \, dt.$$

Writing $A_{0,\varepsilon}^{-\gamma} = A_0^{-1}(A_0 A_{0,\varepsilon}^{-\gamma})$ and noting that $A_0 A_{0,\varepsilon}^{-\gamma}$ is a bounded operator (for any $\varepsilon > 0$), it follows that $A_{0,\varepsilon}^{-\gamma}$ is completely continuous. Since also $\|A_{0,\varepsilon}^{-\gamma} - A_0^{-\gamma}\| \to 0$ as $\varepsilon \to 0$, the complete continuity of $A_0^{-\gamma}$ follows.

We shall finally need the assumption:

(F$_5$) $u_0 \in D(A_0^\beta)$ for some $\beta > \alpha$ and

$$\|A_0^\alpha u_0\| < R. \quad (16.6)$$

THEOREM 16.1. Let the assumptions (F$_1$)–(F$_5$) *hold. Then there exists a number* t*, $0 < $ t* $\le t_0$, *such that there exists at least one continuously differentiable solution of* (16.1) *for* $0 < t \le$ t* *that is continuous for* $0 \le t \le$ t* *and satisfies* (16.2).

THEOREM 16.2. Let the assumptions (F$_1$), (F$_3$)–(F$_5$) *hold with* $\rho = 1$. *Then the assertion of Theorem 16.1 is valid and the solution is unique.*

Proof of Theorem 16.1. We first outline the proof. We shall introduce a set S of functions $v(t)$ $(0 \le t \le t^*)$ and a transformation $w_v = Tv$ defined by $w_v = A_0^\alpha w$, where w is the unique solution of

$$\begin{cases} \dfrac{dw}{dt} + A(t, A_0^{-\alpha}v(t))w = f(t, A_0^{-\alpha}v(t)), \\[2mm] w(0) = u_0 . \end{cases} \quad (16.7)$$

We then show that T has a fixed point—that is, there is a function $\tilde{u} \in S$ such that $T\tilde{u} = \tilde{u}$. $u = A_0^{-\alpha}\tilde{u}$ is the desired solution of (16.1), (16.2).

In proving that T has a fixed point, we shall make use of the following theorem of Schauder [1]:

THEOREM 16.3.   *Let S be a closed convex set in a Banach space* Y *and let* T *be a continuous operator from* S *into* Y *such that* TS *is contained in* S *and such that the closure of* TS *is compact. Then* T *has a fixed point in* S.

To define Y, S we first introduce sets $Q(t^*, K, \eta)$. Here η is any number satisfying $0 < \eta < \beta - \alpha$ and K is any positive number. A function $v(t)$, defined for $0 \le t \le t^*$, is said to belong to $Q(t^*, K, \eta)$ if

$$v(0) = A_0^\alpha u_0 \tag{16.8}$$

and if for any t, τ in $[0, t^*]$,

$$\|v(t) - v(\tau)\| \le K|t - \tau|^\eta. \tag{16.9}$$

From (16.6), (16.9) it follows that if t^* is sufficiently small (depending on $K, \eta, R - \|A_0^\alpha u_0\|$), then

$$\|v(t)\| < R \qquad \text{for } t \in [0, t^*]. \tag{16.10}$$

Hence the operator

$$A_v(t) = A(t, A_0^{-\alpha}v(t)) \tag{16.11}$$

is well defined for $t \in [0, t^*]$ and, by (16.4), (16.9),

$$\|[A_v(t) - A_v(\tau)]A_0^{-1}\| \le C|t - \tau|^\mu \qquad (\mu = \min(\sigma, \rho\eta)), \tag{16.12}$$

where C is a generic constant independent of t^* and of the particular v in $Q(t^*, K, \eta)$. Noting that

$$A_v(0) = A_0, \tag{16.13}$$

it easily follows from (16.3), (16.12) that, if t^* is sufficiently small, then

$$\|(\lambda I - A_v(t))^{-1}\| \le \frac{C}{1 + |\lambda|} \qquad \text{if Re } \lambda \le 0, t \in [0, t^*]. \tag{16.14}$$

From (16.4), (16.9) we also get

$$\|[A_v(t) - A_v(\tau)]A_v^{-1}(s)\| \le C|t - \tau|^\mu \qquad \text{if } t, \tau, s \in [0, t^*]. \tag{16.15}$$

By Theorem 3.1 there exists a fundamental solution $U_v(t, \tau)$ corresponding to $A_v(t)$ and all the estimates for fundamental solutions derived in previous sections hold uniformly with respect to v in $Q(t^*, K, \eta)$. In particular, from (14.15) we get for $0 \leq \alpha < \beta \leq 1, 0 \leq t \leq t + \Delta t \leq t^*$,

$$\|A_0^\alpha [U_v(t + \Delta t, 0) - U_v(t, 0)] A_0^{-\beta}\| \leq C(\Delta t)^{\beta - \alpha}. \tag{16.16}$$

Setting $f_v(t) = f(t, A_0^{-\alpha} v(t))$, it follows from (16.5), (16.9) that

$$\|f_v(t) - f_v(\tau)\| \leq C|t - \tau|^\mu. \tag{16.17}$$

Since $f_v(0) = f(0, A_0^{-\alpha} v(0)) = f(0, u_0)$ is independent of v, (16.17) implies that

$$\|f_v(t)\| \leq C. \tag{16.18}$$

From (14.20) it follows that

$$\left\| A_0^\alpha \left[\int_0^{t+\Delta t} U_v(t + \Delta t, s) f_v(s) \, ds - \int_0^t U_v(t, s) f_v(s) \, ds \right] \right\| \leq C(\Delta t)^{1-\alpha}(|\log (\Delta t)| + 1). \tag{16.19}$$

We set $S = Q(t^*, K, \eta)$ and define a transformation $w_v = Tv$ for $v \in S$ as before. In view of Theorem 3.2, w_v is given by

$$w_v(t) = A_0^\alpha U_v(t, 0) u_0 + A_0^\alpha \int_0^t U_v(t, s) f_v(s) \, ds. \tag{16.20}$$

Writing $u_0 = A_0^{-\beta} A_0^\beta u_0$ and using (16.16), (16.19), we find that

$$\|w_v(t + \Delta t) - w_v(t)\| \leq (\Delta t)^\eta C[(t^*)^{\beta - \alpha - \eta} + (t^*)^\gamma (\Delta t)^{1 - \alpha - \eta - \gamma}(|\log (\Delta t)| + 1)]$$

for any $\gamma > 0, \gamma < 1 - \alpha - \eta$. Hence, if t^* is sufficiently small,

$$\|w_v(t + \Delta t) - w_v(t)\| \leq K(\Delta t)^\eta. \tag{16.21}$$

Since also

$$w_v(0) = A_0^\alpha u_0,$$

it follows that $w_v(t) \in Q(t^*, K, \eta)$—that is, T maps S into itself.

We now consider S as a subset of the Banach space $Y \equiv C([0, t^*]; X)$ consisting of all the continuous functions $v(t)$ from $[0, t^*]$ into X with norm

$$\| v \| = \sup_{0 \leq t \leq t^*} \|v(t)\|.$$

We shall prove that T is a continuous operator in S (with the topology induced by Y). Let v_1, v_2 belong to S and set $z_1 = A_0^{-\alpha} w_{v_1}$, $z_2 = A_0^{-v} w_{v_2}$. Thus, for $i = 1, 2$,

$$
\begin{cases}
\dfrac{dz_j}{dt} + A_{v_i}(t)z_i = f_{v_i}(t), \\[2mm]
z_i(0) = u_0.
\end{cases}
\tag{16.22}
$$

It follows that

$$
\frac{d}{dt}(z_1 - z_2) + A_{v_1}(t)(z_1 - z_2) = [A_{v_2}(t) - A_{v_1}(t)]z_2 + [f_{v_1}(t) - f_{v_2}(t)].
\tag{16.23}
$$

By Lemmas 14.3, 14.5, $A_{v_2}(t)z_2(t)$ is uniformly Hölder continuous for $\tau \le t \le t^*$ ($\tau > 0$). Noting (cf. Lemma 8.1) that $A_0 A_{v_2}^{-1}(t)$ is Hölder continuous in the uniform topology, and writing $A_0 z_2(t) = [A_0 A_{v_2}^{-1}(t)]A_{v_2}(t)z_2(t)$, it follows that $A_0 z_2(t)$ is uniformly Hölder continuous in $[\tau, t^*]$, $\tau > 0$. The same is therefore true of the right-hand side of (16.23). Hence

$$
z_1(t) - z_2(t) = U_{v_1}(t, \tau)[z_1(\tau) - z_2(\tau)]
$$

$$
+ \int_\tau^t U_{v_1}(t, s)\{[A_{v_2}(s) - A_{v_1}(s)]z_2(s) + [f_{v_1}(s) - f_{v_2}(s)]\}\, ds.
\tag{16.24}
$$

Since $A_0 \int_0^t U_{v_2}(t, s)f_{v_2}(s)\, ds$ is a bounded function, it easily follows that

$$
\|A_0 z_2(t)\| \le Ct^{\beta-1}.
\tag{16.25}
$$

Hence we can take $\tau \to 0$ in (16.24) and obtain

$$
z_1(t) - z_2(t) = \int_0^t U_{v_1}(t, s)\{[A_{v_2}(s) - A_{v_1}(s)]z_2(s) + [f_{v_1}(s) - f_{v_2}(s)]\}\, ds.
$$

Using (16.4), (16.5), (16.25), and (14.12), we find that

$$
\|w_{v_1}(t) - w_{v_2}(t)\| \le \int_0^t C|t - s|^{-\alpha}\{C\|v_1(s) - v_2(s)\|^\rho s^{\beta-1}\, ds + C\|v_1(s) - v_2(s)\|^\rho\}\, ds.
$$

Hence

$$
\|w_{v_1} - w_{v_2}\| \le C(t^*)^{\beta-\alpha}\|v_1 - v_2\|^\rho.
\tag{16.26}
$$

It follows that T is a continuous operator.

We now claim that the set TS is contained in a compact subset of Y. Indeed, the functions $w(t)$ of S are uniformly bounded (by (16.10)) and equicontinuous (by (16.9)). If we can show that for each t the set $\{w_v(t); v \in S\}$ is contained in a compact subset of X, then the standard proof of the lemma of Ascoli–Arzela extends with obvious changes, and the assertion that TS is contained in a compact set of Y follows.

We can write, for each $t \in [0, t^*]$, $w_v(t) = A_0^{-\gamma} A_0^\gamma w_v(t)$ $(0 < \gamma < \beta - \alpha)$. Since $\{A_0^\gamma w_v(t); v \in S\}$ is a bounded subset of X, and since $A_0^{-\gamma}$ is completely continuous, it follows that $\{w_v(t); v \in S\}$ is indeed contained in a compact subset of X.

We can now apply Theorem 16.3 and deduce that T has a fixed point z— that is, $z \in Q(t^*, K, \eta)$ and

$$z(t) = A_0^\alpha U_z(t, 0)u_0 + A_0^\alpha \int_0^1 U(t, s)f_z(s)\, ds. \tag{16.27}$$

Then $u(t) = A_0^{-\alpha}z(t)$ satisfies

$$u(t) = U_{A^\alpha o u}(t, 0)u_0 + \int_0^t U_{A^\alpha o u}(t, s)f_{A^\alpha o u}(s)\, ds. \tag{16.28}$$

By Theorem 3.2, $u(t)$ is a solution of (16.1), (16.2).

Proof of Theorem 16.2. If $\rho = 1$, then (16.26) shows that for t^* sufficiently small T is a contraction—that is, $\|Tv_1 - Tv_2\| \leq \theta\|v_1 - v_2\|$ for some $\theta < 1$. We then use the following theorem:

THEOREM 16.4. Let T *be a map from a closed subset* S *of a Banach space* X *into* S. *Assume that* $\|Tx - Ty\| \leq \theta\|x - y\|$ *for all* x, y *in* S *and for some* $0 \leq \theta < 1$. *Then there exists precisely one point z in* S *such that* Tz = z.

Applying this theorem, we conclude that there is precisely one solution $z(t)$ in $Q(t^*, K, \eta)$. If $\tilde{z}$ is another solution, then it must clearly belong to $Q(t^*, \tilde{K}, \eta)$ for some $\tilde{K}$. From the proof of (16.26) it follows that if λ_1 is sufficiently small, then

$$\sup_{0 \leq t \leq \lambda_1} \|z(t) - \tilde{z}(t)\| \leq \tfrac{1}{2} \sup_{0 \leq t \leq 1} \|z(t) - \tilde{z}(t)\|.$$

Hence $\tilde{z}(t) = z(t)$ for $0 \leq t \leq \lambda_1$. We can now proceed to show that $\tilde{z}(t) = z(t)$ for $\lambda_1 \leq t \leq \lambda_2$, $\lambda_2 \leq t \leq \lambda_3$, and so on. We can take the λ_i so that $\lambda_i - \lambda_{i-1} = c$, where c is a positive number independent of i. Hence $\tilde{z}(t) = z(t)$ for all $t \in [0, t^*]$.

Having constructed the solution (in case $\rho = 1$) for $0 \leq t \leq t^*$, we can now proceed to extend it into an interval $t^* \leq t \leq t^{**}$, then to a third interval, and so on. From the previous proof one sees that the lengths of these intervals remain bounded from below by a fixed positive constant (except for the one that ends at t_0) provided the solution $u(t)$ is a priori known to satisfy:

$$\|A^\alpha(t, u(t))u(t)\| \leq R' < R, \tag{16.29}$$

$$\|A^\beta(t, u(t))u(t)\| \leq C', \tag{16.30}$$

$$\|R(\lambda; A(t, u(t))\| \leq \frac{C'}{|\lambda| + 1} \quad (\operatorname{Re} \lambda \leq 0), \tag{16.31}$$

(F_3) and (F_4) hold with A_0 replaced by $A(s, u(s))$, uniformly in s, (16.32)

where R', C' are constants. We sum up:

THEOREM 16.5. *Let the assumptions* (F_1), (F_3)–(F_5) *hold with* $\rho = 1$ *and assume that any possible solution in* $[0, \sigma]$ $(0 < \sigma < t_0)$ *satisfies* (16.29)–(16.32). *Then there exists a unique solution of* (16.1) *for* $0 < t \leq t_0$ *satisfying* (16.2).

Note that the solution of (16.1), (16.2) satisfies a uniform Hölder condition with exponent min (σ, ρ).

We proceed to consider in more detail the question of existence *in the large*—that is, for $0 \leq t \leq t_0$, restricting ourselves to equations of the form

$$\frac{du}{dt} + A(t)u = f(t, u). \tag{16.33}$$

In view of Theorem 16.5 our purpose is to find an a priori bound on $\|A^\rho(t)u(t)\|$ for some $0 < \rho < 1$ (and, hopefully, for *any* $0 < \rho < 1$).

THEOREM 16.6. *Let the assumption* (B_1)–(B_3) *hold and assume that if* u(t) *is a solution of* (16.33) *for* $0 \leq t \leq \sigma$, *then*

$$\|f(s, u(s))\| \leq C(1 + \|A^\rho(s)u(s)\|) \qquad (0 \leq s \leq \sigma), \tag{16.34}$$

where C, ρ *are independent of* u, σ, *and* $0 < \rho < 1$. *Then*

$$\|A^\rho(t)u(t)\| \leq \frac{C'}{t^\rho} \qquad (0 \leq t \leq \sigma), \tag{16.35}$$

where C' *is a constant independent of* u, σ.

Proof. In view of Theorem 3.2 we can write

$$u(t) = U(t, 0)u_0 + \int_0^t U(t, s)f(s, u(s))\, ds.$$

Applying $A^\rho(t)$ to both sides and using (16.34) and (14.12) with $\gamma = \rho$, $\beta = 0$, we get

$$\|A^\rho(t)u(t)\| \leq \frac{C}{t^\rho} + C \int_0^t (t-s)^{-\rho}(1 + \|A^\rho(s)u(s)\|)\, ds.$$

From this the inequality (16.35) follows.

We conclude with a result on the behavior at $t = \infty$ of solutions of (16.33).

THEOREM 16.7. *Let the assumptions* (i), (ii) *of Section* 13 *hold and assume that* u(t) *is a solution of* (16.33) *for* $0 < t < \infty$ *and*

$$\|f(t, u(t))\| \leq C\|A^\rho(t)u(t)\| + \varepsilon(t) \tag{16.36}$$

for some $\rho \in (0, 1)$, *where* $\varepsilon(t) \to 0$ *as* $t \to 0$. *Then*

$$\lim_{t \to \infty} \|A^\rho(t)u(t)\| = 0. \tag{16.37}$$

Proof. From (13.18), (13.19) and Theorem 14.1 it follows that there exists a positive number θ, $\theta < \delta$, such that, for any $0 \leq \gamma \leq 1$,

$$\|A^\gamma(t)U(t, \tau)\| \leq \frac{C}{(t-\tau)^\gamma} e^{-\theta(t-\tau)} \qquad \text{if } t > \tau \geq 0. \tag{16.38}$$

Now write

$$u(t) = U(t, \tau)u(\tau) + \int_\tau^t U(t, s)f(s, u(s))\, ds.$$

Applying $A^\rho(t)$ to both sides, and using (16.38), (16.36), we get, for the function $\phi(t) = e^{\theta(t-\tau)}\|A^\rho(t)u(t)\|$, the inequality

$$\phi(t) \leq \left\{ C(t-\tau)^{-\rho}\|u(\tau)\| + C \int_\tau^t (t-s)^{-\rho}e^{\theta(s-\tau)}\varepsilon(s)ds \right\} + C \int_\tau^t (t-s)^{-\rho}\phi(s)\, ds. \tag{16.39}$$

Denoting by $\psi(t)$ the expression in braces, we see that

$$\psi(t) \leq C(t-\tau)^{-\rho}\|u(\tau)\| + Ce^{\theta(t-\tau)} \, \underset{\tau \leq s < \infty}{\text{l.u.b.}} \, \varepsilon(s). \tag{16.40}$$

From (16.39), by iteration, we get

$$\phi(t) \le \psi(t) + \int_\tau^t \left[\sum_{j=1}^\infty \frac{(t-s)^{j-1-j\rho}(\Gamma(1-\rho))^j C^j}{\Gamma(j(1-\rho))} \right] \psi(s)\, ds.$$

Observing that the last series is bounded by $C(t-s)^{-\rho} \exp[C(t-s)^{1-\rho}]$ and using (16.40), we get, for $t - \tau \ge 1$ and for any $\lambda > 0$,

$$\phi(t) \le C e^{\lambda(t-\tau)} \|u(\tau)\| + C e^{\theta(t-\tau)} \underset{\tau \le s < \infty}{\text{l.u.b.}} \ \varepsilon(s).$$

Hence, for any $0 < \theta_0 < \theta$,

$$\|A^\rho(t)u(t)\| \le C e^{-\theta_0(t-\tau)} \|u(\tau)\| + C \underset{\tau \le s < \infty}{\text{l.u.b.}} \ \varepsilon(s). \tag{16.41}$$

This proves (16.37).

PROBLEM. (1) Prove that if in Theorem 16.7 $\varepsilon(t) \le C_0 t^{-h}$ for some $C_0 > 0$, $h > 0$, then $\|A^\rho(t)u(t)\| \le C e^{-h}$ for some $C > 0$. Also, if $\varepsilon(t) \le C_0 e^{-ht}$ for some $C_0 > 0, 0 < h < 1$, then $\|A^\rho(t)u(t)\| \le C e^{-ht}$

17 | NONLINEAR PARABOLIC EQUATIONS

We shall apply the results of Section 16 to nonlinear parabolic equations. We begin with some preliminaries.

Let A be a closed linear operator in a Banach space X with a dense domain D_A, and assume that

$$\|R(\lambda; A)\| \le \frac{C}{1 + |\lambda|} \qquad \text{for Re } \lambda \le 0. \tag{17.1}$$

LEMMA 17.1. Let B *be a closed linear operator in* X *with domain* $D_B \supseteq D_A$. *Let* $0 < \gamma < \alpha < 1$ *and suppose that, for some constant* K > 0,

$$\|Bx\| \le \delta^{1-\gamma}\|Ax\| + \frac{K}{\delta^\gamma} \|x\| \tag{17.2}$$

for all $x \in D_A$ *and for all* δ *sufficiently small (say* $0 < \delta \le \delta_0$). *Then there is a constant* C *such that*

$$\|Bx\| \le C\|A^\alpha x\| \qquad \text{for all } x \in D_A. \tag{17.3}$$

Proof. Since B is closed, if $A^{-\alpha}x \in D_B$, then

$$BA^{-\alpha}x = \frac{1}{\Gamma(\alpha)} \int_0^\infty t^{\alpha-1} Be^{-tA}x \, dt,$$

provided the integral is convergent. Hence

$$\|BA^{-\alpha}x\| \le \frac{1}{\Gamma(\alpha)} \int_0^{\delta_0} t^{\alpha-1}\|Be^{-tA}x\| \, dt + \frac{1}{\Gamma(\alpha)} \int_{\delta_0}^\infty t^{\alpha-1}\|Be^{-tA}x\| \, dt.$$

We estimate the first integral on the right by using (17.2) with $\delta = t$ and the second integral by using (17.2) with $\delta = \delta_0$. Making use of (13.7), (13.8), we easily find that $\|BA^{-\alpha}x\| \le C\|x\|$. Hence, if $x \in D_A$,

$$\|Bx\| = \|BA^{-\alpha}(A^\alpha x)\| \le C\|A^\alpha x\|.$$

Consider now any derivative D^β. In every space $L^p(\Omega)$ ($1 < p < \infty$) we can associate with D^β an operator T_β as follows:

(i) The domain of T_β is the set of all functions $u(x)$ in $L^p(\Omega)$ that have a βth weak derivative $D^\beta u$ in $L^p(\Omega)$.

(ii) $(Tu)(x)$ is the weak derivative $D^\beta u(x)$.

It is easily verified that this operator is closed. For simplicity of notation we denote T_β also by D^β.

Now let $(A, \{B_j\}, \Omega)$ be a regular elliptic boundary value problem satisfying all the assumptions of Theorem I.19.4 with $\pi/2 \le \theta \le 3\pi/2$. Then, for some $k \ge 0$, $A + kI$ satisfies (17.1) in $L^p(\Omega)$ ($1 < p < \infty$). For simplicity we assume that $k = 0$.

LEMMA 17.2. Let $B = D^\beta$, $|\beta| = 2m - 1$. Then for any $p \in (1, \infty)$ B and A *satisfy the assumptions of Lemma 17.1 in* $X = L^p(\Omega)$ *for any* $\gamma = (2m - 1)/2m + \varepsilon$, *where* $\varepsilon > 0$ *and is sufficiently small.*

Proof. From the definition of D_A in Section I.19 it is clear that $D_B \supset D_A$. It remains to prove (17.2). From Theorem I.10.1 we find that

$$|D^\beta u|_{0,p} \le C(|u|_{2m,r})^a (|u|_{0,p})^{1-a}$$

$$\le C\delta^{1-a}|Au|_{0,r} + \frac{C}{\delta^a}|u|_{0,p}$$

$$\le C\delta^{1-a}|Au|_{0,p} + \frac{C}{\delta^a}|u|_{0,p}, \tag{17.4}$$

provided $1 < r < p$, $(2m - 1)/2m < a < 1$, and

$$\frac{1}{p} = \frac{2m - 1}{n} + a\left(\frac{1}{r} - \frac{2m}{n}\right) + (1 - a)\frac{1}{p}.$$

Taking $a = (2m - 1)/2m + \varepsilon$, $\varepsilon > 0$, the last relation reduces to

$$\left(\frac{2m - 1}{2m} + \varepsilon\right)\left(\frac{1}{r} - \frac{1}{p}\right) = \frac{2m}{n}\varepsilon.$$

Clearly, if ε is sufficiently small, then one can choose $r > 1$ so that the last relation holds. (17.4) then gives (17.2) with $\gamma = a$.

The assertion of Lemma 17.2 holds also for D^β with $0 \le |\beta| \le 2m - 1$ and with $\gamma = |\beta|/(2m) + \varepsilon$, $\varepsilon > 0$ and sufficiently small. The proof is similar to the proof above.

Combining this result with Lemma 17.1, we obtain

LEMMA 17.3.   *Let* A *be a strongly elliptic operator as in Lemma* 17.2. *Then, for any* $\alpha \in ((2m - 1)/2m, 1)$, $p > 1$,

$$|D^\beta A^{-\alpha}u|_{0,p} \le C|u|_{0,p} \qquad (0 \le |\beta| \le 2m - 1) \qquad (17.5)$$

for all $u \in H^{2m, p}(\Omega; \{B_j\})$.

We shall show that Theorems 16.1, 16.2, 16.5–16.7 can be applied to nonlinear parabolic equations

$$\frac{\partial u}{\partial t} + \sum_{|\alpha| = 2m} a_\alpha(x, t, u, Du, \ldots, D^{2m-1}u)D^\alpha u = f(x, t, u, Du, \ldots, D^{2m-1}u) \quad (17.6)$$

in a cylinder $Q_T = \Omega \times (0, T)$ with coefficients in $\bar{Q}_T$. Parabolicity here means that for any real vector $\xi \ne 0$ and for arbitrary values of $u, Du, \ldots, D^{2m-1}u$,

$$(-1)^m \operatorname{Re}\left\{\sum_{|\alpha| = 2m} a_\alpha(x, t, u, Du, \ldots, D^{2m-1}u)\xi^\alpha\right\} \ge c|\xi|^{2m} \qquad (c > 0).$$

The initial and boundary conditions are taken to be (9.6), (9.5), or, more generally, (9.6), (10.9). If $u_0(x) \in C^{2m-1}(\bar{\Omega})$, then

$$A_0 u \equiv \sum_{|\alpha| = 2m} a_\alpha(x, 0, u_0(x), Du_0(x), \ldots, D^{2m-1}u_0(x))D^\alpha u$$

is a strongly elliptic operator with continuous coefficients. Without loss of generality we may assume that (F_1) holds, since otherwise we first perform a transformation $u = e^{kt}v$ which gives a new parabolic equation for which A_0 is replaced by $A_0 + kI$.

We take X to be $L^p(\Omega)$ $(1 < p < \infty)$. By Section I.19, A_0^{-1} maps bounded subsets of $L^p(\Omega)$ into bounded subsets of $W^{2m, p}(\Omega)$. In view of Theorem I.11.2, A_0^{-1} is a completely continuous operator in $L^p(\Omega)$. Furthermore, by Lemma 17.3, if $(2m - 1)/2m < \alpha < 1$, then

$$|D^\beta A_0^{-\alpha} u|_{0, p}^\Omega \le C|u|_{0, p}^\Omega \qquad \text{for } 0 \le |\beta| \le 2m - 1, \tag{17.7}$$

where C depends only on a bound on the coefficients A_0, on a module of strong ellipticity, and on a modulus of continuity of the leading coefficients. It follows that if f and the a_α are continuously differentiable in all their variables, then (F_3), (F_4) hold with $\sigma = \rho = 1$. Hence Theorems 16.1, 16.2, and 16.5–16.7 can be applied.

PROBLEMS   (1) Prove the following converse of Lemma 17.1: Let A be as in Lemma 7.1 and let B be a closed linear operator with domain $D_B \supseteq D_A$. Suppose that (17.3) holds. Then (17.2) holds for $\gamma = \alpha$, $x \in D_A$ and all δ sufficiently small.

(2) Consider the parabolic equation (17.6) in Q_∞ with the boundary conditions (9.5), and suppose that $a_\alpha = a_\alpha(x, t)$ if $|\alpha| = 2m$ and that

$$\text{Re} \left\{ \int_\Omega \sum_{|\alpha| = 2m} a_\alpha(x, t) D^\alpha u \cdot \bar u \, dx \right\} \ge c \int_\Omega |u|^2 \, dx \qquad (c > 0)$$

for all $u \in H^{2m}(\Omega) \cap H_0^m(\Omega)$. Prove that if for any solution u, in some domain Q_σ,

$$\text{Re} \{ \bar u f(x, t, u, Du, \ldots, D^{2m-1} u) \} \le C_1 |u|^2 + C_2,$$

where C_1, C_2 are independent of σ, then the following a priori inequality holds:

$$\int_\Omega |u(x, t)|^2 \, dx \le C, \qquad C \text{ independent of } \sigma.$$

(3) Prove, under the assumptions of the previous problem, that if

$$|f(t, x, u, \ldots, D^{2m-1} u)| \le C \left(1 + \sum_{j=1}^{2m-1} |D^j u|^{r_j} \right)$$

for all values of u, Du, $\ldots$, $D^{2m-1} u$, where $0 \le r_j \le (2m - 1 + n/2)/(j + n/2)$ then the condition (16.34) holds for any $\rho \in ((2m - 1/2m), 1)$.

(4) Prove, under the assumptions of the previous problem, that if for any $R > 0$

$$|f(t, x, u, \ldots, D^{2m-1} u)| \le C(R) \sum_{j=1}^{2m-1} |D^j u|^{r_j} + \varepsilon_R(t) \qquad \left(0 \le r_j \le \frac{2m - 1 + n/2}{j + n/2} \right)$$

for $|u| \leq R$ and for all values of $Du, \ldots, D^{2m-1}u$, and if

$$\text{Re } \{\bar{u}f(x, t, u, \ldots, D^{2m-1}u)\} \leq c'|u|^2 + (|u|^2 + 1)\delta(t),$$

where $c' < c$ and $\varepsilon_R(t) \to 0$, $\delta(t) \to 0$ as $t \to \infty$, then

$$\sum_{|\gamma| \leq 2m-1} \int_\Omega |D^\gamma u(x, t)|^2 \, dx \to 0 \qquad \text{as } t \to \infty.$$

18 | UNIQUENESS FOR BACKWARD EQUATIONS

In previous sections we have proved existence and uniqueness theorems for evolution equations with a given initial condition. The backward equation

$$\frac{du}{dt} - A(t)u = f(t)$$

(with A as in Section 3) does not always have a solution satisfying a given condition

$$u(0) = u_0 .$$

We shall prove, however, that, under some further restrictions on $A(t)$, the solution, if existing, is unique.

Let $u(t)$ be a solution of

$$\frac{du}{dt} + A(t)u = 0 \qquad (0 < t < T), \tag{18.1}$$

continuous in the closed interval $0 \leq t \leq T$. We shall assume:

(i) $A(t)$ is a closed operator with domain $D(A(t))$ dense in a Hilbert space X, and $u(t)$ belongs to $D(A(t))$ as well as to the domain $D(A^*(t))$ of the adjoint $A^*(t)$ of $A(t)$.

(ii) $A(t)u(t)$ is continuously differentiable for $0 < t < T$, and there exist positive constants c, k such that, for $0 < t < T$,

$$c \text{ Re } ((A(t) + kI)u(t), u(t)) - \text{Re } \frac{d}{dt}(A(t)u(t), u(t)) \geq \tfrac{1}{2} \|(A(t) + A^*(t))u(t)\|^2.$$

$$\tag{18.2}$$

PROBLEM. (1) Consider the parabolic operator (9.1) and assume that $A(x, t, D)$ is formally self-adjoint with sufficiently smooth coefficients. Prove that if u is a smooth solution of $Lu = 0$ in Q_T satisfying (9.5), then the assumptions (i), (ii) hold.

THEOREM 18.1.   *Let* u(t) *be a solution of* (18.1) *satisfying the assumptions* (i), (ii). *Then the function* $\log \|e^{-kt}u(t)\|$ *is a convex function of the variable* $s = e^{ct}$.

The convexity of a function $\phi(s)$ implies that if $s_0 < s_1 < s_2$, then

$$\phi(s_1) \le \frac{s_2 - s_1}{s_2 - s_0} \phi(s_0) + \frac{s_1 - s_0}{s_2 - s_0} \phi(s_2).$$

It easily follows that if $0 < t_1 < t$, then

$$e^{-kt} \|u(t)\| \ge \|u(0)\| \left[\frac{\|u(t_1)\|}{\|u(0)\|} e^{-kt_1}\right]^{\gamma}, \qquad \gamma = \frac{e^{ct} - 1}{e^{ct_1} - 1}. \qquad (18.3)$$

Hence, with t_1 fixed,

$$\|u(t)\| e^{-kt} \ge \rho \|u(0)\| \rho^{-a^t} \qquad (a = e^c), \qquad (18.4)$$

where ρ is a number depending on the particular solution.

COROLLARY.   *Let* u(t) *be a solution of* (18.1) *satisfying* (i), (ii). *If* u(T) = 0, *then* u(t) ≡ 0 *for* 0 ≤ t ≤ T.

Proof of Theorem 18.1.   By assumption, $u(t)$ is twice continuously differentiable. Set $v(t) = e^{-kt}u(t), q = (v, v)$. If we prove that

$$q \frac{d^2q}{ds^2} \ge \left(\frac{dq}{ds}\right)^2,$$

then the assertion follows. The last inequality is equivalent to

$$\ddot{q} \ge c\dot{q} + \frac{1}{q} \dot{q}^2, \qquad (18.5)$$

where d/dt is denoted by a dot. To prove (18.5), note that v satisfies

$$\dot{v} + A(t)v + kv = 0.$$

Hence

$$\dot{q} = 2 \operatorname{Re} (\dot{v}, v) = -2 \operatorname{Re} (Av, v) - 2k(v, v)$$

$$= -2 \operatorname{Re} e^{-2kt}(Au, u) - 2kq, \tag{18.6}$$

$$\ddot{q} = 4k \operatorname{Re} (Av, v) - 2e^{-2kt} \frac{d}{dt} \operatorname{Re} (Au, u) - 2k\dot{q}. \tag{18.7}$$

We thus get

$$\ddot{q} - \frac{1}{q} \dot{q}^2 - c\dot{q} = 4k \operatorname{Re} (Av, v) - 2e^{-2kt} \frac{d}{dt} \operatorname{Re} (Au, u) - 2k\dot{q}$$

$$- \frac{((A + A^*)v, v)^2}{(v, v)} - 8k \operatorname{Re} (Av, v) - 4k^2(v, v) - c\dot{q}$$

$$\geq -2e^{-2kt} \frac{d}{dt} \operatorname{Re} (Au, u) - \|(A + A^*)v\|^2 - c\dot{q},$$

where in deriving the last inequality we have made use of (18.6) and of Schwarz's inequality. If we now use (ii), then (18.5) follows.

From (18.6) it follows that if $((A + kI)u, u) \geq 0$, then $e^{-kt} \|u(t)\|$ is a monotone-decreasing function.

PROBLEMS. (2) Let $u(t)$ be a twice continuously differentiable solution of $du/dt = Au$, where $A = e^{i\theta}B$ and B is self-adjoint in a Hilbert space X. Prove that $\log \|u(t)\|$ is a convex function of t.

(3) Let $u(t)$ be a solution in a Hilbert space X of

$$\frac{1}{i} \frac{du}{dt} - Au = 0, \qquad 0 \leq t \leq T,$$

and assume that $u(T) = 0$ and that on some ray $\arg \lambda = \theta_0$, $0 < \theta_0 < \pi$, $\|\operatorname{Re} (\lambda; A)\| \leq C \exp \{\alpha \operatorname{Im} (\lambda)\}$ for some $\alpha > 0$. Prove: If $\alpha < T$, then $u(t) \equiv 0$ for $\alpha \leq t \leq T$. [*Hint:* $\sqrt{2\pi} \, i\hat{u}(\lambda) = R(\lambda; A)u(0)$ is an entire function. Show that $\hat{u}(\lambda) = 0(\exp \{\alpha \operatorname{Im} (\lambda)\})$.]

19 | LOWER BOUNDS ON SOLUTIONS AS $t \to \infty$

In Section 13 we have derived upper bounds, as $t \to \infty$, on the fundamental solutions of evolution equations satisfying the conditions (B_1)–(B_3) of Section 3 (cf. (13.18), (13.19), and (13.26), (13.27)). These immediately yield upper bounds on the solutions of $du/dt + A(t)u = f(t)$ as $t \to \infty$. We shall now derive lower bounds, as $t \to \infty$, on solutions of

$$\frac{du}{dt} + Au = 0, \tag{19.1}$$

assuming that A is a closed densely defined linear operator in a Banach space X, satisfying

$$\|R(\lambda; A)\| \le \frac{C}{1 + |\lambda|} \quad \text{if } \frac{\pi}{2} - \frac{\pi\alpha}{2} \le \arg \lambda \le \frac{3\pi}{2} + \frac{\pi\alpha}{2}. \tag{19.2}$$

THEOREM 19.1.   *Let* A *satisfy the foregoing assumptions. For any* $y \in X$ *there exist positive constants* d_0, d *such that*

$$\|e^{-tA}y\| \ge d_0 e^{-dt^{1/\alpha}} \|y\| \qquad \text{for } 0 \le t < \infty. \tag{19.3}$$

Proof.   By continuity, (19.3) holds if $0 \le t \le \delta$ for some $\delta > 0$. Thus it remains to establish (19.3) for $\delta < t < \infty$. Choose β such that $\alpha < \beta < 1$ and such that (19.2) holds with α replaced by β. By Theorem 2.1, e^{-tA} is analytic and bounded in the sector Σ_β: $|\arg t| \le \pi\beta/2$, $t \ne 0$. Let $\|e^{-tA}\| \le M$ in Σ_β. For any bounded linear functional f in X, consider the function

$$\phi(t) = f(e^{-tA}y). \tag{19.4}$$

$\phi(t)$ is a complex-valued bounded analytic function in Σ_β.
For any $c > 0, 0 < \gamma < 1$, introduce the region

$$G(c, \gamma) = \left\{ t = \sigma + i\tau; \sigma \ge c \text{ and } |\tau| \le (\sigma - c) \tan\left(\frac{\pi\gamma}{2}\right) \right\}.$$

If $t \in G(c, \beta)$, then $|\arg (t - c)| \le \pi\beta/2$. Hence

$$|f(e^{-tA}y)| = |f(e^{-(t-c)A}e^{-cA}y)| \le \|f\| M \|e^{-cA}y\|$$

—that is,

$$|\phi(t)| \le \|f\| M \|e^{-cA}y\| \qquad \text{if } t \in G(c, \beta). \tag{19.5}$$

The function

$$\psi(t) = \phi(t^\alpha) \tag{19.6}$$

is analytic in t for Re $t > 0$. Consider $\psi(t)$ in a strip $S: 0 \leq$ Re $t \leq c$. Clearly

$$|\psi(t)| \leq M \|f\| \|y\| \equiv N_1 \qquad \text{in } S. \tag{19.7}$$

We shall next estimate $|\psi(t)|$ on Re $t = c$. First we show that

$$\text{if Re } t = c, \qquad \text{then } t^\alpha \in G(bc^\alpha, \beta) \tag{19.8}$$

for some positive constant b independent of c. Indeed, write $t = c + i\sigma = Re^{i\theta}$, where $R \cos \theta = c, 0 \leq |\theta| < \pi/2$. Then

$$t^\alpha = (c + i\sigma)^\alpha = R^\alpha e^{i\alpha\theta} = R^\alpha(\cos \alpha\theta + i \sin \alpha\theta).$$

Therefore, t^α belong to $G(bc^\alpha, \beta)$ if

$$R^\alpha |\sin \alpha\theta| \leq (R^\alpha \cos \alpha\theta - bR^\alpha \cos^\alpha \theta) \tan\left(\frac{\pi\beta}{2}\right)$$

—that is, if

$$|\tan \alpha\theta| \leq \tan\left(\frac{\pi\beta}{2}\right)\left[1 - \frac{b \cos^\alpha \theta}{\cos \alpha\theta}\right] \qquad \left(0 \leq |\theta| < \frac{\pi}{2}\right).$$

Since $\alpha < \beta < 1$, this inequality holds if b is positive and sufficiently small. From (19.8), (19.5), (19.6) it follows that

$$|\psi(t)| \leq \|f\| M \|e^{-bc^\alpha A}y\| \equiv N_2 \qquad \text{if Re } t = c. \tag{19.9}$$

We shall need the following lemma, which is a special case of the three-lines theorem.

LEMMA 19.1.   *Let $\psi(t)$ be a continuous bounded function in S, analytic in the interior of S, and let*

$$|\psi(t)| \leq N_1 \text{ on Re } t = 0, \qquad |\psi(t)| \leq N_2 \text{ on Re } t = c.$$

Then for any t_0, $0 \leq t_0 \leq c$,

$$|\psi(t_0)| = N_1^{1 - t_0/c} N_2^{t_0/c}.$$

Proof. The function

$$h(t) = \frac{\psi(t)}{N_1} \left(\frac{N_1}{N_2}\right)^{t/c}$$

is analytic in the interior of S, continuous and bounded in S, and $|h(t)| \leq 1$ on Re $t = 0$ and on Re $t = c$. For any $\varepsilon > 0$, consider the function

$$h_\varepsilon(t) = h(t)e^{\varepsilon(t^2 - c^2)}.$$

$|h_\varepsilon(t)| \leq 1$ on Re $t = 0$ and on Re $t = c$, and $|h_\varepsilon(t)| \to 0$ if $|t| \to \infty$, $t \in S$. By the maximum modulus theorem, $|h_\varepsilon(t)| \leq 1$ for all $t \in S$. Letting $\varepsilon \to 0$, we conclude that $|h(t)| \leq 1$ in S. This yields the assertion of the lemma.

Recalling (19.7), (19.9) and using the lemma, it follows that

$$|f(e^{-t_0^\alpha A}y)| \leq \|f\| \, M \, \|y\|^{1 - t_0/c} \, \|e^{-bc^\alpha A}y\|^{t_0/c}.$$

Thus,

$$\|e^{-t_0^\alpha A}y\| \leq M \, \|y\|^{1 - t_0/c} \, \|e^{-bc^\alpha A}y\|^{t_0/c}. \tag{19.10}$$

Since $\|e^{-\delta A}y\| \geq c_1\|y\|$ $(c_1 > 0)$, taking $t_0^\alpha = \delta$ in (19.10) and setting $bc^\alpha = t$, (19.3) follows.

PROBLEMS. (1) Prove that if A is an elliptic operator with real coefficients with domain $H^{2m}(\Omega) \cap H_0^m(\Omega)$, then (19.3) holds for any $\alpha < 1$. [*Hint:* Use Problem I.18.5.]

(2) Let $p(z)$ be a polynomial in z of degree m, and consider the Cauchy problem:

$$p\left(\frac{d}{dt}\right)u + Au = f(t) \qquad (0 < t \leq t_0),$$

$$\frac{d^j u(0)}{dt^j} = u_j \qquad (0 \leq j \leq m - 1),$$

where $f(t)$, $u(t)$ and the u_j belong to a Banach space X and A is a closed operator in X. Assume that whenever this problem has a solution for given u_j and $f(t)$ (continuous), the inequality

$$\|u(t_0)\| \leq C \sum_{j=0}^{m-1} \|u_j\| + C \int_0^{t_0} \|f(s)\| \, ds$$

holds for some constant C (independent of u_j, f). Prove: If $m \geq 3$ and if $R(\lambda; A)$ exists for some $\lambda = \lambda_0$ with $|\lambda_0|$ sufficiently large, then A is a bounded operator.

20 | PROBLEMS

Consider the operator

$$Lu \equiv \sum_{i,\,j=1}^{n} a_{ij}(x,\,t)\frac{\partial^2 u}{\partial x_i\,\partial x_j} + \sum_{i=1}^{n} b_i(x,\,t)\frac{\partial u}{\partial x_i} + c(x,\,t)u - \frac{\partial u}{\partial t} \qquad (20.1)$$

in an $(n + 1)$-dimensional domain Q. We shall assume:

(A) The coefficients of L are real-valued and continuous in $\bar{Q}$, and (a_{ij}) is a positive definite matrix in $\bar{Q}$.

Given a point $P^0 = (x^0,\,t^0)$ in Q, denote by $S(P^0)$ the set of all points P in Q that can be connected to P^0 by a continuous curve in Q along which the t-coordinate is nondecreasing from P to P^0. Denote by $C(P^0)$ the connected components (in $t = t^0$) of $Q \cap \{t = t^0\}$, which contains P^0. Note that $S(P^0) \supset C(P^0)$.

We write $u \in \tilde{C}^2(Q)$ if $u, D_x u, D_x^2 u, D_t u$ are continuous in Q.

(1) Let (A) hold and let $u \in \tilde{C}^2(Q)$. Assume that at each point of Q, either $Lu \geq 0$, $c < 0$ or $Lu > 0$, $c \leq 0$. Then u cannot take a positive maximum in Q.

(2) Consider the function

$$h(x,\,t) = \exp\left\{-\alpha\left[\sum_{i=1}^{n} \lambda_i(x_i - x_i^*)^2 + \lambda_0(t - t^*)^2\right]\right\} - \exp\{-\alpha R^2\}$$

in the ellipsoid

$$E: \sum_{i=1}^{n} \lambda_i(x_i - x_i^*)^2 + \lambda_0(t - t^*)^2 < R^2. \qquad (20.2)$$

Set $x^* = (x_1^*,\,\ldots,\,x_n^*)$ and let $\bar{P} = (\bar{x},\,\bar{t})$ be a point on ∂E with $\bar{x} \neq x^*$. Let B be a ball with center $\bar{P}$ and radius $< |\bar{x} - x^*|$. Assume, finally, that $\bar{B} \subset Q$, that (A) holds, and that $c \leq 0$. Prove: If α is a sufficiently large constant, then $Lh > 0$ in $\bar{B}$.

(3) Let E be the ellipsoid (20.2) and assume that $\bar{E} \subset Q$, that (A) holds, and that $c \leq 0$. Let u be a function in $\tilde{C}^2(Q)$ satisfying $Lu \geq 0$ in Q, and suppose that $M \equiv \sup_Q u > 0$. Prove: If, for some $\bar{P} = (\bar{x},\,\bar{t})$ on ∂E, $u(\bar{P}) = M$ and $u(\bar{P}) > u(P)$ for all $P \in E$, then $\bar{x} = x^*$.

[*Hint:* We may assume that $u(\bar{P}) > u(P)$ for all $P \in \bar{E}$, $P \neq \bar{P}$. If $\bar{x} \neq x^*$, consider $u + \varepsilon h$ in B.]

(4) Assume that L satisfies (A), that $c \leq 0$, and let $u \in \tilde{C}^2(Q)$, $Lu \geq 0$ in Q, $M \equiv \sup_Q u > 0$. Prove: If for some $P^0 = (x^0,\,t^0)$ in Q, $u(P^0) = M$, then $u(P) = u(P^0)$ for all $P \in C(P^0)$.

[*Hint:* If not, there are points $\bar{P}$, P^* in $C(P^0)$ such that $u(\bar{P}) = M$, $u(P^*) < M$, and $|\bar{P} - P^*|$ is sufficiently small. Consider the ellipsoids $|x - x^*|^2 + \lambda |t - t^0|^2 < R^2$, $\lambda > 0$, $R^2 = \lambda \varepsilon^2$ and derive a contradiction to the result of Problem 3.]

(5) Let R be a rectangle $|x_i - x_i^0| \leq a_i$, $t^0 - a^0 \leq t \leq t^0$ in Q. Assume that (A) holds, that $c \leq 0$, and let u be a function in $\tilde{C}^2(Q)$ satisfying: $Lu \geq 0$ in Q, $M \equiv \sup_Q u > 0$ and $u(P^0) = M$, where $P^0 = (x^0, t^0)$, $x^0 = (x_1^0, \ldots, x_n^0)$. Prove that $u(P) \equiv u(P^0)$ in R.

[*Hint:* If the assertion is false, deduce, by using the previous problem, that there exists a smaller rectangle (denote it again by R) such that $u(x, t) < u(P^0)$ for all $(x, t) \in R$, $t < t^0$. Introduce

$$k(x, t) = t^0 - t - K |x - x^0|^2 \qquad (K > 0)$$

in the subset R' of R, where $t^0 - t > K |x - x^0|^2$, and consider $u + \varepsilon k$.]

(6) Prove the *strong maximum principle*, which asserts the following: Assume that L satisfies (A) and that $c \leq 0$ in Q. Let $u \in \tilde{C}^2(Q)$, $Lu \geq 0$ in Q. If u takes a positive maximum in Q at a point P^0 of Q, then $u(P) \equiv u(P^0)$ in $S(P^0)$.

(7) Prove that the strong maximum principle for parabolic equations implies the strong maximum principle for elliptic equations.

(8) Let $Q = \Omega \times (0, T)$, $S = \partial\Omega \times (0, T]$, $B = \bar{Q} \cap \{t = 0\}$ and assume that L satisfies (A), that $c \leq \alpha$, and let u be a solution of $Lu = f$ in Q, with u, f continuous in $\bar{Q}$. Prove:

$$\max_{\bar{Q}} |u| \leq e^{\alpha T} \left\{ \max_{B \cup S} |u| + \gamma \max_{\bar{Q}} |f| \right\},$$

where γ is a constant independent of f.

(9) Let L satisfy (A) with $Q = R^n \times (0, T)$, $c \leq 0$ in Q, and let $u \in \tilde{C}^2(Q)$, $Lu \leq 0$ in Q. Assume further that u is continuous in $\bar{Q}$, that $u(x, 0) \geq 0$ on R^n, and that

$$\liminf_{|x| \to \infty} u(x, t) \geq 0$$

uniformly with respect to t in bounded sets of $[0, T)$. Prove that $u \geq 0$ in Q.

(10) Let L, Q be as in Problem 9 and assume that

$$|a_{ij}(x, t)| \leq M, \quad |b_i(x, t)| \leq M(1 + |x|), \quad c(x, t) \leq M(1 + |x|^2). \qquad (20.3)$$

Prove: If $u \in \tilde{C}^2(Q)$, $Lu \leq 0$ in Q, u continuous in $\bar{Q}$, and

$$u(x, t) \geq -Be^{\beta |x|^2} \qquad \text{in } Q,$$

where B, β are positive constants, then $u \geq 0$ in Q.

[*Hint:* Consider

$$k(x, t) = \exp\left[\frac{k\,|x|^2}{1 - \mu t} + vt\right] \qquad \text{in } 0 \le t \le \frac{1}{2\mu}.$$

Show that $Lh/h \le 0$ and apply Problem 9 to $v = u/h$.]

(11) Let L satisfy (A) in $Q = R^n \times (0, T)$, and let (20.3) hold. Show that there exists at most one solution to the Cauchy problem:

$$\begin{cases} Lu = f & \text{in } Q, \\ u(x, 0) = \phi & \text{on } \bar{Q} \cap \{t = 0\}, \end{cases} \tag{20.4}$$

that satisfies the condition:

$$|u(x, t)| \le B e^{\beta |x|^2} \qquad \text{in } Q, \tag{20.5}$$

for some positive constants B, β.

(12) Let $f(t)$ be a nonzero C^∞ function that vanishes for $t < 0$ and for $t > 1$ and that satisfies

$$|f^{(k)}(t)| \le C^k k^{k(1+\varepsilon)} \qquad (k = 0, 1, 2, \ldots),$$

for some positive constants C, ε. It is well known [see S. Mandelbrojt, *Series de Fourier et classes quasi-analytique de fonctions* (Paris: Gauthier-Villars, 1935)] that such a function exists for any $\varepsilon > 0$. Prove that the function

$$u(x, t) = \sum_{j=0}^{\infty} \frac{f^{(j)}(t)}{\alpha^j (2mj)!} x^{2mj} \qquad (\alpha \text{ positive constant})$$

is a solution of

$$\frac{\partial u}{\partial t} = \alpha \frac{\partial^{2m} u}{\partial x^{2m}} \qquad \text{in } R^n \times (0, 1),$$

$$u(x, 0) = 0 \qquad \text{in } R^n,$$

provided $1 + \varepsilon < 2m$, and

$$|u(x, t)| \le B \exp\{\beta\,|x|^{2m/(2m-1)+\delta}\},$$

where C, β, δ are positive constants and $\delta \to 0$ if $\varepsilon \to 0$. This example shows that the condition (20.5) cannot be essentially improved.

(13) Let $F(x, t, p, p_i, p_{ij})$ $(1 \le i, j \le n)$ be a continuous function together with its first derivatives $\partial F/\partial p_{ij}$ in a domain E. Let u, v be two functions in $\bar{C}^2(Q)$, where $Q = \Omega \times (0, T)$, Ω a bounded domain. Denote by S the set of

points (x, t, p, p_i, p_{ij}) with $(x, t) \in Q$, p in the interval with end-points $v(x, t)$, $u(x, t)$, p_i in the interval with end-points $\partial v(x, t)/\partial x_i$, $\partial u(x, t)/\partial x_i$, and p_{ij} in the interval with end-points $\partial^2 v(x, t)/\partial x_i\, \partial x_j$, $\partial^2 u(x, t)/\partial x_i\, \partial x_j$. Assume that $\bar{S} \subset E$ and that $(\partial F/\partial p_{ij})$ is a positive semidefinite matrix on S. Assume finally that

$$\frac{\partial v}{\partial t} > F(x, t, v, v_{x_i}, v_{x_i x_j}) \qquad \text{in } Q,$$

$$\frac{\partial u}{\partial t} \leq F(x, t, u, u_{x_i}, u_{x_i x_j}) \qquad \text{in } Q,$$

and that $v > u$ on $\bar{Q} \cap \{t = 0\}$ and on $\partial\Omega \times \{0 \leq t < T\}$. Prove that $v > u$ in Q.

[*Hint:* If $v > u$ on $Q \cap \{0 \leq t < t_0\}$ but not on $Q \cap \{t = t_0\}$, then evaluate $\partial(v - u)/\partial t$ at a point (x^0, t^0), where $v = u$.]

(14) Generalize the result of Problem 13 to general bounded domains Q.

(15) Let L satisfy (A) in $Q = \Omega \times (0, \infty)$, where Ω is a bounded domain in R^n, and let $c \leq 0$. Let $f(u)$ be a positive monotone increasing in u, $u_0 \leq u < \infty$, and assume that

$$\int_{u_0}^{\infty} \frac{du}{f(u)} < \infty,$$

$$\frac{f(\mu u)}{\mu f(u)} \geq K(\mu) \qquad \text{for } u \geq u_0, \mu \geq \mu_0; \quad K(\mu) \to 0 \text{ if } \mu \to \infty.$$

Then for any $0 < T_0 < T$ and for any closed subdomain Ω_0 of Ω there exists a positive constant N (depending only on $f, \Omega, \Omega_0, T, T_0, L$) and function z satisfying

$$Lz > -f(z) \qquad \text{in } \Omega \times (0, T_0),$$

$$z < N \qquad \text{on } \bar{\Omega} \cap \{t = 0\} \text{ and on } \partial\Omega \times (0, T_0),$$

$$z(x, t) \to 0 \qquad \text{if } x \in \Omega_0, \ t \to T_0.$$

[*Hint:* Let $G(v) = \int_v^{\infty} (1/f(v))\, dv$ and let g be the inverse of G. $g(w)$ is defined for $0 < w < \varepsilon_0$, $g(w) > 0$, $g'(w) < 0$, $g''(w) \geq 0$, $g(w) \to \infty$ if $w \to 0$. Let $\zeta(x) \in C_0^{\infty}(\Omega)$, $\zeta = \varepsilon_0/T_0$ on Ω_0, $0 \leq \zeta \leq \varepsilon_0/T_0$ elsewhere, and take

$$z(x, t) = \mu g(\varepsilon_0 - t\zeta(x)), \quad N = \tfrac{1}{2}\mu g(\varepsilon_0).]$$

(16) Let $L, f, \Omega, T, \Omega_0, T_0$ be as in the previous problem. Consider the boundary value problem:

$$Lu = -f(u) \qquad \text{in } \Omega \times (0, T),$$

$$u = h \qquad \text{on } \bar{Q} \cap \{t = 0\} \text{ and on } \partial\Omega \times \{0 \leq t < T\}.$$

Prove: If $h \geq N$, then this problem has no solutions.

(17) Let u be a solution of

$$Lu = f(x, t, u, D_x u) \qquad \text{in } \Omega \times (0, t_0),$$

$$u(x, 0) = 0 \qquad \text{on } \partial\Omega \times (0, t_0),$$

for some $t_0 > 0$, where L satisfies (A) in $\Omega \times (0, \infty)$. Assume that

$$-wf(x, t, w, 0) \leq \gamma w^2 + \delta \qquad (\gamma, \delta \text{ constants})$$

for any $x \in \Omega$, $0 < t < \infty$, $-\infty < w < \infty$. Prove: For any $\gamma' > \gamma$, $|u(x, t)| \leq M\{\exp \gamma' t\}$, where M is a constant *independent* of t_0.

[*Hint:* Use Problem 13.]

(18) Let u be a solution of the heat equation $\Delta u - u_t = 0$ in $Q = R^n \times (a, b)$, and let u be continuous in $\bar{Q}$ and satisfy (20.5). Prove:

$$u(x, t) = \int_{R^n} \Gamma(x, t; \xi, a)u(\xi, a) \, d\xi$$

if $t - a < 1/4\beta$; Γ is defined by (10.11).

[*Hint:* Use Problem 2 of Section 10 and Problem 2 of Section 12.]

(19) Prove the following Liouville-type theorem for the heat equation: If $\Delta u - u_t = 0$ in $Q = R^n \times \{-\infty < t < 0\}$ and if u is bounded in Q, then $u \equiv$ const.

(20) Let u be a solution of the heat equation in $Q = R^n \times (a, b)$. Prove that the function

$$w(x, t) = \frac{1}{\Omega_n r^{n-1}} \int_{|\xi|=r} u(\xi, t) \, dS_\xi \qquad (|x| = r),$$

where Ω_n is the surface area of the unit sphere in R^n, is also a solution of the heat equation in Q. Note that w is symmetric—that is, $w(x, t) = w(y, t)$ if $|x| = |y|$.

(21) Set

$$K_n(s) = \sum_{k=0}^{\infty} \frac{(s/2)^{2k}}{k! \, \Gamma(k + n/2)}.$$

Prove that for any $\xi > 0$, the function

$$w(x, t) = e^{\xi^2 t} K_m(|x| \xi)$$

is a symmetric nonnegative solution of the heat equation in n dimensions— that is, $\Delta w - w_t = 0$, where Δ is the Laplacian in R^n.

(22) Let $\phi(\xi)$ be a monotone nondecreasing function with $\int_0^\infty d\phi(\xi) < \infty$. Prove that the function

$$w(x, t) = \int_0^\infty e^{\xi^2 t} K_n(|x| \, \xi) \, d\phi(\xi) \qquad \text{in } R^n \times (-\infty, 0) \tag{20.6}$$

is a symmetric nonnegative solution of the heat equation.

(23) Let u be a nonnegative solution of the heat equation in $Q = R^n \times (a, b)$, and let $a < s < t < b$, $R > 0$. Prove that

$$u(x, t) \geq (4\pi(t - s))^{-n/2} \int_{|x| < R} u(\xi, s) \exp\left\{-\frac{|x - \xi|^2}{4(t - s)}\right\} d\xi.$$

(24) Let Γ be defined by (10.11) and let u be a solution of the heat equation in $Q = \Omega \times (a, b)$, where Ω is a bounded domain in R^n with smooth boundary. Assume that u is continuous in $\bar{Q}$ and that $D_x u$ is continuous in $\partial\Omega \times [a, b]$. Prove: For any $(x, t) \in Q$,

$$u(x, t) = \int_\Omega \Gamma(x, t; \xi, a) u(\xi, a) \, d\xi$$

$$+ \int_a^t \int_{\partial\Omega} \left[u(\xi, \tau) \frac{\partial}{\partial \nu} \Gamma(x, t; \xi, \tau) - \Gamma(x, t; \xi, \tau) \frac{\partial u(\xi, \tau)}{\partial \nu} \right] dS_\xi \, d\tau.$$

It follows that solutions of the heat equations belong to C^∞.

[*Hint:* Use the relation

$$\int_a^t \int_\Omega [v(u_\tau - \Delta u) + u(v_\tau + \Delta v)] \, d\xi \, d\tau$$

$$= \int_a^t \int_\Omega (uv)_\tau \, d\xi \, d\tau - \int_a^t \int_{\partial\Omega} \left(v \frac{\partial u}{\partial \nu} - u \frac{\partial v}{\partial \nu} \right) dS_\xi \, d\tau.] \tag{20.7}$$

(25) Let u be a solution of the heat equation in $Q = R^n \times (a, b)$, continuous in $\bar{Q}$ and satisfying

$$\int_a^b \int_{R^n} |u(x, t)| \, e^{-b|x|^2} \, dx \, dt < \infty \tag{20.8}$$

for some positive constant b. Prove: If $u(x, a) \equiv 0$ on R^n, then $u \equiv 0$ in Q.

[*Hint:* Use (20.7) with $v = \zeta\Gamma$, $\zeta(x) = 1$ if $|x| < R$, $\zeta(x) = 0$ if $|x| \geq R + 1$ $|D^\alpha\zeta| \leq \text{const.}$ for $|\alpha| \leq 2$.]

(26) Let u be a nonnegative solution of the heat equation in $Q = R^n \times (a, c)$, continuous in $\bar{Q}$. Prove that for any $a < b < c$, u satisfies (20.8).

[*Hint:* Integrate the inequality

$$u(x, t) \geq (4\pi(t - s))^{-n/2} \int_{R^n} u(\xi, s) \exp\left\{-\frac{|x - \xi|^2}{4(t - s)}\right\} d\xi \qquad (20.9)$$

for $x = 0$, $a < s < \eta$ $(\eta < t)$.]

(27) Let u be a nonnegative solution of the heat equation in $Q = R^n \times (a, b)$, continuous in $\bar{Q}$. Prove: If $u(x, a) \equiv 0$ on R^n, then $u \equiv 0$ in Q.

(28) Let u be as in Problem 26. Use the inequality (20.9) with $s = a$ to conclude that the right-hand side of (20.9), for $s = a$, is a solution of the heat equation in Q and is continuous in $R^n \times [a, c)$.

(29) Let u be a nonnegative solution of the heat equation in $Q = R^n \times (a, b)$, continuous in $\bar{Q}$. Prove: If $(x, t) \in Q$,

$$u(x, t) = (4\pi(t - a))^{-n/2} \int_{R^n} u(\xi, a) \exp\left\{-\frac{|x - \xi|^2}{4(t - a)}\right\} d\xi. \qquad (20.10)$$

(30) Let u be a nonnegative solution of the heat equation in $Q = R^n \times (-\infty, 0)$. Then $u(x, t_1) \geq u(x, t_2)$ if $t_2 \leq t_1 < 0$.
[*Hint:* Show that

$$u(x, t_1) \geq u(x, t_2)\left(\frac{t_2 - s}{t_1 - s}\right)^{n/2}, \qquad t_1 < s < t_2.]$$

(31) Let u be a nonnegative solution of the heat equation in $Q = R^n \times (-\infty, 0)$. Show that $\partial^m u/\partial t^m \geq 0$ for all $m \geq 0$.

(32) Let $u(r, t)$ be a symmetric nonnegative solution of the heat equation in $Q = R^n \times (-\infty, 0)$. Use (20.10) to show that

$$u(r, t) = \sum_{m=0}^{\infty} a_m(t)r^{2m}, \qquad a_m(t) = \frac{1}{(2m)!} \frac{\partial^{2m}}{\partial r^{2m}} u(0, t),$$

and the series represents an entire analytic function (for r in the complex plane). Derive the similar expansion

$$u_t(r, t) = \sum_{m=0}^{\infty} b_m(t)r^{2m}, \qquad b_m(t) = \frac{1}{(2m)!} \frac{\partial^{2m}}{\partial r^{2m}} u_t(0, t),$$

and verify that $b_m(t) = da_m(t)/dt$.

(33) Let u be as in the previous problem. Show that

$$a_m(t) = \frac{\Gamma(n/2)}{4^m m! \Gamma(m + n/2)} \frac{\partial^m}{\partial t^m} u(0, t).$$

(34) A theorem of Hausdorff–Bernstein–Widder states that if a C^∞ function $v(t)$ satisfies: $v^{(n)}(t) \geq 0$ for $-\infty < t < 0$, $n = 0, 1, 2, \ldots$, then there exists a nondecreasing function $\phi(s)$ ($0 < s < \infty$) such that

$$v(t) = \int_0^\infty e^{st} \, d\phi(s).$$

Use this theorem and the previous problem to derive the following theorem:

If w is a symmetric nonnegative solution of the heat equation in $Q = R^n \times (-\infty, 0)$, then w has the form (20.6).

(35) Let u be a positive solution of the heat equation in $Q = R^n \times (-\infty, 0)$. Prove that

$$\gamma \equiv \lim_{t \to -\infty} \frac{\log u(0, t)}{t} \quad \text{exists.}$$

(36) Let w be a positive symmetric solution of the heat equation in $Q = R^n \times (-\infty, 0)$, and define

$$h(\alpha) = \overline{\lim_{t \to -\infty}} \frac{\log w(|\alpha| t, t)}{t}.$$

Prove: If α_0 is the distance from 0 to the support of ϕ (cf. (20.6)), then

$$h(\alpha) = \alpha_0(\alpha_0 - \alpha) \quad \text{if } 0 \leq \alpha \leq \alpha_0.$$

(37) Let w, α_0 be as in the previous problem. Prove that

$$h(\alpha) \leq \alpha_0(\alpha_0 - \alpha) \quad \text{if } \alpha_0 < \alpha < \infty.$$

(38) Let u, γ be as in Problem 35. Set $K_\alpha = \{(x, t); |x| < |\alpha| t, -\infty < t < 0\}$. Prove: If $\gamma > 0$, then u is unbounded in K_α for any $\alpha > \gamma^{1/2}$; more generally, there does not exist a positive function $\phi(t)$ such that

$$\lim_{t \to -\infty} \frac{\log \phi(t)}{t} = 0, \quad u(x, t) \leq \phi(t) \text{ in } K_\alpha \quad (\alpha > \gamma^{1/2}).$$

(39) A domain Q in $R^n \times (-\infty < t < \infty)$ is said to have the property P if for any two points in Q there exists a continuous curve connecting them along which the t-coordinate varies monotonically. *Harnack's inequality* for the heat equation states the following:

If u is a nonnegative solution of the heat equation in a domain Q_0, then for any bounded domain Q having the property P and satisfying $\bar{Q} \subset Q_0$ (more precisely, with $[\bar{Q} \cap \{t < \lambda\}] \subset [Q_0 \cap \{t < \lambda\}]$ for any λ),

$$u(x, t) \geq cu(x', t')$$

for all (x, t), (x', t') in Q provided $t \geq t'$; here c is a positive constant independent of u.

Prove this inequality in the case Q is a rectangular domain.

[*Hint:* Represent u in terms of Green's function (cf. Problem 3, Section 10) and show that $G(x, t; \xi, 0)/G(x', t'; \xi, 0) \geq c > 0$ for $(\xi, 0)$ at the bottom base of Q, and so on.]

(40) Prove Harnack's inequality in the general case.

(41) Let Ω be a bounded domain in R^{n+1} and let $P^0 = (x^0, t^0)$ be a point on its boundary $\partial\Omega$. We say that P^0 has the *inside sphere property* if there exists a closed ball B with center $(\bar{x}, \bar{t})$ such that $B \cap \{t \leq t^0\}$ lies in $\bar{\Omega}$, $\bar{Q} \cap \partial B \cap \{t \leq t^0\} = \{P^0\}$ and $\bar{x} \neq x^0$. Prove the following theorem:

Let L satisfy (A), and let $c \leq 0$. Let $u \in \tilde{C}^2(Q)$, $Lu \geq 0$ in Q, u continuous in $\bar{Q}$, and assume that u takes its positive maximum M (in $\bar{Q}$) at a point P^0 having the inside sphere property. Assume finally that there exists a neighborhood V of P^0 such that $u < M$ in $\Omega \cap V$. Then, for any nontangential inward direction τ lying on $t = t^0$ (or even pointing downward with respect to t),

$$\frac{\partial u}{\partial \tau} \equiv \lim_{\Delta\tau \to 0} \frac{\Delta u}{\Delta\tau} < 0 \qquad \text{at } P^0.$$

[*Hint:* Suppose $B \subset V$, and let π be a hyperplane separating P^0 and $\bar{P}$ and denote by B^* the subset of B lying on the same side as P^0. Choose π such that the set $B_0 = B^* \cap \{t < t^0\}$ has nonempty interior. Introduce

$$h(x, t) = \exp\{-\alpha[|x - \bar{x}|^2 + |t - \bar{t}|^2]\} - \exp\{-\alpha R^2\} \qquad (R = \text{radius of } B)$$

and consider $u + \varepsilon h$ in B_0. Show that this function attains its positive maximum at P^0.]

(42) Give an example to show that if the assumption $\bar{x} \neq x^0$ is omitted in the definition of the inside sphere property, then the assertion of the previous theorem is generally false.

(43) Let Ω be a bounded domain in R^n and let $x^0 \in \partial\Omega$. If there exists a closed ball B such that $B \subset \bar{\Omega}$, $B \cap \bar{\Omega} = \{x^0\}$, then we say that x^0 has the *inside sphere property*. Prove the analog of the theorem of Problem 41 for elliptic operators—that is, if L is a second-order elliptic operator with real continuous coefficients in $\bar{\Omega}$ and with $c \leq 0$, and if $Lu \geq 0$ in Ω, u continuous in $\bar{\Omega}$ and u takes its positive maximum at a point $x^0 \in \partial\Omega$ having the inside sphere property, then $\partial u/\partial\tau < 0$.

(44) Consider the initial-boundary value problem

$$
\begin{cases}
Lu = f & \text{in } Q = \Omega \times (0, T), \\
u = h & \text{on } \bar{Q} \cap \{t = 0\}, \\
\dfrac{\partial u}{\partial v} + \mu u = g & \text{on } \partial \Omega \times (0, T),
\end{cases}
\tag{20.11}
$$

where v is the outward normal and μ is a continuous nonnegative function. Assume that each point of $\partial\Omega$ has the inside sphere property (this is the case, for example, if $\partial\Omega$ is in C^2), and that L satisfies (A). Prove: If u_1, u_2 are two solutions of (20.11), then $u_1 - u_2 \equiv \text{const.}$ (If $\mu \not\equiv 0$, then $u_1 \equiv u_2$.)

(45) Prove a uniqueness theorem, similar to the one of the previous problem, for the elliptic boundary value problem

$$
\begin{cases}
Lu = f & \text{in } \Omega, \\
\dfrac{\partial u}{\partial v} + \mu u = g & \text{on } \partial \Omega,
\end{cases}
\tag{20.12}
$$

where L is as in Problem 43.

(46) Let u be a solution of (20.11), where L satisfies (A), $c \leq 0$, and $\mu \geq \mu_0 > 0$, μ_0 a constant. Prove that there exists a constant K depending only on L, Ω, μ_0 such that

$$
\underset{Q}{\text{l.u.b.}} \, |u| \leq K \left\{ \underset{Q}{\text{l.u.b.}} \, |f| + \underset{\partial\Omega \times (0,T)}{\text{l.u.b.}} \, |g| + \underset{\Omega}{\text{l.u.b.}} \, |h| \right\}.
\tag{20.13}
$$

(47) Derive a bound analogous to (20.13) for solutions of the elliptic boundary value problem (20.12).

(48) Let Ω be a cube in R^n. Solve the equation

$$
\begin{cases}
\Delta u - u_t = 0 & \text{in } \Omega \times (0, \infty), \\
u(x, 0) = f(x) & \text{on } \Omega, \\
u = 0 & \text{on } \partial\Omega \times (0, \infty)
\end{cases}
\tag{20.14}
$$

by the method of separation of variables, assuming f to be sufficiently smooth and, say, to belong to $C_0^\infty(\Omega)$.

(49) Let Ω be a bounded domain in R^n and let $f(x) \, (\not\equiv 0)$ be a sufficiently smooth nonnegative function in Ω having a compact support. Prove that the solution of (20.14) (its existence was considered in Section 10) satisfies, for any $1 < p < \infty$:

$$
\left\{ \int_\Omega |u(x, t)|^p \, dx \right\}^{1/p} \geq c e^{-bt} \qquad (c > 0, \, b > 0).
$$

Thus, in the assertion of Theorem 19.1 the exponent $1/\alpha$ (in (19.3)) cannot generally be replaced by $1/(\alpha - \varepsilon)$, $\varepsilon > 0$.

[*Hint:* Compare the solution u with the solution u' of the corresponding problem when Ω is replaced by a cube Ω', $\overline{\Omega}' \subset \Omega$.]

(50) Prove that if A is as in Problem 9, Section 14, then for any $0 < \alpha < 1$,

$$\|A^\alpha x\| \leq C \|Ax\|^\alpha \|x\|^{1-\alpha} \qquad \text{for all } x \in D(A),$$

where C is a constant.

Part 3 | SELECTED TOPICS

We shall refer to Section m (and to Theorem $m.n$, and so on) of Parts I and II by Section I.m (and Theorem I.$m.n$) and Section II.m (Theorem II.$m.n$), respectively.

We shall deal with several independent topics. Sections 1–3 are concerned with analyticity of solutions and depend upon results derived in Parts I and II. Section 4 is independent of any previous material, but it is related to Section II.19. Sections 5–7 form one subject and employ results of Part I. Section 8 extends some of the results of Part II. In Section 9 we use some of the basic theorems of Part II.

1 | ANALYTICITY OF SOLUTIONS OF ELLIPTIC EQUATIONS

Let

$$A(x, D) = \sum_{|\alpha| \leq 2m} a_\alpha(x)D^\alpha \tag{1.1}$$

be an elliptic operator with coefficients defined in the closure $\overline{\Omega}$ of a bounded domain Ω. By results of Part I, if f and the a_α belong to $C^\infty(\Omega)$, then any solution of

$$A(x, D)u = f \tag{1.2}$$

in Ω also belongs to $C^\infty(\Omega)$. Suppose further that $\partial\Omega$ is of class C^∞ and that f and the a_α belong to $C^\infty(\bar\Omega)$. If A is strongly elliptic in $\bar\Omega$ and u is a solution of (1.2) in Ω with zero Dirichlet data on $\partial\Omega$, then $u \in C^\infty(\bar\Omega)$. In this section we shall prove similar results with C^∞ replaced by analyticity.

Denote by B_R the ball $\{x; \sum_{i=1}^n x_i^2 < R\}$ and by G_R the hemiball $B_R \cap \{x; x_n > 0\}$. Our first result is the following:

THEOREM 1.1.   *Let* u *be a solution of the strongly elliptic equation* (1.2) *with analytic coefficients in* $\bar G_R$, $R < 1$. *If the Dirichlet data of* u *on* $x_n = 0$ *are zero, then* u *is analytic in some* $\bar G_R$-*neighborhood of the origin.*

Proof.   Without restriction of generality we may assume in what follows that the first $4m$ derivatives of u are continuous in $\bar G_R$. We introduce the following notation: $x' = (x_1, \ldots, x_{n-1})$, $y = x_n$, $D_y = \partial/\partial y$ and $D_{x'}^\alpha$ is an αth derivative with respect to components of x'. We introduce the notation

$$\int_\Omega |D^{2m}D_{x'}^p u|^2 \, dx = \sup_{|\gamma| = 2m} \sup_{|\beta| = p} \int_\Omega |D^\gamma D_{x'}^\beta u|^2 \, dx, \tag{1.3}$$

$$\int_\Omega |D^q u|^2 \, dx = \sup_{|\beta| = q} \int_\Omega |D^\beta u|^2 \, dx. \tag{1.4}$$

PROBLEMS.   (1) Prove that a function $u(x) \in C^\infty(\Omega)$ is analytic in Ω if and only if for any subdomain Ω_0 of Ω with $\bar\Omega_0 \subset \Omega$ there exist constants H_0, H such that

$$|D^\beta u(x)| \le H_0 H^{|\beta|}\beta! \qquad \text{for } x \in \Omega_0, 0 \le |\beta| < \infty.$$

(2) Prove that a function $u(x) \in C^\infty(\bar\Omega)$ is analytic in $\bar\Omega$ if and only if there exist constants H_0, H such that

$$\left\{ \int_\Omega |D^p u|^2 \, dx \right\}^{1/2} \le H_0 H^p p! \qquad (1 \le p < \infty).$$

Set $\sigma_R = \bar G_R \cap \{y = 0\}$, and denote by Σ_R the part of the boundary of G_R that lies in $y > 0$. From Theorem I.18.1 it follows that if

$$\begin{cases} A(x, D)v = g & \text{in } G_R, \\ D_y^j v = 0 & \text{on } \sigma_R \, (0 \le j \le m - 1), v = 0 \text{ in a neighborhood of } \Sigma_R, \end{cases}$$
$$\tag{1.5}$$

then

$$\int_{G_R} |D^{2m}v|^2 \, dx \le C \int_{G_R} |g|^2 \, dx + C \int_{G_R} |v|^2 \, dx \qquad (C \text{ constant}). \tag{1.6}$$

In fact, we can extend v by zero into a domain G^* lying in $y > 0$ and having a C^{2m} boundary and then apply Theorem I.18.1 in G^*.

Let $h(t)$ be a C^∞ function satisfying: $h(t) = 1$ if $t \le 0$, $h(t) = 0$ if $t \ge 1$. Set

$$\zeta(x) = h\left(\frac{|x| - r}{\delta}\right). \tag{1.7}$$

LEMMA 1.1.   *Let* u *be as in Theorem 1.1. There exists a constant* K_1 *such that if* $0 < r < r + \delta < R, r > \delta$, *then*

$$\int_{G_r} |D^{2m}u|^2 \, dx \le K_1 \left\{ \int_{G_{r+\delta}} |f|^2 \, dx + \sum_{q=1}^{2m} \left[\int_{G_{r+\delta}} |D^{2m-q}u|^2 \, dx \right] \delta^{-2q} \right\}. \tag{1.8}$$

Proof.   The function $v = \zeta u$ satisfies (1.5) with

$$g = \zeta f + \sum_{0 < |\beta| \le 2m} D^\beta \zeta \cdot A_\beta(x, D)u,$$

where A_β is an operator of order $2m - |\beta|$. Now apply (1.6).

LEMMA 1.2.   *Under the assumptions of Theorem 1.1,* u *satisfies the following inequalities:*

$$\int_{G_r} |D_{x'}^p D^{2m}u|^2 \, dx \le [M\lambda^p p! (R - r)^{-p}]^2 \quad \left(\frac{R}{2} \le r < R, 0 \le p < \infty\right), \tag{1.9}$$

where M, λ *are constants.*

Proof.   (1.9) is a consequence of the inequalities

$$\int_{G_r} |D_{x'}^\beta D^{2m}u|^2 \, dx \le [M\lambda^{|\beta|}\beta! (R - r)^{-|\beta|}]^2 \quad \left(\frac{R}{2} \le r < R, 0 \le |\beta| < \infty\right). \tag{1.10}$$

We shall prove (1.10) by induction on $|\beta|$. First we choose $M \ge 1$ such that (1.10) holds for $|\beta| \le 2m$, $\lambda = 1$. From now on M is fixed. We now assume that (1.10) holds for all β with $|\beta| \le p - 1$ $(p > 2m)$ and we shall prove it for any β with $|\beta| = p$. In the process of this proof we shall determine λ independently of β.

Applying $D_{x'}^\beta$ to both sides of (1.2), where $|\beta| = p$, we obtain

$$A(x, D)(D_{x'}^\beta u) = D_{x'}^\beta f - \sum_{|\alpha| \le 2m} \sum_{\substack{\gamma \le \beta \\ |\gamma| > 0}} \binom{\beta}{\gamma} D_{x'}^\gamma a_\alpha \cdot D_{x'}^{\beta - \gamma} D^\alpha u \equiv D_{x'}^\beta f - F_0 \equiv F.$$

(1.11)

We can apply Lemma 1.1 to the function $D_{x'}^\beta u$. We obtain

$$\int_{G_r} |D_{x'}^\beta D^{2m} u|^2 \, dx \le K_1 \left\{ \int_{G_{r+\delta}} |F|^2 \, dx + \sum_{q=1}^{2m} \left[\int_{G_{r+\delta}} |D_{x'}^\beta D^{2m-q} u|^2 \, dx \right] \delta^{-2q} \right\}.$$

(1.12)

Since the a_α are analytic functions in $\bar{G}_R$, the result of Problem 1 implies that, for some constant c,

$$|D^\gamma a_\alpha| \le c^{|\gamma|+1} \gamma!$$

(1.13)

for all $x \in \bar{G}_R$, $0 \le |\gamma| < \infty$, $0 \le |\alpha| \le 2m$. Hence

$$\left\{ \int_{G_\rho} \left| \sum_{|\alpha| \le 2m} \sum_{\substack{\gamma \le \beta \\ |\gamma| > 0}} \binom{\beta}{\gamma} D_{x'}^\gamma a_\alpha \cdot D_{x'}^{\beta - \gamma} D^\alpha u \right|^2 dx \right\}^{1/2}$$

$$\le \sum_{|\alpha| \le 2m} \sum_{\substack{\gamma \le \beta \\ |\gamma| > 0}} c^{|\gamma|+1} \binom{\beta}{\gamma} \gamma! \left\{ \int_{G_\rho} |D_{x'}^{\beta - \gamma} D^\alpha u|^2 \, dx \right\}^{1/2}$$

$$\le M \sum_{|\alpha| \le 2m} \sum_{\substack{\gamma \le \beta \\ |\gamma| > 0}} c^{|\gamma|+1} \binom{\beta}{\gamma} \gamma! \, \lambda^{|\beta|-|\gamma|} (R - \rho)^{|\beta|+|\gamma|} (\beta - \gamma)!,$$

where the inductive assumption has been used. Hence, if $\lambda > 2cR$, then

$$\left\{ \int_{G_\rho} |F_0|^2 \, dx \right\}^{1/2} \le M c_1 p! \, \lambda^{p-1} (R - \rho)^{-p+1},$$

(1.14)

where c_1 is a constant depending only on c, R.
    Next, if $1 \le q \le 2m$,

$$\left\{ \int_{G_\rho} |D_{x'}^\beta D^{2m-q} u|^2 \, dx \right\}^{1/2} \le \sup_{|\gamma| = |\beta| - q} \left\{ \int_{G_\rho} |D^{2m} D_{x'}^\gamma u|^2 \, dx \right\}^{1/2}$$

$$\le c_2 M \frac{\lambda^{p-q}}{p^q} p! (R - \rho)^{-p+q},$$

where c_2 is a constant independent of β. Substituting this inequality and (1.14) into (1.12), we get, after using the fact that

$$|D^\gamma f| \leq c_3^{|\gamma|+1} \gamma! \qquad (c_3 \text{ constant, } 0 \leq |\gamma| < \infty), \qquad (1.15)$$

and taking $\lambda > Rc_3$,

$$\left\{ \int_{G_r} |D_{x'}^\beta D^{2m} u|^2 \, dx \right\}^{1/2}$$

$$\leq c_4 Mp! \lambda^{p-1} (R - r + \delta)^{-p+1} \left\{ 1 + \sum_{q=1}^{2m} \frac{1}{\lambda^q p^q} (R - r + \delta)^{q-1} \delta^{-q} \right\},$$

where c_4 is a constant independent of p. Choosing $\delta = (R - r)/p$, we get

$$\left\{ \int_{G_r} |D_{x'}^\beta D^{2m} u|^2 \, dx \right\}^{1/2} \leq c_5 Mp! \lambda^{p-1} (R - r)^{-p},$$

provided $\lambda \geq 1$, where c_5 is a constant depending only on c_4, R. Hence, if we take $\lambda \geq c_5$, then the proof is complete.

LEMMA 1.3.   Under the assumptions of Theorem 1.1, u satisfies the following inequalities:

$$\left\{ \int_{\sigma_r} |D_y^q D_{x'}^p u|^2 \, dx' \right\}^{1/2} \equiv \sup_{|\beta|=p} \left\{ \int_{\sigma_r} |D_y^q D_{x'}^\beta u|^2 \, dx' \right\}^{1/2}$$

$$\leq \bar{M} \bar{\lambda}^p p! (R - r)^{-p} \qquad \left(\frac{R}{2} \leq r < R, 0 \leq p < \infty \right) \quad (1.16)$$

for $0 \leq q \leq 2m - 1$, where $\bar{M}$, $\bar{\lambda}$ are constants.

Proof.   From (1.9) it follows that for $0 \leq q \leq 2m - 1, \frac{1}{2} R \leq r < R$, $0 \leq p < \infty$,

$$\int_{G_r} |D_y^q D_{x'}^p u|^2 \, dx \leq [M\lambda^p p! (R - r)^{1-p}]^2,$$

$$\int_{G_r} |D_y^{q+1} D_{x'}^p u|^2 \, dx \leq [M\lambda^p p! (R - r)^{-p}]^2,$$

$$\int_{G_r} |D_y^q D_{x'}^{p+1} u|^2 \, dx \leq [M\lambda^p p! (R - r)^{-p}]^2.$$

Using the inequality (for any $v \in C^1(\bar{G}_R)$)

$$\int_{\sigma_r} |v|^2 \, dx' \le \text{const.} \left[\frac{1}{r} \int_{G_r} |v|^2 \, dx + r \int_{G_r} |Dv|^2 \, dx \right], \tag{1.17}$$

(1.16) now follows (with $\bar{\lambda} = \lambda$). The proof of (1.17) is left to the reader.

We shall now prove by induction on q that

$$\sup_{\sigma_{R/2}} |D_y^q D_{x'}^\beta u(x', 0)| \le \bar{M} \hat{\lambda}^p \Lambda^q p! \, q!$$

$$\text{for } |\beta| = p, 0 \le p < \infty, 0 \le q < \infty. \tag{1.18}$$

In view of (1.16) and Theorem I.9.1, (1.18) holds for all $q \le 2m - 1$ with $\Lambda = 1$ and, say, $\hat{\lambda} = \lambda^*$. We now assume that (1.18) holds for all $q < k$ ($k \ge 2m$) and prove it for $q = k$.

We write the elliptic equation (1.2) more explicitly:

$$a D_y^{2m} u + \sum_{|\alpha| \le 2m}' a_\alpha D^\alpha u = f, \tag{1.19}$$

where in $\sum'$ there occurs no D^α with $\alpha = (0, \ldots, 0, 2m)$. The ellipticity implies that $a \ne 0$ in $\bar{G}_R$.

Applying D_y^{k-2m} to both sides of (1.19), we get

$$a D_y^k u = - \sum_{|\alpha| \le 2m}' \sum_{j=0}^{k-2m} \binom{k-2m}{j} D_y^j a_\alpha \cdot D_y^{k-2m-j} D^\alpha u$$

$$- \sum_{j=1}^{k-2m} \binom{k-2m}{j} D_y^j a \cdot D_y^{k-j} u + D_y^{k-2m} f.$$

Next we apply $D_{x'}^\beta$ ($|\beta| = p$) to both sides and obtain

$$a D_y^k D_{x'}^\beta u = - \sum_{|\alpha| \le 2m}' \sum_{j=0}^{k-2m} \sum_{\gamma \le \beta} \binom{k-2m}{j} \binom{\beta}{\gamma} D_y^j D_{x'}^\gamma a_\alpha \cdot D_y^{k-2m-j} D_{x'}^{\beta-\gamma} D^\alpha u$$

$$- \sum_{j=1}^{k-2m} \sum_{\gamma \le \beta} \binom{k-2m}{j} \binom{\beta}{\gamma} D_y^j D_{x'}^\gamma a \cdot D_y^{k-j} D_{x'}^{\beta-\gamma} u$$

$$- \sum_{\substack{\gamma \le \beta \\ |\gamma| > 0}} \binom{\beta}{\gamma} D_{x'}^\gamma a \cdot D_y^k D_{x'}^{\beta-\gamma} u + D_y^{k-2m} D_{x'}^\beta f. \tag{1.20}$$

Using (1.13), (1.15), and the inductive assumption, we can evaluate the right-hand side of (1.20) and obtain

$$\sup_{\sigma_{R/2}} |a D_y^k D_{x'}^\beta u| \le \bar{c} \bar{M} \hat{\lambda}^{p+1} \Lambda^{k-1} p! \, k!,$$

where $\bar{c}$ is some constant (independent of $\overline{M}, \hat{\lambda}, \Lambda, \beta, k$) provided $\hat{\lambda}, \Lambda$ are larger than some constant (independent of k, β) and provided $\Lambda \geq \hat{\lambda}$—that is, provided $\hat{\lambda} \geq \bar{c}, \Lambda \geq \hat{\lambda}$. We now fix $\hat{\lambda}$ such that $\hat{\lambda} \geq \lambda^*, \hat{\lambda} \geq \bar{c}$, and then fix Λ such that $\Lambda \geq \hat{\lambda}, \Lambda \geq \bar{c}\hat{\lambda}/(\min |a|)$. Then (1.18) follows for $q = k$.

The assertion of Theorem 1.1 now follows from (1.18).

Our next result is concerned with analyticity in the interior.

THEOREM 1.2. *Let* u *be a solution of* (1.2) *in a domain* Ω, *where* A *is elliptic in* Ω *and all the coefficients are analytic in* Ω. *Then* u *is also analytic in* Ω.

Proof. It suffices to consider the case $\Omega = B_R$. If $A(x, D)v = f$ and v vanishes in a neighborhood of ∂B_R, then we have (cf. (1.6))

$$\int_{B_R} |D^{2m}v|^2 \, dx \leq K_1 \int_{B_R} |f|^2 \, dx. \tag{1.21}$$

The proof of (1.21) is similar to the proof of Theorem I.16.3 if one uses Fourier transforms instead of Fourier series. Taking a function ζ as in (1.7) and applying (1.21) to $v = \zeta u$, we get the inequality (1.8) but with G_ρ replaced by B_ρ. This inequality can now be used to prove by induction on p that

$$\left\{ \int_{B_r} |D^p u|^2 \, dx \right\}^{1/2} \leq M_0 \lambda_0^p \, p! (R - r)^{-p} \qquad \left(\frac{R}{2} \leq r \leq R \right).$$

The details are left to the reader.

PROBLEM. (3) Let $y = f(x)$ be a C^{2m} map from a neighborhood of $x^0 \in R^n$ into a neighborhood of $y^0 = f(x^0)$ in R^n and suppose that the Jacobian is $\neq 0$. Then this map transforms elliptic (strongly elliptic) operators into elliptic (strongly elliptic) operator.

Combining Theorems 1.1, 1.2, and the result of Problem 3, and noting that an analytic function of analytic functions is analytic, we get the following theorem.

THEOREM 1.3. *Assume that* u *is a solution of the strongly elliptic equation* (1.2) *with coefficients analytic in* $\overline{\Omega}$ *and let* $\partial \Omega$ *be analytic. If the Dirichlet data of* u *vanish on* $\partial \Omega$, *then* u *is analytic in* $\overline{\Omega}$.

PROBLEM. (4) Extend Theorem 1.3 by showing that if the Dirichlet data of u on $\partial \Omega$ are analytic functions then u is analytic in $\overline{\Omega}$.

2 | ANALYTICITY OF SOLUTIONS OF EVOLUTION EQUATIONS

In this section we shall prove analyticity of solutions $u(t)$ of the evolution equation, in a Banach space X,

$$\frac{du}{dt} + A(t)u = f(t). \tag{2.1}$$

Recall that $u(t)$ with values in X is said to be analytic at $t = \sigma$ if, for some $\delta > 0$, $u(t) = \sum_{m=0}^{\infty} (t - \sigma)^m u_m$ for $|t - \sigma| < \delta$, where the u_m belong in X and the series is convergent (or absolutely convergent).

PROBLEM. (1) A function $u(t)$ in $C^{\infty}((a, b); X)$ is analytic in the interval (a, b) if and only if, for any $a < \alpha < \beta < b$,

$$\left\| \frac{d^m u(t)}{dt^m} \right\| \leq c_0 c^m m! \qquad \text{for } \alpha \leq t \leq \beta, 0 \leq m < \infty,$$

where c_0, c are constants.

Let Δ be a bounded open convex neighborhood in the complex plane of the real interval $(0, t_0]$. Denote by $\bar{\Delta}$ the closure of Δ. We shall make the following assumptions:

(E_1) For $t \in \bar{\Delta}$, $A(t)$ is a closed linear operator in X with domain D_A dense in X and independent of t.

(E_2) $R(\lambda; A(t))$ exists for all $t \in \bar{\Delta}$ and λ in a sector $Q_\phi = \{\lambda; \lambda \neq 0, |\pi - \arg \lambda| \leq \pi/2 + \phi\}$, and

$$\|R(\lambda; A(t))\| \leq \frac{C}{1 + |\lambda|} \qquad (t \in \bar{\Delta}, \lambda \in Q_\phi), \tag{2.2}$$

where C is a constant.

(E_3) $A(t)A^{-1}(0)$ is analytic in $\bar{\Delta}$.

Denote by Σ_ϕ the sector $\{\lambda; \lambda \neq 0, |\arg \lambda| < \phi\}$. We write $t > s \pmod{\phi}$ if $t - s \in \Sigma_\phi$. If either $t > s \pmod{\phi}$ or $t = s$, then we write $t \geq s \pmod{\phi}$.

PROBLEM. (2) Prove that, under the assumptions (E_1)–(E_3), the function $A(0)A^{-1}(t)$ is analytic in $\bar{\Delta}$.

THEOREM 2.1.   If the conditions (E_1)–(E_3) hold, then the fundamental solution $U(t, s)$ can be extended as an analytic function in (t, s) for $t \in \Delta$, $s \in \Delta$ if $t > s \pmod \phi$. This extended function can be defined at $t = s$ so that it remains strongly continuous for $t \geq s \pmod \phi$.

Proof.   Recall, by (II.2.5), that

$$e^{-tA(s)} = \int_\Gamma e^{\lambda t}(\lambda I + A(s))^{-1} \, d\lambda, \tag{2.3}$$

and write

$$(\lambda I + A(s))^{-1} = A^{-1}(0)(\lambda A^{-1}(0) + A(s)A^{-1}(0))^{-1}. \tag{2.4}$$

From (E_3) it follows that the right-hand side of (2.4) is analytic in s ($s \in \bar\Delta$). From the proof of Theorem II.2.1 it then follows that $e^{-tA(s)}$ is analytic in $s \in \bar\Delta$ for any $t \in \Sigma_\phi$. From Theorem II.2.1 we also know that $e^{-tA(s)}$ is analytic in t ($t \in \Sigma_\phi$) for any $s \in \bar\Delta$. We now use the fact that if $F(t, s)$ is a bounded function that is analytic separately in t and in s, then it is also analytic in the variable (t, s). We then conclude that $e^{-(t-s)A(s)}$ is analytic in (t, s) for $t \in \bar\Delta$, $s \in \bar\Delta$, provided $t > s \pmod \phi$. This function is also uniformly bounded with respect to t and s, and strongly continuous in (t, s) for t, s in $\bar\Delta$, $t \geq s \pmod \phi$.

LEMMA 2.1.   Let $P(t, s)$ and $Q(t, s)$ be two bounded linear operators defined and analytic for $t > s \pmod \phi$, t and s in Δ. Assume that they are uniformly bounded with respect to t and s. Then the integral

$$\int_s^t P(t, \tau)Q(\tau, s) \, d\tau \tag{2.5}$$

is defined in the same domain and it is analytic and uniformly bounded in (t, s).

Proof.   The integral in (2.5) is defined along any contour for which $t > \tau > s \pmod \phi$, and the integral then exists as an improper integral. The value of the integral is independent of the contour. If we take the contour to be a straight segment, then we get

$$\left\| \int_s^t P(t, \tau)Q(\tau, s) \, d\tau \right\| \leq \sup \|P(t, \tau)\| \cdot \sup \|Q(\tau, s)\| \cdot |t - s|.$$

This proves the uniform boundedness of the integral.

To prove analyticity in (t, s), it suffices to prove analyticity separately in t and in s. We shall only prove analyticity in t, since the proof of analyticity in s is similar. We shall prove analyticity in some closed disc W_ε with center t_0 and radius ε.

Fix τ_0 such that $t > \tau_0 > s \pmod{\phi}$ for $t \in W_{\varepsilon_0}$ (ε_0 sufficiently small) and write

$$\int_s^t P(t, \tau)Q(\tau, s)\, d\tau = \int_s^{\tau_0} P(t, \tau)Q(\tau, s)\, d\tau + \int_{\tau_0}^t P(t, \tau)Q(\tau, s)\, d\tau \equiv I(t, s) + J(t, s).$$

Clearly $I(t, s)$ is analytic in t. To consider $J(t, s)$, substitute $t - \tau = \sigma$. Then

$$J(t, s) = \int_0^{t - \tau_0} P(t, t - \sigma)Q(t - \sigma, s)\, d\sigma \equiv \int_0^{t - \tau_0} S(t, \sigma, s)\, d\sigma.$$

The integrand $S(t, \sigma, s)$ is analytic in (t, σ, s) if $\sigma > 0 \pmod{\phi}$, $t - \tau_0 > 0 \pmod{\phi}$ and $t - \sigma > s \pmod{\phi}$. If $\xi > 0 \pmod{\phi}$ and $|\xi|$ is sufficiently small, then the integral

$$\int_\xi^{t_0 - \tau_0} S(t, \sigma, s)\, d\sigma$$

is well defined and analytic in t. Indeed this follows by noting that $t - (t_0 - \tau_0) > s \pmod{\phi}$ if ε_0 is sufficiently small, and thus $t - \sigma > s \pmod{\phi}$. Since

$$\left\| \int_0^\xi S(t, \sigma, s)\, d\sigma \right\| \to 0$$

as $\xi \to 0$, uniformly in t, it follows that

$$\int_0^{t_0 - \tau_0} S(t, \sigma, s)\, d\sigma$$

is analytic in t. Thus it remains to consider the integral

$$K(t, s) \equiv \int_{t_0 - \tau_0}^{t - \tau_0} S(t, \sigma, s)\, d\sigma.$$

If h is sufficiently small and $t \in W_\varepsilon$ ($\varepsilon \leq \varepsilon_0$), then

$$\frac{1}{h}(K(t + h, s) - K(t, s)) = \frac{1}{h} \int_{t - \tau_0}^{t + h - \tau_0} S(t + h, \sigma, s)\, d\sigma$$

$$+ \frac{1}{h} \int_{t_0 - \tau_0}^{t - \tau_0} [S(t + h, \sigma, s) - S(t, \sigma, s)]\, d\sigma.$$

We take ε such that $t - \sigma > s \pmod{\phi}$ for σ on the segment connecting $t_0 - \tau_0$ to $t - \tau_0$. Then $\partial K(t, s)/\partial t$ exists and equals

$$S(t, t - \tau_0, s) + \int_{t_0 - \tau_0}^{t - \tau_0} \frac{\partial}{\partial t} S(t, \sigma, s) \, d\sigma.$$

Since each of these terms is a bounded operator (in fact, even analytic in t), the proof is completed.

We have proved before that $e^{-(t-s)A(s)}$ is analytic in (t, s) if t and s belong to $\bar{\Delta}$ and $t > s \pmod{\phi}$. Hence

$$\frac{\partial}{\partial t} e^{-(t-s)A(s)} = -A(s)e^{-(t-s)A(s)} \tag{2.6}$$

is also analytic in (t, s). Write

$$[A(t) - A(s)]e^{-(t-s)A(s)}$$
$$= [A(t)A^{-1}(0) - A(s)A^{-1}(0)][A(s)A^{-1}(0)]^{-1}A(s)e^{-(t-s)A(s)}. \tag{2.7}$$

Using the result of Problem 2 and the analyticity of the right-hand side of (2.6), we conclude that the right-hand side of (2.7) is analytic in (t, s)—that is, $\phi_1(t, s)$ (cf. (II.4.5)) is analytic in (t, s).

We now wish to apply Lemma 2.1 to conclude that

$$\phi_2(t, s) = \int_s^t \phi_1(t, \tau)\phi_1(\tau, s) \, d\tau$$

is analytic in t, s, for $t \in \Delta$, $s \in \Delta$, $t > s \pmod{\phi}$. Since, however, $\phi_1(t, s)$ is not uniformly bounded, we cannot use Lemma 2.1. Nevertheless, since ϕ_1 satisfies (II.4.20), the proof of Lemma 2.1 can easily be extended to this case. Thus, we find that $\phi_2(t, s)$ is analytic in (t, s) if t and s vary in Δ and $t > s \pmod{\phi}$, and

$$|\phi_2(t, s)| \leq \frac{C}{|t - s|^{1 - 2\alpha}}.$$

We can proceed in this way step by step and thus conclude that each $\phi_k(t, s)$ is analytic and satisfies (II.4.23). It follows that $\Phi(t, s)$ is also analytic and satisfies (II.4.21). The proof of Theorem 2.1 is now completed by using (II.4.8) and applying once again the argument used in the proof of Lemma 2.1.

COROLLARY. Let the assumptions of Theorem 2.1 hold and let f(t) *be analytic in Δ. Then any solution* u(t) *of (2.1) is analytic in* t *if* $t \in \Delta$, t > 0 (mod ϕ).

Proof.   For any $\varepsilon > 0$, we have

$$u(t) = U(t, \varepsilon)u(\varepsilon) + \int_{\varepsilon}^{t} U(t, s)f(s) \, ds.$$

Now apply Theorem 2.1 and Lemma 2.1.

We shall now prove another analyticity theorem in a Hilbert space, under weaker assumptions on $A(t)$, namely:

($\mathbf{F_1}$) For $0 \leq t \leq t_0$ ($t_0 < \infty$), $A(t)$ is a closed linear operator in X with a domain D_A dense in X and independent of t.

($\mathbf{F_2}$) The resolvent $R(\lambda; A(t))$ exists for all $0 \leq t \leq t_0$ and λ pure imaginary with $|\lambda| \geq N(t)$, and

$$\|R(\lambda; A(t))\| \leq \frac{C(t)}{1 + |\lambda|} \qquad (\lambda \text{ pure imaginary, } |\lambda| \geq N(t)),$$

where $N(t)$, $C(t)$ are constants (depending on t).

($\mathbf{F_3}$) $A^{-1}(t)$ and $A(t)A^{-1}(s)$ are bounded operators for all t, s in $[0, t_0]$; the derivatives $d^k(A(t)A^{-1}(s)x/dt^k \equiv A^{(k)}(t)A^{-1}(s)x$ exist for all $x \in X$ and $k \geq 0$, and

$$\|A^{(k)}(t)A^{-1}(s)\| \leq C_1^{k+1}k! \qquad (0 \leq k < \infty, 0 \leq t, s \leq t_0), \qquad (2.8)$$

where C_1 is a constant.

($\mathbf{F_4}$) $f(t)$ is infinitely differentiable in $[0, t_0]$ and

$$\|f^{(k)}(t)\| \leq C_2^{k+1}k! \qquad (0 \leq k < \infty, 0 \leq t \leq t_0), \qquad (2.9)$$

where C_2 is a constant.

In view of the result of Problem 1, the conditions (2.8), (2.9) assert in effect that $A(t)A^{-1}(s)$ and $f(t)$ are analytic in $t, 0 \leq t \leq t_0$.

Note also that the assumption ($\mathbf{F_2}$) appears also in Theorem II.11.2.

THEOREM 2.2.   Let the assumptions ($\mathbf{F_1}$)–($\mathbf{F_4}$) *hold. Then any solution* u(t) *of* (2.1) *is analytic in the interval* (0, t_0].

Proof.   It suffices to prove analyticity in a neighborhood of any point of $(0, t_0]$. We take this neighborhood to be an interval (a, b) with $b - a$ sufficiently small (to be determined later on). We take $s = (a + b)/2$. Set

$$|v(t)|_{m, \delta} = \left\{ \int_{a+\delta}^{b-\delta} \|v^{(m)}(t)\|^2 \, dt \right\}^{1/2}.$$

We shall prove by induction that

$$\|A(s)u(t)\|_{m-1, \delta} + \|u(t)\|_{m, \delta} \leq M\lambda^m m! \delta^{-m} \qquad (2.10)$$

for $m = 0, 1, \ldots; 0 < \delta \leq (b - a)/2$. In view of Problem 1, this would complete the proof of Theorem 2.2.

We first fix $M \geq 1$ such that (2.10) holds with $\lambda = 1$, $m = 1$. We shall now assume that (2.10) holds for all $m \leq p$ and prove it for $m = p + 1$. Let $\zeta(t)$ be real-valued C^∞ function such that $\zeta(t) = 1$ if $a + \delta < t < b - \delta$ and $\zeta(t) = 0$ if $t < a + \delta'$ or if $t > b - \delta'$, where $0 < \delta' < \delta$, and, finally $|\zeta(t)| \leq C$, $|\zeta'(t)| \leq C/(\delta - \delta')$, where C is independent of δ, δ'. Consider the function $v = \zeta u^{(p)}$. It satisfies

$$\frac{dv}{dt} + A(t)v = \zeta F_p + \zeta' u^{(p)},$$

where

$$F_p \equiv f^{(p)} - \sum_{q=1}^{p} \binom{p}{q} [A^{(q)}(t)A^{-1}(s)]A(s)u^{(p-q)}.$$

Applying the inequalities (II.11.9), (II.11.11) with $m = 0$, and using (2.8), (2.9) and the inductive assumption we obtain, after choosing $\delta = \delta'(1 + 1/p)$,

$$\|A(s)u(t)\|_{p, \delta} + \|u(t)\|_{p+1, \delta} \leq CM\lambda^p (p + 1)! \delta^{-p-1},$$

provided λ is sufficiently large (depending only on the constants in (2.8), (2.9)). Taking $\lambda \geq C$, (2.10) follows for $m = p + 1$.

PROBLEM. (3) Let u be a solution of (II.9.2), (II.9.5), where L is uniformly parabolic with continuous coefficients, $D_t^j f(x, t)$, $D_t^j a_\alpha(x, t)$ exist and are continuous in $\bar{Q}_T$ for all $j \geq 0$, and

$$|D_t^j f(x, t)| + \sum_{|\alpha| \leq 2m} |D_t^j a_\alpha(x, t)| \leq C^{j+1} j! \qquad (0 \leq j < \infty).$$

Let Ω_0 be an open subset of Ω. Prove: If $u(x, \sigma) \equiv 0$ for $x \in \Omega_0$ and some $\sigma \in (0, T)$, then $u(x, t) \equiv 0$ for $x \in \Omega_0$ and $0 < t < T$.

3 | ANALYTICITY OF SOLUTIONS OF PARABOLIC EQUATIONS

We consider uniformly parabolic equations

$$\frac{\partial u}{\partial t} + A(x, t, D)u = f(x, t) \qquad \text{in } Q_T, \tag{3.1}$$

where $Q_T = \Omega \times (0, T)$, Ω is a bounded domain in R^n, and

$$A(x, t, D) = \sum_{|\alpha| \leq 2m} a_\alpha(x, t)D^\alpha. \tag{3.2}$$

The solutions u are subject to the Dirichlet boundary conditions

$$\frac{\partial^j u}{\partial v^j} = 0 \qquad \text{on } S_T = \partial\Omega \times (0, T), \qquad 0 \leq j \leq m - 1. \tag{3.3}$$

We shall prove the following theorem.

THEOREM 3.1. *If* f *and the* a$_\alpha$ *are analytic in* $\bar{\Omega}_T$ *and if* $\partial\Omega$ *is analytic, then any solution of* (3.1), (3.3) *is analytic in* $\bar{\Omega} \times \{0 < t < T\}$.

Proof. In view of Theorem II.10.2, u is in $C^\infty(\bar{\Omega} \times \{0 < t < T\})$. Thus we can proceed to estimate the successive derivatives of u. We shall derive the necessary estimates by combining the methods of proof of Theorems 1.1–1.3 and 2.2.

We denote by G_R the hemiball $\{x; \sum_{i=1}^n x_i^2 < R^2, x_n > 0\}$ and set $\sigma_R = \bar{G}_R \cap \{x_n = 0\}$, $\Sigma_R = \partial G_R \cap \{x; x_n > 0\}$, where ∂G_R is the boundary of G_R. We also use the notation $x' = (x_1, \ldots, x_{n-1})$, $y = x_n$, and

$$\int_{G_r} |D_t^p D_{x'}^j D_y^k D^i u|^2 \, dx = \sup_{|\beta| = j} \sup_{|\alpha| = i} \int_{G_r} |D_t^p D_{x'}^\beta D_y^k D^\alpha u|^2 \, dx.$$

The integral

$$\int_{\sigma_r} |D_t^p D_{x'}^j D_y^k D^i u|^2 \, dx'$$

is defined in a similar way.

LEMMA 3.1. *Let* u *be a solution of* (3.1) *with* $\Omega = G_R$ *and let* $\partial^j u/\partial v^j = 0$
on $\sigma_R \times (0, T)$ *for* $0 \le j \le m - 1$. *Assume that the equation* (3.1) *is parabolic in*
$\bar{Q}_T = \bar{G}_R \times \{0 \le t \le T\}$ *with* $C^\infty(\bar{Q}_T)$ *coefficients and that* u, f *are in* $C^\infty(\bar{G}_R \times$
$\{0 < t < T\})$. *Then for any* $0 < \eta < \sigma < T$, $0 < \delta < r < r + \delta < R, \sigma < T/3$,
$R/2 \le r < R$,

$$
\int_\sigma^{T-\sigma} \int_{G_r} |D_t u|^2 \, dx \, dt + \int_\sigma^{T-\sigma} \int_{G_r} |D^{2m} u|^2 \, dx \, dt
$$

$$
\le K_1 \int_{\sigma-\eta}^{T-\sigma+\eta} \int_{G_{1+\delta}} |f|^2 \, dx \, dt
$$

$$
+ \sum_{q=1}^{2m} \left[\int_{\sigma-\eta}^{T-\sigma+\eta} \int_{G_{r+\delta}} |D^{2m-q} u|^2 \, dx \, dt \right] \delta^{-2q}
$$

$$
+ \frac{1}{\eta^2} \int_{\sigma-\eta}^{T-\sigma+\eta} \int_{G_{r+\delta}} |u|^2 \, dx \, dt \Bigg\}, \tag{3.4}
$$

where K_1 *is a constant.*

Proof. Let $\zeta(x)$ be the function introduced in (1.7) and let $\zeta_0(t)$ be a C^∞
function satisfying $\zeta_0(t) = 0$ if $t < \sigma - \eta$ or if $t > T - \sigma + \eta$, $\zeta_0(t) = 1$ if
$\sigma < t < T - \sigma$, and $|\zeta_0(t)| \le C, |D_t \zeta_0(t)| \le C/\eta$ for all t, where C is a constant
independent of σ, η. The function $v = \zeta_0 \zeta u$ satisfies

$$
\frac{\partial v}{\partial t} + A(x, t, D)v = g \qquad \text{in } G_R \times \{0 < t < T\},
$$

$$
\frac{\partial^j v}{\partial v^j} = 0 \qquad \text{on } \partial G_R \times \{0 < t < T\}, 0 \le j \le m - 1,
$$

$$
v(x, t) = 0 \qquad \text{if } t < \sigma - \eta \qquad \text{or if } t > T - \sigma + \eta,
$$

where

$$
g = \zeta_0 \zeta f + D_t \zeta_0 \cdot \zeta f + \zeta_0 \sum_{0 < |\beta| \le 2m} D^\beta \zeta \cdot A_\beta(x, t, D)u,
$$

where A_β is an operator of order $2m - |\beta|$. Now apply the inequalities (II.11.9),
(II.11.11) with $m = 0$ to $u = v$ (with $X = L^2(G_R)$), and use Theorem I.18.1.

LEMMA 3.2. *Let the assumptions of Lemma* 3.1 *hold and assume that*

$$
\int_\sigma^{T-\sigma} \int_{G_R} |D_t^{p+1} u|^2 \, dx \, dt + \int_\sigma^{T-\sigma} \int_{G_R} |D_t^p D^{2m} u|^2 \, dx \, dt \le [C^{p+1} p! \sigma^{-p-1}]^2
$$

$$
(0 \le p < \infty, 0 < \sigma < T/3) \tag{3.5}
$$

for some constant C. *Then, for* $0 < \sigma < T/3$, $R/2 \leq r < R$,

$$\int_\sigma^{T-\sigma} \int_{G_r} |D_t^{p+1} D_x^k u|^2 \, dx \, dt + \int_\sigma^{T-\sigma} \int_{G_r} |D_t^p D_x^k D^{2m} u|^2 \, dx \, dt$$

$$\leq [M\lambda^p \Lambda^k p! \, k! \, (R-r)^{-k} \sigma^{-p-1}]^2 \qquad (0 \leq p < \infty, 0 \leq k < \infty), \qquad (3.6)$$

where M, λ, Λ *are constants.*

Proof. We shall prove (3.6) by induction on k. If $k = 0$, then (3.6) is a consequence of (3.5). We shall now prove (3.6) for all $p \geq 0$, assuming that (3.6) holds for all $p \geq 0$ with k replaced by k' and $0 \leq k' \leq k - 1$; here $k \geq 1$. The proof is by induction on p. We assume that (3.6) holds with p replaced by p', and $0 \leq p' \leq p - 1$, and then prove (3.6). We shall assume that $p \geq 1$, since the argument that follows below can easily be modified to yield a proof for the case $p = 0$. In the proof below we shall determine the constants λ, Λ independently of k, p (M is already fixed).

Applying $D_t^p D_{x'}^\gamma$ ($|\gamma| = k$) to both sides of (3.1), we get

$$\frac{\partial}{\partial t} (D_t^p D_{x'}^\gamma u) + A(x, t, D)(D_t^p D_{x'}^\gamma u) = F, \qquad (3.7)$$

where

$$F = D_t^p D_{x'}^\gamma f - \sum_{|\alpha| \leq 2m} \sum_{0 \leq \beta \leq \gamma}' \sum_{q=0}^p{}' \binom{\gamma}{\beta} \binom{p}{q} D_t^q D_{x'}^\beta a_\alpha \cdot D_t^{p-q} D_{x'}^{\gamma-\beta} D^\alpha u, \qquad (3.8)$$

where the prime in the last two sums indicates that only terms with $q + |\beta| > 0$ may appear.

By Lemma 3.1,

$$\int_\sigma^{T-\sigma} \int_{G_r} |D_t^{p+1} D_{x'}^k u|^2 \, dx \, dt + \int_\sigma^{T-\sigma} \int_{G_r} |D_t^p D_{x'}^k D^{2m} u|^2 \, dx \, dt$$

$$\leq K_1 \left\{ \left[\int_{\sigma-\eta}^{T-\sigma+\eta} \int_{G_{r+\delta}} |F|^2 \, dx \, dt \right. \right.$$

$$+ \sum_{q=1}^{2m} \left[\int_{\sigma-\eta}^{T-\sigma+\eta} \int_{G_{r+\delta}} |D_t^p D_{x'}^k D^{2m-q} u|^2 \, dx \, dt \right] \delta^{-2q}$$

$$+ \left[\int_{\sigma-\eta}^{T-\sigma+\eta} \int_{G_{r+\delta}} |D_t^p D_{x'}^k u|^2 \, dx \, dt \right] \eta^{-2} \right\}. \qquad (3.9)$$

Using the inductive assumption and the analyticity of f and the a_α, we find that

$$\int_{\sigma-\eta}^{T-\sigma+\eta} \int_{G_{r+\delta}} |F|^2 \, dx \, dt$$

$$\le [cM(\lambda^{p-1}\Lambda^k + \lambda^p\Lambda^{k-1})p!\,k!\,(R-r-\delta)^{-k}(\sigma-\eta)^{-p-1}]^2, \quad (3.10)$$

where c is a constant independent of p, k, λ, Λ, provided λ, Λ are larger than some constant (depending only on A, f). Substituting (3.10) into (3.9) and taking $\delta = (R-r)/k, \eta = \sigma/p$, we find that the right-hand side of (3.9) is bounded by

$$\left[c'M\lambda^p\Lambda^k\left(\frac{1}{\lambda} + \frac{1}{\Lambda}\right)p!\,k!\,(R-r)^{-k}\sigma^{-p-1}\right]^2,$$

where c' is a constant independent of p, k, λ, Λ. Taking λ, Λ sufficiently large, (3.6) follows.

Having proved Lemma 3.2, we can now obtain (cf. the proof of Lemma 1.3) the following inequalities:

$$\int_\sigma^{T-\sigma} \int_{\sigma_r} |D_t^p D_{x'}^k \cdot D_y^q u|^2 \, dx \, dt \le [K^{p+k+1}p!\,k!\,(R-r)^{-k}\sigma^{-p-1}]^2$$

$$(0 \le p < \infty, 0 \le k < \infty) \quad (3.11)$$

for $0 < \sigma < T/3, R/2 \le r < R, 0 \le q \le 2m - 1$.

Now write the equation (3.1) in the form

$$aD_y^{2m}u = -\sum_{|\alpha|\le 2m}' a_\alpha D^\alpha u - D_t u + f,$$

where the prime in Σ indicates that the coefficient a_α with $\alpha = (0, \ldots, 0, 2m)$ is missing. By induction on q one can then prove that

$$|D_y^q D_{x'}^\beta \cdot D_t^p u(x', 0, t)| \le N\mu^q\mu_1^{|\beta|}\mu_2^p q!\,\beta!\,p! \quad (3.12)$$

for x' in some neighborhood $|x'| \le \delta_0$ (for any $\delta_0 < R$) and t in some interval $t_1 \le t \le t_2$ $(t_1 > 0, t_2 < T)$; here N, μ, μ_1, μ_2 are constants. In fact, for $0 < q \le 2m - 1$ this follows from (3.11), whereas the proof for any $q \ge 2m$ (assuming it to hold for all values $\le q - 1$) follows by calculations similar to those which follow (1.19).

From (3.12) it follows that u is analytic in a neighborhood of any point of $\sigma_r \times (0, T), r < R$.

We shall now complete the proof of Theorem 3.1. From Theorem 2.2 it follows that (3.5) holds with G_R replaced by Ω. Now consider u in a Q_T-neighborhood $V = \Omega_0 \times (0, T)$ of a point (ξ^0, t^0) on S_T. We perform a local analytic transformation with nonvanishing Jacobian. The image of V contains a cylinder $G_R \times (s_1, s_2)$ such that the image of (ξ^0, t^0) is the center of the flat part of ∂G_R. The inequalities (3.5) obviously hold in the new variables. Hence we can apply Lemma 3.2. The analyticity of u at (ξ^0, t^0) thus follows.

To prove analyticity at an interior point (ξ^*, t^*) of Q_T we modify the previous arguments. Thus we replace G_R by a ball B_R lying in Ω (and also do not use any local analytic mapping). (3.6) is modified, replacing D_x^k by D^k. Once the modified form of (3.6) is proved, the analyticity of u at (ξ^*, t^*) follows.

4 | LOWER BOUNDS FOR SOLUTIONS OF EVOLUTION INEQUALITIES

Consider the evolution operator

$$Lu = \frac{du}{dt} - Au \tag{4.1}$$

in a Hilbert space X. We shall assume that A is symmetric (that is, $(Ax, y) = (x, Ay)$ for all x, y in the domain of A. Let $\phi(t)$ be a given nonnegative function. Consider solutions u of the inequality

$$\|Lu\| \leq \phi(t) \|u(t)\| \tag{4.2}$$

for $0 \leq t < \infty$, where $u(t)$ is continuous, du/dt is piecewise continuous, and $Au(t)$ is continuous. We shall derive lower bounds on $\|u(t)\|$.

THEOREM 4.1.   Let A be symmetric and let u *be a solution of* (4.2) *with* $u(0) \neq 0$. (i) *If* $\phi \in L^p(0, \infty)$ *for some* p, $2 \leq p \leq \infty$, *then*

$$\|u(t)\| \geq \|u(0)\| \exp \{\lambda t - \mu(t + 1)^{2 - 2/p}\}. \tag{4.3}$$

(ii) *If* $\phi(t) \leq K(t + 1)^\alpha$, $\alpha > 0$, *then*

$$\|u(t)\| \geq \|u(0)\| \exp \{\lambda t - \mu(t + 1)^{2\alpha + 2}\}. \tag{4.4}$$

In each case, λ *is a constant depending on* u *and* μ *is a constant depending only on* ϕ.

Proof.   Suppose first that $u(t) \neq 0$ for all $t > 0$. Clearly $d\|u\|^2/dt = 2(Au, u) + 2 \operatorname{Re}(Lu, u)$. By taking finite differences and using the fact that A is symmetric, one finds that (Au, u) is differentiable and

$$\frac{d}{dt}(Au, u) = 2 \operatorname{Re}\left(Au, \frac{du}{dt}\right).$$

It follows that

$$\|u\|^4 \frac{d}{dt}\frac{(Au, u)}{\|u\|^2} = 2\|u\|^2 (Au, Au + Lu) - 2(Au, u)\operatorname{Re}(Au + Lu, u)$$

$$= 2\|Au + \tfrac{1}{2}Lu\|^2 \|u\|^2 - \tfrac{1}{2}\|Lu\|^2 \|u\|^2$$
$$- 2[\operatorname{Re}(Au + \tfrac{1}{2}Lu, u)]^2 + \tfrac{1}{2}[\operatorname{Re}(Lu, u)]^2.$$

Using Schwarz's inequality and (4.2), we get

$$\frac{d}{dt}\frac{(Au, u)}{\|u\|^2} \geq -\tfrac{1}{2}\phi^2. \tag{4.5}$$

Consider now the case (i) with $2 < p < \infty$. Then

$$\int_0^t \phi^2 \, ds \leq \left(\int_0^t \phi^p \, ds\right)^{2/p}\left(\int_0^t ds\right)^{1-2/p} \leq Mt^{1-2/p}.$$

Hence, integrating both sides of (4.5), we get

$$\frac{(Au, u)}{\|u\|^2} \geq \lambda - Mt^{1-2/p} \quad \left(\lambda = \frac{(Au(0), u(0))}{\|u(0)\|^2}\right), \tag{4.6}$$

where M depends only on ϕ. If $p = 2$ or if $p = \infty$, then (4.6) is clearly also valid. We now integrate the relation

$$\frac{d}{dt}\log\|u\|^2 = \frac{2(Au, u) + 2\operatorname{Re}(Lu, u)}{\|u\|^2} \tag{4.7}$$

and make use of (4.6) and (4.2). We get

$$\log\|u(t)\| \geq \log\|u(0)\| + \lambda t - \frac{p}{2p-2}Mt^{2-2/p} - Nt^{1-1/p},$$

where N depends only on ϕ. From this (4.3) follows.

In case (ii) holds, we again integrate (4.5). We get

$$\frac{(Au, u)}{\|u\|^2} \geq -M(t + 1)^{2\alpha+1}.$$

Using this relation and (4.2) in (4.7), (4.4) follows.

We can now show that the assumption $u(t) \neq 0$ for all $t > 0$ is valid. Indeed, if not, then there exists a value t_0 such that $u(t) \neq 0$ if $0 \leq t < t_0$ and $u(t_0) = 0$. Since (4.3), (4.4) hold for all $t \in (0, t_0)$, it follows (by taking $t \nearrow t_0$) that $u(t_0) \neq 0$—a contradiction.

We shall give another approach to the problem of deriving lower bounds on $\|u(t)\|$.

THEOREM 4.2. Let u(t) *be defined for* a $\leq$ t $\leq$ b *and belong to the domain of the self-adjoint operator* A, *and assume that du/dt is piecewise continuous and* Au(t) *is continuous. Then the following inequality holds:*

$$\sup_{a \leq t \leq b} \|u(t)\|^2 \leq 2(\|u(a)\|^2 + \|u(b)\|^2) + 4\left(\int_a^b \left\| \frac{du}{dt} - Au \right\| dt\right)^2. \quad (4.8)$$

Proof. Let E be the projection operator $\int_0^\infty dE_\lambda$, where $\{E_\lambda\}$ is the spectral family associated to A. Set $u_1 = Eu$, $u_2 = (I - E)u$. If $du/dt - Au = f$, then $du_1/dt - Au_1 = Ef = f_1$ and $du_2/dt - Au_2 = (I - E)f \equiv f_2$. Hence

$$\frac{d}{dt}(u_i, u_i) = 2 \operatorname{Re}(Au_i, u_i) + 2 \operatorname{Re}(f_i, u_i) \qquad (i = 1, 2).$$

Since $(Au_1, u_1) \geq 0$, $(Au_2, u_2) \leq 0$, we conclude that

$$\frac{d}{dt}(u_1, u_1) \geq 2 \operatorname{Re}(f_1, u_1), \qquad \frac{d}{dt}(u_2, u_2) \leq 2 \operatorname{Re}(f_2, u_2).$$

Integrating, we get

$$\|u_1(b)\|^2 - \|u_1(t)\|^2 \geq 2 \operatorname{Re} \int_t^b (f_1(s), u_1(s))\, ds,$$

so that

$$\|u_1(t)\|^2 \leq \|u_1(b)\|^2 + 2M \int_t^b \|f(t)\|\, dt,$$

where $M = \sup_{a \le t \le b} \|u(t)\|$. Similarly,

$$\|u_2(t)\|^2 \le \|u_2(a)\|^2 + 2M \int_a^t \|f(t)\| \, dt.$$

Adding the last two inequalities, we find that

$$\|u(t)\|^2 \le \|u_2(a)\|^2 + \|u_1(b)\|^2 + 2M \int_a^b \|f(t)\| \, dt.$$

Hence

$$M^2 \le \|u_2(a)\|^2 + \|u_1(b)\|^2 + \frac{M^2}{2} + 2\left(\int_a^b \|f(t)\| \, dt\right)^2,$$

and (4.8) follows.

THEOREM 4.3.   *Let* A *be a self-adjoint operator and let* u *be a solution of* (4.2) *in* a $\le$ t $\le$ b *with*

$$\int_a^b \phi(t) \, dt \le \frac{1}{2\sqrt{2}}. \tag{4.9}$$

Then

$$\|u(t)\| \le 2\sqrt{2} \, \|u(a)\|^{(b-t)/(b-a)} \|u(b)\|^{(t-a)/(b-a)}. \tag{4.10}$$

Proof.   Set $w(t) = e^{\sigma t} u(t)$, σ real. Then

$$\left\| \frac{dw}{dt} - (A + \sigma I)w \right\| = e^{\sigma t} \left\| \frac{du}{dt} - Au \right\|. \tag{4.11}$$

Applying Theorem 4.2 to $A + \sigma I$, we find, after using first (4.9) and then (4.11), that

$$\|e^{\sigma t} u(t)\|^2 \le 4\|e^{\sigma a} u(a)\|^2 + 4\|e^{\sigma b} u(b)\|^2.$$

Choosing σ such that the two terms on the right become equal, (4.10) follows.

Theorem 4.3 gives a convexity-like inequality that can be used to derive results similar to Theorem 4.1.

PROBLEMS. (1) Suppose there is a sequence $\{t_n\}$ with $t_0 = 0$, $t_n < t_{n+1}$, $t_n \to \infty$ as $n \to \infty$, such that

$$\int_{t_{n-1}}^{t_n} \phi(t)\, dt = \frac{1}{6\sqrt{2}},$$

and set $\rho_n = t_{n+1} - t_n$. Prove: If

$$\sum_{0}^{n} \frac{1}{\rho_j} \leq K \left(1 + \sum_{j=0}^{n} \rho_j \right)^k \qquad (n = 1, 2, \ldots),$$

then any solution u of (4.2), with A self-adjoint, satisfies

$$\|u(t)\| \geq \|u(0)\| \exp \{ -\mu(t + 1)^{k+1} \} \beta^t \qquad (0 < t < \infty),$$

where μ is a constant independent of u and β is a constant depending on the solution u. [*Hint*: Set $\sigma_j = \log \|u(t_j)\|$ and estimate $(\sigma_j - \sigma_{j-1})/\rho_{j-1}$ successively, using Theorem 4.3.]

(2) Let u be a solution of (4.2) with A self-adjoint. If $\phi \in L^p(0, \infty)$ $(1 \leq p \leq 2)$, then

$$\|u(t)\| \geq \|u(0)\| \exp \{ -\mu(t + 1) \} \beta^t.$$

[*Hint*: Use Problem 1].

If $\phi \in L^p(0, \infty)$ $(2 < p \leq \infty)$ or if $\phi(t) \leq K(1 + t)^\alpha$ $(\alpha \geq 0)$, then one gets the same lower bounds as asserted in Theorem 4.1.

5 | WEIGHTED ELLIPTIC EQUATIONS

In this and in the following two sections we use the notation: $D_j = (1/i)\partial/\partial x_j$, $D_t = (1/i)\partial/\partial t$, and $D_x^\alpha = D_1^{\alpha_1} \cdots D_n^{\alpha_n}$, where $\alpha = (\alpha_1, \ldots, \alpha_n)$. We consider differential operators in $(n + 1)$ variables

$$A(x, D_x, D_t) = A_l(x, D_x) + A_{l-1}(x, D_x)D_t + \cdots + A_0 D_t^l, \qquad (5.1)$$

where $A_0 = \text{const.} \neq 0$ and $A_j(x, D_x)$ (for $j \geq 1$) is a linear differential operator of order s_j with coefficients in a bounded domain Ω of R^n. Set

$$Q_a^b = \Omega \times \{a < t < b\}, \qquad Q = Q_{-\infty}^\infty,$$
$$S_a^b = \partial\Omega \times \{a < t < b\}, \qquad S = S_{-\infty}^\infty.$$

We say that $A(x, D_x, D_t)$ is of *order type* $(2m, l)$ if

$$s_l = 2m, \qquad s_j \le \frac{2m}{l} j \qquad (1 \le j \le l - 1). \qquad (5.2)$$

Denote by $A_j^0(x, D_x)$ the sum of terms in $A_j(x, D_x)$ that are of order $2mj/l$ $(A_j^0(x, D_x) \equiv 0$ if there are no such terms). The operator

$$A_0(x, D_x, D_t) = \sum_{j=0}^{l} A_{l-j}^0(x, D_x) D_t^j \qquad (5.3)$$

is called the *weighted principal part* of $A(x, D_x, D_t)$.

DEFINITION. An operator $A(x, D_x, D_t)$ of order type $(2m, l)$ is said to be *weighted elliptic* in Q $(\bar{Q})$ if

$$A_0(x, \xi, \tau) \neq 0 \qquad (5.4)$$

for all real $(\xi, \tau) \neq 0$ and all x in Q $(\bar{Q})$.

Observe that a weighted elliptic operator of order type $(2m, 2m)$ is elliptic, and of order type $(2m, 1)$ is parabolic. If $A(x, D_x, D_t)$ is weighted elliptic of order type $(2m, l)$, then $A_l(x, D_x)$ is elliptic of order $2m$.

If $n \le 2$, we always impose on $A(x, D_x, D_t)$ the *root condition*, namely, that for every real $\tau \neq 0$, real linearly independent vectors ξ, η, and $x \in \partial\Omega$, the polynomial $A_0(x, \xi + z\eta, \tau)$ has exactly m roots with positive imaginary parts.

We consider solutions $u(x, t)$ of the system

$$\begin{cases} A(x, D_x, D_t)u = f(x, t) & \text{in } Q_a^b, \\ B_j(x, D_x)u = 0 & \text{on } S_a^b \quad (1 \le j \le m), \end{cases} \qquad (5.5)$$

where $A(x, D_x, D_t)$ is a weighted elliptic operator of order type $(2m, l)$. One refers to the triple $(A, \{B_j\}, Q)$ as a *weighted elliptic boundary value problem*.

We shall now extend the results of Section I.19 to such boundary value problems.

COMPLEMENTARY CONDITION. At any point x of $\partial\Omega$ let ν be the normal to $\partial\Omega$ and let (ξ, τ) be a nonzero vector in R^{n+1} parallel to S at x. Then the polynomials in z: $B_{j0}(x, \xi + z\nu)$ $(j = 1, \ldots, m)$ are linearly independent modulo the polynomial

$$\prod_{k=1}^{m} (z - z_k^+(\xi, \tau)),$$

where $z_k^+(\xi, \tau)$ $(k = 1, \ldots, m)$ are the roots of $A_0(x, \xi + z\nu, \tau)$ with positive imaginary parts. (B_{j0} is the principal part of B_j.)

DEFINITION. A weighted elliptic boundary value problem $(A, \{B_j\}, Q)$ of order type $(2m, l)$ is called *regular* if:

(i) the complementary condition holds;

(ii) $\partial\Omega$ is of class C^{2m}, the coefficients of $A(x, D_x, D_t)$ are bounded measurable functions in Ω, and the coefficients of $A_0(x, D_x, D_t)$ are continuous in $\overline{\Omega}$;

(iii) B_j is of order $m_j \leq 2m - 1$, the coefficients of $B_j(x, D_x)$ are of class C^{2m-m_j} on $\partial\Omega$, and the system $\{B_j\}$ is normal.

We can now state:

THEOREM 5.1.   Let $(A, \{B_j\}, Q)$ be a regular weighted elliptic boundary value problem of order type $(2m, \ell)$ and let $1 < p < \infty$. Set $d = 2m/\ell$. Then for all functions $u \in C^{2m}(\overline{\Omega})$ satisfying the boundary conditions

$$B_j(x, D_x)u = 0 \qquad \text{on } \partial\Omega \qquad (1 \leq j \leq m), \tag{5.6}$$

and for all real λ, $|\lambda| \geq N_0$,

$$\sum_{j=0}^{2m} |\lambda|^{(2m-j)/d} |u|_{j, p}^{\Omega} \leq C |A(x, D_x, \lambda)u|_{0, p}^{\Omega}, \tag{5.7}$$

where C, N_0 are constants depending only on $(A, \{B_j\}, Q)$ and p.

Proof. If $l = 2m$, then $A(x, D_x, D_t)$ is an elliptic operator in (x, t) of order $2m$, and the proof is similar to the proof of Theorem I.18.2. Suppose then that $l \neq 2m$. Set $d = a/b$, where a, b are relatively prime positive integers. Since $A_j^0 \equiv 0$ if $2jm/l$ is not an integer, we can write

$$A_0(x, \xi, \tau) = \sum_{j=0}^{h} A_{l-bj}^0(x, \xi)\tau^{bj} \qquad \left(h = \frac{l}{b}\right), \tag{5.8}$$

where $A_0(x, D_x, D_t)$ is the weighted principal part of $A(x, D_x, D_t)$. Define

$$L^+(x, \xi, \tau) = \sum_{j=0}^{h} A_{l-bj}^0(x, \xi)\tau^{aj},$$

$$L^-(x, \xi, \tau) = \sum_{j=0}^{h} (-1)^{bj} A_{l-bj}^0(x, \xi)\tau^{aj}.$$

Clearly, if $\tau \geq 0$, then $\tau^b \geq 0$ and

$$\begin{cases} L^+(x, \xi, \tau) = A_0(x, \xi, \tau^d), \\ L^-(x, \xi, \tau) = A_0(x, \xi, -\tau^d). \end{cases} \tag{5.9}$$

Also, for fixed x, L^+ and L^- are homogeneous polynomials in (ξ, τ) of degree $2m$. From (5.9) and the weighted ellipticity of A it follows that $L^+(x, \xi, \tau) \neq 0$ for all real $(\xi, \tau) \neq 0$, $\tau \geq 0$. By the homogeneity of L^+ in (ξ, τ) we then conclude that $L^+(x, \xi, \tau) \neq 0$ for all real $(\xi, \tau) \neq 0$. Similarly $L^-(x, \xi, \tau) \neq 0$ for all real $(\xi, \tau) \neq 0$. One can verify that L^+ and L^- satisfy the root condition (if $n \leq 2$) and that the $\{B_j\}$ satisfy the complementary condition with respect to L^+ and with respect to L^-. Thus $(L^+, \{B_j\}, \Omega)$ and $(L^-, \{B_j\}, \Omega)$ are regular weighted elliptic boundary value problems in (x, t) of order-type $(2m, 2m)$. We can therefore apply the special case $l = 2m$ and, upon using (5.9), get

$$\sum_{j=0}^{2m} |\lambda|^{l(2m-j)/2m} |u|^{\Omega}_{j,p} \leq C |A_0(x, D_x, \lambda)u|^{\Omega}_{0,p}$$

if λ is real, $|\lambda| \geq N_0$. Now use the fact that

$$|A_0(x, D_x, \lambda)u - A(x, D_x, \lambda)u|^{\Omega}_{0,p} \leq \frac{K}{|\lambda|^{1/2m}} \sum_{j=0}^{2m} |\lambda|^{l(2m-j)/2m} |u|^{\Omega}_{j,p}$$

for some constant K depending only on bounds on the coefficients of $A - A_0$.

We set $A^\lambda(x, D_x) = A(x, D_x, \lambda)$ and call $(A^\lambda, \{B_j\}, \Omega)$ the *reduced* weighted elliptic boundary value problem. The following theorem, due to Agmon [4], is analogous to Theorem I.19.3.

THEOREM 5.2.   *Let the assumptions of Theorem 5.1 hold and let* A^λ *be the operator defined (as in Section I.19) by the reduced weighted elliptic boundary value problem* $(A^\lambda(x, D_x), \{B_j\}, \Omega)$. *Then for any* $1 < p < \infty$, A *maps its domain* $H^{2m,p}(\Omega; \{B_j\})$ *onto* $L^p(\Omega)$ *for all* λ *real with* $|\lambda|$ *sufficiently large.*

From now on we make the assumption:

$$d = \frac{2m}{l} \quad \text{is an integer} \tag{5.10}$$

For simplicity we also assume that

$$A_0(x) \equiv 1. \tag{5.11}$$

Introducing the functions $u_j = D_t^j u$ ($0 \leq j \leq l - 1$), we can write (5.1) in the form

$$D_t u_j - u_{j+1} = 0 \quad (j = 0, 1, \ldots, l - 2),$$

$$D_t u_{l-1} + \sum_{j=0}^{l-1} A_{l-j} u_j = f. \tag{5.12}$$

Introducing the vectors $U = (u_0, \ldots, u_{l-1})$, $F = (f_0, \ldots, f_{l-1})$, we can write (5.12) in the matrix form

$$D_t U - AU = F, \tag{5.13}$$

where

$$AU = \left(u_1, u_2, \ldots, u_{l-1}, -\sum_{j=0}^{l-1} A_{l-j} u_j \right) \tag{5.14}$$

and $F = (0, 0, \ldots, 0, f)$.

PROBLEM. (1) Prove that for general F, the equation (5.13) is the same as the system

$$
\begin{aligned}
u_j &= D_t^j u_0 - \sum_{k=1}^{j} D_t^{j-k} f_{k-1} \qquad (0 \le j \le l - 1), \\
A u_0 &= f_{l-1} + \sum_{k=1}^{l-1} \left[D_t^{l-k} + \sum_{j=k}^{l-1} D_t^{j-k} A_{l-j} \right] f_{k-1}.
\end{aligned}
\tag{5.15}
$$

We now introduce two Banach spaces:

$$X = W^{2m,\,p}(\Omega) \times W^{2m-d,\,p}(\Omega) \times \cdots \times W^{d,\,p}(\Omega),$$

$$Y = W^{2m-d,\,p}(\Omega) \times W^{2m-2d,\,p}(\Omega) \times \cdots \times W^{0,\,p}(\Omega).$$

We take for the domain of the operator A of (5.13) the subset

$$D_A = H^{2m,\,p}(\Omega; \{B_j\}) \times W^{2m-d,\,p}(\Omega) \times \cdots \times W^{d,\,p}(\Omega)$$

of X. Then AU is in Y and A is a closed linear operator from X into Y. By a *solution* of (5.13) we mean a function $U(t)$ in an interval (a, b) with values in D_A such that (i) $U(t)$ is continuous in the X-topology, (ii) the derivative $D_t U$ exists and is piecewise continuous in the Y-topology, and (iii) (5.13) holds.

We denote by S the subspace of Y consisting of all vectors of the form $(0, 0, \ldots, 0, f)$, $f \in L^p(\Omega)$. We denote by $R_S(\lambda; A)$ the restriction of $R(\lambda; A)$ to S. We write $|R(\lambda; A)|_X$ ($|R(\lambda; A)|_Y$) for the norm of $R(\lambda; A)$ considered as a map from Y into X (Y). The norms $|R_S(\lambda; A)|_X$, $|R_S(\lambda; A)|_Y$ are defined similarly.

THEOREM 5.3.   *Let the assumptions of Theorem 5.1 hold. The resolvent* $R(\lambda; A)$ *considered as a mapping of* Y *into* X (Y) *exists and is a bounded (completely continuous) operator for all complex* λ *except for a discrete set whose points are* λ_k, *the eigenvalues of* A. *As a function of* λ, $R(\lambda; A)$ *is a meromorphic* *function with poles at the points* λ_k. *Furthermore, there exists numbers* $0 < \delta < \pi/2$, $N > 0$, *such that the region* Σ: $|\arg(\pm\lambda)| \leq \delta$, $|\lambda| \geq N$, *is free of poles, and such that*

$$|R(\lambda; A)|_X + |\lambda| \, |R(\lambda; A)|_Y \leq C|\lambda|^{\ell-1} \qquad in \ \Sigma, \qquad (5.18)$$

$$|R_s(\lambda; A)|_X + |\lambda| \, |R_s(\lambda; A)|_Y \leq C \qquad in \ \Sigma, \qquad (5.19)$$

where C *is a constant, and*

$$\left| \frac{d}{d\lambda} R_s(\lambda; A) \right|_X + \left| \lambda \frac{d}{d\lambda} R_s(\lambda; A) \right|_Y = 0\left(\frac{1}{\lambda}\right) \qquad as \ \lambda \in \Sigma, \ |\lambda| \to \infty. \quad (5.20)$$

Proof. The existence of the resolvent means that for every $F \in Y$ there is a unique $U \in D_A$ such that

$$\lambda U - AU = F. \qquad (5.21)$$

This equation is equivalent to the system (cf. (5.15))

$$A(x, D_x, \lambda)u_0 = f_{l-1} + \sum_{k=1}^{l-1} \left(\lambda^{l-k} + \sum_{j=k}^{l-1} \lambda^{j-k} A_{l-j} \right) f_{k-1}, \qquad (5.22)$$

$$u_j = \lambda^j u_0 - \sum_{k=1}^{j} \lambda^{j-k} f_{k-1} \qquad (1 \leq j \leq l-1). \qquad (5.23)$$

This system has a unique solution if and only if (5.22) has a unique solution u_0 in $H^{2m,p}(\Omega; \{B_j\})$. By Theorems 5.1, 5.2 this is in fact the case for all real λ with $|\lambda|$ sufficiently large. Since A is a closed linear operator from X into Y, $R(\lambda; A)$ not only exists for such values of λ but is also a bounded operator from Y into X. By Theorem I.11.2, $R(\lambda; A)$ is a completely continuous operator from Y into Y.

Set $R_0 = R(\lambda_0; A)$ for some λ_0 in the resolvent set $\rho(A)$ of A. For any $\lambda \in \rho(A)$ with $1/(\lambda_0 - \lambda)$ in $\rho(R_0)$, we have

$$R(\lambda; A) = \frac{1}{\lambda_0 - \lambda} R_0 R\left(\frac{1}{\lambda_0 - \lambda}; R_0\right). \qquad (5.24)$$

Since R_0 is completely continuous (as a map from Y into Y), the resolvent $R(\mu; R_0)$ exists for all complex μ except for a discrete set $\{\mu_k\}$, and $R(\mu, R_0)$ has poles at these points. In view of (5.24), a similar conclusion follows for $R(\lambda; A)$ (as a map from Y into Y)—that is, $R(\lambda; A)$ exists for all λ except for a discrete set $\{\lambda_k\}$, it is holomorphic in λ for $\lambda \neq \lambda_k$ for all k, and at each λ_k it has a pole. From this it follows that, as a mapping from Y into X, $R(\lambda; A)$ also exists for all $\lambda \neq \lambda_k (1 \leq k < \infty)$. From the resolvent equation it further follows that $dR(\lambda; A)/d\lambda$ exists. Thus, $R(\lambda; A)$ is holomorphic in λ for $\lambda \neq \lambda_k$ for all k. At each point λ_k it has an isolated singularity. This singularity is a pole, since the Laurent expansions about $\lambda = \lambda_k$ of $R(\lambda; A)$ considered as a map from Y into Y and considered as a map from Y into X must coincide.

Consider now a family of boundary value problems

$$P_\theta = (A(x, D_x, e^{i\theta}D_t), \{B_j(x, D_x)\}, \Omega)$$

depending on a real parameter θ. Since P_0 is a given regular weighted elliptic boundary value problem, it follows by continuity that each P_θ with $|\theta| < \delta$ is also a regular weighted elliptic boundary value problem, provided δ is sufficiently small. Furthermore, the assertions of Theorems 5.1, 5.2 hold uniformly with respect to θ.

It follows that in some region Σ: $|\arg(\pm\lambda)| \leq \delta, |\lambda| \geq N$ there are no poles λ_k, and

$$\sum_{j=0}^{2m} |\lambda|^{(2m-j)/d} |u|_{j, p}^{\Omega} \leq K |A(x, D_x, \lambda)u|_{0, p}^{\Omega},$$

where K is a constant independent of u, λ. Using this relation and (5.22), (5.23), the inequalities (5.18), (5.19) easily follow. Finally, (5.20) follows from (5.19) and the Cauchy integral representation theorem. The details are left to the reader.

6 | ASYMPTOTIC EXPANSIONS OF SOLUTIONS OF EVOLUTION EQUATIONS

We shall consider evolution equations

$$Lu \equiv \frac{1}{i} \frac{du}{dt} - A(t)u = f, \tag{6.1}$$

where $A(t)$ has an asymptotic expansion

$$A(t) = A_0 + \sum_{j=1}^{m} e^{-\alpha_j t} A_j + B(t), \qquad B(t) \text{ "small,"}$$

and similarly f has an asymptotic expansion. We shall prove that the solution u then also has an asymptotic expansion. The assumptions on A will be such that our results will apply to elliptic equations. Applications will be given in Section 7.

We are given two Hilbert spaces X and Y, with $X \subset Y$, $| \ |_X \geq | \ |_Y$. $A(t)$ is a closed linear operator in Y with a domain $D(A(t))$ in X. By a *solution* $u(t)$ of (6.1) we mean a function $u(t)$ defined in some interval (a, b) with values in $D(A(t))$ such that $u(t)$ is continuous as a function from (a, b) into X, du/dt exists and is piecewise continuous as a function from (a, b) into Y, and (6.1) holds. If instead of (6.1), $|Lu - f| = 0(e^{-\delta t})$, then we call u a δ-*asymptotically approximate solution.*

We shall need the following assumptions:

(A_1) The operator $A(t)$ $(0 \leq t < \infty)$ has the form

$$A(t) = A_0 + A_1(t) + B(t), \tag{6.2}$$

where

$$A_1(t) = \sum_{j=1}^{m} e^{-\alpha_j t} A_j. \tag{6.3}$$

The operators A_j $(0 \leq j \leq m)$ and $B(t)$ are closed operators in Y with domains $D(A_j)$, $D(B(t))$ in X satisfying: $D(A_j) \supset D(A_0)$ $(1 \leq j \leq m)$, $D(B(t)) \supset D(A_0)$, and

$$|B(t)u|_Y + \sum_{j=0}^{m} |A_j u|_Y \leq K |u|_X \qquad \text{for all } u \in D(A_0), 0 \leq t < \infty, \tag{6.4}$$

where K is a constant.

(A_2) The integers α_i satisfy

$$0 < \alpha_1 < \cdots < \alpha_m. \tag{6.5}$$

(A_3) The range of the A_j $(1 \leq j \leq m)$ and of $B(t)$ lie in a closed subspace S of Y. We set

$$L_0 u = \frac{1}{i} \frac{du}{dt} - A_0 u,$$

$$L_1 u = \frac{1}{i} \frac{du}{dt} - (A_0 + A_1)u.$$

DEFINITION. Let S be a closed subspace of Y. Suppose there exists an operator ζ_T defined for all continuous functions $u(t)$ $(0 < t < \infty)$ with $L_0 u(t)$ in S for all $t > 0$, such that:

(i) $v(t) \equiv (\zeta_T \cdot u)(t) = u(t)$ if $t \geq T + 1$;

(ii) $v(t) \equiv 0$ if $t \leq T$;

(iii) $L_0 v(t) \in S$ for all $t > 0$;

(iv) $|L_0 v|_Y \leq c(|L_0 u|_Y + |u|_Y)$ and $|v|_X \leq c |u|_X$ for $T \leq t \leq T + 1$, where c is a constant.

Then we say that L_0 satisfies the ζ_T *hypothesis* (with respect to S).

If $S = Y$, then this hypothesis certainly holds, for we can take $(\zeta_T \cdot u)(t) = g(t)u(t)$, where $g(t)$ is a suitable C^∞ function.

PROBLEM. (1) Let $(A_1), (A_3)$ hold and let L_0 satisfy the ζ_T hypothesis. Prove: if $v = \zeta_T \cdot u$ and $Lu \in S$, then $Lv \in S$.

For a real positive number α, and numbers $\alpha_1, \ldots, \alpha_m$ we define sets

$$\Gamma_{N-l} = \left\{ x; x = \sum_{k=1}^{l} \alpha_{j_k}, \quad 1 \le j_k \le m \right\}$$

for $l = N, N - 1, \ldots, 1$, where $N = [\alpha/\alpha_1] + 1$. Finally, we set $\Gamma_N = \{0\}$ and

$$\Gamma = \bigcup_{i=0}^{N} \Gamma_i, \quad \Gamma(\alpha; \alpha_1, \ldots, \alpha_m) = \Gamma \cap \{x; x < \alpha\}.$$

THEOREM 6.1.   Let u be a solution of (6.1) in $(0, \infty)$ with $|u|_X \in L^2(0, \infty)$. Let the assumptions $(A_1)–(A_3)$ hold and let L_0 satisfy the ζ_0 hypothesis. Assume further that

(i) $R_S(\lambda; A_0)$ is a meromorphic function in the strip $0 \le \mathrm{Im}\, \lambda < \alpha$ and $|R_S(\lambda; A_0)|_X = O(1)$ as $|\lambda| \to \infty$ in the strip.

(ii) For every $\varepsilon > 0$ there is a constant K_ε such that $|B(t)u|_Y \le K_\varepsilon e^{-(\alpha-\varepsilon)t}|u|_X$.

(iii) $f(t) \in S$ for every $t \ge 0$ and $|f(t)|_Y = O(e^{-(\alpha-\varepsilon)t})$ for every $\varepsilon > 0$.

Then for every $\varepsilon > 0$ sufficiently small there exists a finite number of exponential functions $u_j(t) = e^{i\lambda_j t}p_j(t)$ with $\mathrm{Im}\, \lambda_j \le \alpha - \varepsilon$, $\lambda_j = \tilde{\lambda}_j + i\gamma$, where $\gamma \in \Gamma(\alpha; \alpha_1, \ldots, \alpha_m)$ and $\tilde{\lambda}_j$ is a pole of $R_S(\lambda; A_0)$ in the strip $0 < \mathrm{Im}\, \lambda \le \alpha - \varepsilon$ and $p_j(t)$ is a polynomial in t with coefficients in $D(A_0)$, such that

$$\int_0^\infty \left| e^{-(\alpha-\varepsilon)t} \left| u(t) - \sum_j u_j(t) \right|_X \right|^2 dt \le C_0 \int_0^\infty |u|_X^2 \, dt + C_0'. \tag{6.6}$$

C_0 is a constant depending only on the A_i, α_i, $B(t)$, α, ε, and C_0' is a constant depending also on f, and $C_0' = 0$ if $f(t) \equiv 0$.

Proof.   We may assume that α/α_1 is not an integer, for otherwise we replace α by $\alpha - \eta$ with η arbitrarily small such that $(\alpha - \eta)/\alpha_1$ is not an integer. Let $v = \zeta_0 \cdot u$ and write, for all real t,

$$Lv = f + h, \tag{6.7}$$

where, by definition, $f(t) = 0$ if $t < 0$. Then the support of h is contained in the interval $[0, 1]$.

Writing (6.7) in the form

$$L_1 v = f + h + B(t)v \equiv g(t) \tag{6.8}$$

and using the result of Problem 1 and the condition (A_3), we see that $g(t) \in S$ for every t. By the assumptions (ii), (iii), $e^{\delta t} |g(t)|_Y \in L^2(0, \infty)$ if $0 \le \delta < \alpha$, and

$$\int_0^\infty |e^{\delta t} |g(t)|_Y|^2 \, dt \le C_1 \int_0^\infty |u|_X^2 \, dt + C_1' \tag{6.9}$$

if $0 \le \delta \le \alpha - \varepsilon/2$. Here C_1 (C_1') is a generic constant depending only on $A_j, \alpha_j, B, \alpha, \varepsilon$ $(A_j, \alpha_j, B, \alpha, \varepsilon, f$ and $= 0$ if $f \equiv 0)$.

Using Plancherel's theorem and (6.9), we get

$$\int_{\text{Im } \lambda = \delta} |\hat{g}(\lambda)|_Y^2 \, d\lambda = \int_0^\infty |e^{\delta t} g(t)|_Y^2 \, dt \le C_1 \int_0^\infty |u|_X^2 \, dt + C_1'. \tag{6.10}$$

If we take the Fourier transform of both sides of (6.8), we obtain

$$\lambda \hat{v}(\lambda) - A_0 \hat{v}(\lambda) - \sum_{j=1}^m A_j \hat{v}(\lambda - i\alpha) = \hat{g}(\lambda). \tag{6.11}$$

This relation holds not only for λ real, but also if $\text{Re } \lambda \le 0$. Now, $\hat{g}(\lambda)$ is an analytic function for $\text{Im } \lambda < \alpha$, whereas $\hat{v}(\lambda)$ is an analytic function for $\text{Im } \lambda < 0$ with values in L^2 on the real axis, and

$$\int_{\text{Im } \lambda = \delta \le 0} |\hat{v}(\lambda)|_X^2 \, d\lambda \le \int_0^\infty |v|_X^2 \, dt \le C_1 \int_0^\infty |u|_X^2 \, dt. \tag{6.12}$$

Consider the relation (6.11). Since $g(t) \in S$, also $\hat{g}(\lambda) \in S$ if $\text{Im } \lambda < \alpha$. The functions $A_j \hat{v}(\lambda - i\alpha_j), A_0 \hat{v}(\lambda)$ also belong to S. Hence, if $\lambda \in \rho_S(A_0)$, then (6.11) takes the form

$$\hat{v}(\lambda) = R_S(\lambda; A_0)\hat{g}(\lambda) + \sum_{j=1}^m R_S(\lambda; A_0)A_j \hat{v}(\lambda - i\alpha_j). \tag{6.13}$$

This formula enables us to extend $\hat{v}(\lambda)$ as a meromorphic function into the domain $\text{Im } \lambda < \alpha$ in $N = [\alpha/\alpha_1] + 1$ steps. It is clear that the poles λ_j of $\hat{v}(\lambda)$ in $\text{Im } \lambda < \alpha$ have the form $\tilde{\lambda}_j + i\gamma$, where $\gamma \in \Gamma(\alpha; \alpha_1, \ldots, \alpha_m)$ and $\tilde{\lambda}_j$ is a pole of $R_S(\lambda; A_0)$.

It is easily verified that for every $\varepsilon > 0$ sufficiently small there is a $\delta = \delta(\varepsilon)$ such that:

(a) $\alpha - \varepsilon < \delta(\varepsilon) < \alpha - \varepsilon/2$, $[\delta(\varepsilon)/\alpha_1] = [\alpha/\alpha_1]$ (recall that α/α_1 is not an integer).

(b) There are no poles of $R_S(\lambda; A_0)$ in the strip $\alpha - \varepsilon < \text{Im } \lambda < \delta(\varepsilon)$.

(c) On the lines $\text{Im } \lambda = \delta(\varepsilon) - \gamma$, where $\gamma \in \Gamma(\alpha; \alpha_1, \ldots, \alpha_m)$ and $\text{Im } \lambda \ge 0$, $R_S(\lambda; A_0)$ has no poles.

Denote by $\Delta(\delta)$ the set of lines Im $\lambda = \delta(\varepsilon) - \gamma$, where $\gamma \in \Gamma(\alpha; \alpha_1, \dots, \alpha_m)$ and $\gamma \le \delta(\varepsilon)$. Then

$$|R_S(\lambda; A_0)|_X \le C_1 \qquad \text{on } \Delta(\delta). \tag{6.14}$$

LEMMA 6.1. The following inequality holds:

$$\int_{\text{Im } \lambda = \delta(\varepsilon)} |\hat{v}(\lambda)|_X^2 \, d\lambda \le C_1 \int_0^\infty |u|_X^2 \, dt + C_1'. \tag{6.15}$$

PROBLEM. (2) Prove, by induction, that for $0 \le k \le N$,

$$\int_{\text{Im } \lambda = \delta(\varepsilon)} |\hat{v}(\lambda - i\gamma)|_X^2 \, d\lambda \le C_1 \int_0^\infty |u|_X^2 \, dt + C_1' \qquad \text{for } \gamma \in \Gamma_k. \tag{6.16}$$

(6.16) for $k = 0$ gives (6.15).

LEMMA 6.2. For any $0 < \delta < \alpha$,

$$\left| \int_s^{s+i\delta} e^{i\lambda t} \hat{v}(\lambda) \, d\lambda \right|_X \to 0 \qquad \text{as } s \to \pm \infty. \tag{6.17}$$

Proof. Since $|e^{\delta t} g(t)|_Y \in L^1(0, \infty)$ if $\delta < \alpha$ and $|e^{\delta t} v(t)|_X \in L^1(0, \infty)$ if $\delta < 0$, the lemma of Riemann–Lebesgue gives

$$\lim_{s \to \pm \infty} |\hat{g}(s + i\eta)|_Y = 0 \qquad \text{if } \eta < \alpha,$$

$$\lim_{s \to \pm \infty} |\hat{v}(s + i\eta)|_X = 0 \qquad \text{if } \eta < 0.$$

By the Lebesgue bounded convergence theorem we then obtain

$$\lim_{s \to \pm \infty} \int_{r_1}^{r_2} |\hat{g}(s + i\eta)|_Y \, d\eta = 0, \qquad 0 \le r_1 < r_2 < \alpha,$$

$$\lim_{s \to \pm \infty} \int_{-r_1}^{-r_2} |\hat{v}(s + i\eta)|_X \, d\eta = 0, \qquad 0 < r_2 < r_1 < \infty.$$

Using these relations in (6.13) and noting that $|R_S(\lambda; A_0)|_X \le C_1$ if $0 \le \text{Im } \lambda < \alpha$, $|\text{Re } \lambda| \ge d$ (for some constant d), one can easily prove, by induction on k, that

$$\lim_{s \to \pm \infty} \int_{(k-1)\alpha_1}^{k\alpha_1} |\hat{v}(s + i\eta)|_X \, d\eta = 0 \qquad (1 \le k \le N).$$

(In the last interval, we define $\hat{v}(s + i\eta) = 0$ for $\eta > \delta$.) Hence,

$$\lim_{s \to \pm \infty} \int_0^\delta |\hat{v}(s + i\eta)|_X \, d\eta = 0.$$

This implies (6.17).

We return to the proof of Theorem 6.1. Recall that

$$v(t) = \frac{1}{\sqrt{2\pi}} \int_{-\infty}^\infty e^{i\lambda t} \hat{v}(\lambda) \, d\lambda.$$

Using Lemma 6.2 and the residue theorem, we conclude that

$$w(t) \equiv v(t) - i\sqrt{2\pi} \sum_i \operatorname*{Res}_{\lambda = \lambda_j} e^{i\lambda t} \hat{v}(\lambda) = \frac{1}{\sqrt{2\pi}} \int_{\operatorname{Im} \lambda = \delta(\varepsilon)} e^{i\lambda t} \hat{v}(\lambda) \, d\lambda,$$

where λ_j are the poles of $\hat{v}(\lambda)$ in the strip $0 \leq \operatorname{Im} \lambda \leq \alpha - \varepsilon$.

Using Lemma 6.1 and Plancherel's theorem, we find that

$$\int_{-\infty}^\infty |e^{\delta t}|w(t)|_X|^2 \, dt = \int_{-\infty}^\infty |\hat{v}(\mu + i\delta)|_X^2 \, d\mu \leq C_1 \int_0^\infty |u|_X^2 \, dt + C_1'.$$

Noting that $v(t) = u(t)$ if $t \geq 1$, we easily find that (6.6) holds with

$$u_j = i\sqrt{2\pi} \operatorname*{Res}_{\lambda = \lambda_j} e^{i\lambda t} \hat{v}(\lambda).$$

Recall that λ_j has the form $\tilde{\lambda}_j + i\gamma$, where $\gamma \in \Gamma(\alpha; \alpha_1, \ldots, \alpha_m)$ and $\tilde{\lambda}_j$ is a pole of $R_S(\lambda; A_0)$. If

$$\hat{v}(\lambda) = \sum_{k=-h}^\infty (\lambda - \lambda_j)^k v_k$$

is the Laurent expansion of $\hat{v}(\lambda)$ near $\lambda = \lambda_j$, then

$$u_j = i\sqrt{2\pi} \, e^{i\lambda_j t} \sum_{k=0}^{h-1} \frac{(it)^k}{k!} v_{-k-1}.$$

This completes the proof of the theorem.

COROLLARY. *If* $\lambda_j = a + ib$ *is a pole of order* n *of* $\hat{v}(\lambda)$ *in the strip* $0 \leq \operatorname{Im} \lambda < \alpha$, *then*

$$|u_j(t)|_X = 0(t^{n-1} e^{-bt}). \tag{6.18}$$

PROBLEM.   (3) Set $u_\lambda = i\sqrt{2\pi}\ \mathrm{Res}_{z=\lambda}\ e^{izt}\hat{v}(z)$. Prove: If $\lambda = \mu + iv$ is a pole of $\hat{v}(z)$ and $v < \alpha$, then

$$\frac{1}{i}\frac{du_\lambda}{dt} - A_0 u_\lambda = \sum_{j=1}^{m} e^{-\alpha_j t} A_j u_{\lambda - i\alpha_j}. \tag{6.19}$$

Define

$$u(t; x) = \sum_{\substack{\mu = x \\ 0 \le v \le \alpha - \varepsilon}} u_\lambda(t), \qquad \lambda = \mu + iv. \tag{6.20}$$

THEOREM 6.2.   *Let the assumptions of Theorem 6.1 hold. Then the function* u(t; x) *satisfies*

$$|Lu(t; x) - f|_Y = 0(e^{-(\alpha - \varepsilon)t}).$$

Proof.   Using (6.19), one finds that

$$\frac{1}{i}\frac{du(t; x)}{dt} - A_0 u(t; x) - A_1 u(t; x) = -\sum_{j=1}^{m} e^{-\alpha_j t} A_j \sum_{\substack{\mu = x \\ \alpha - \varepsilon - \alpha_j < v \le \alpha - \varepsilon}} u_\lambda(t) \equiv g(t; x).$$

If we use the corollary to Theorem 6.1, we get

$$|g(t; x)|_Y = 0(t^{n-1}e^{-(\alpha + \alpha_1 - \varepsilon)t}) = 0(e^{-(\alpha - \varepsilon)t}).$$

Hence

$$|Lu(t; x) - f|_Y = |Lu(t; x)|_Y + |f(t)|_Y$$

$$\le |L_1 u(t; x)|_Y + |B(t)u(t; x)|_Y + |f(t)|_Y$$

$$= 0(e^{-(\alpha - \varepsilon)t})(1 + |u(t; x)|_X).$$

Since $|u|_X \in L^2(0, 1)$, $\hat{v}(\lambda)$ has no poles on the real axis. Consequently, if $u_\lambda(t) \ne 0$, then Im $\lambda > 0$. By the corollary to Theorem 6.1 we then conclude that $|u(t; x)|_X \le C_1$ for $t \ge 0$. This completes the proof.

From Theorems 6.1 and 6.2 we conclude:

COROLLARY.   *Let* u(t) *be a solution of* Lu = 0 *with* $|u|_X \in L^2(0, \infty)$, *and let all the assumptions of Theorem 6.1 hold. Then, for any* $0 < \varepsilon < \alpha$ *there is a finite number of functions* $u_h(t)$ $(1 \le h \le h_0)$ *of the form*

$$u_h(t) = e^{i\mu t}\left[\sum_{\gamma \in \Gamma(\alpha; \alpha_1, \ldots, \alpha_m)} e^{-(v+\gamma)t} p_\gamma(t)\right],$$

where $\lambda = \mu + i\nu$ *is a pole of* $R_S(\lambda; A_0)$ *in the strip* $0 < \text{Im}\,\lambda \leq \alpha - \varepsilon$,

$$|Lu_h|_Y = 0(e^{-(\alpha-\varepsilon)t}), \tag{6.21}$$

and $u(t) = \sum_{h=1}^{h_0} u_h(t) + w(t)$, *where*

$$\int_0^\infty |e^{-(\alpha-\varepsilon)t}|w(t)|_X|^2 \, dt \leq C_1 \int_0^\infty |u|_X^2 \, dt. \tag{6.22}$$

So far we have assumed that $A(t)$ has an asymptotic expansion and f is "small." We shall now extend these results by assuming that also f has an asymptotic expansion, and $f \in S$. First we consider the case that $A(t) \equiv A_0$ is a closed linear operator in Y with domain $D(A_0)$ dense in X. We further assume that L_0 satisfies the ζ_0 hypothesis.

LEMMA 6.3. Assume that $R_S(\lambda; A_0)$ *is analytic in a neighborhood of a point* $\lambda = i\beta$ (β *real) except at* $\lambda = i\beta$, *where it has at most a pole of order* j. *Then, for any positive integer* q *and* $f \in S$ *there exists a solution* u(t) *of*

$$L_0 u \equiv \frac{1}{i}\frac{du}{dt} - A_0 u = e^{-\beta t}t^q f \qquad (t \geq 0)$$

having the form

$$u(t) = e^{-\beta t}p(t),$$

where p(t) *is a polynomial in* t *with coefficients in* $D(A_0)$ *of degree at most* q + j.

PROBLEM. (4) Prove that $u(t) = \text{Res}_{\lambda=i\beta}\, w(\lambda, t)$, where

$$w(\lambda, t) = ie^{i\lambda t}q!\,\frac{R_S(\lambda; A_0)f}{(\beta + i\lambda)^{q+1}},$$

satisfies the assertions of the lemma.

LEMMA 6.4. Let u *be a solution of* $L_0 u = f$ *for* $t \geq 0$. *Assume that* $R_S(\lambda; A_0)$ *is meromorphic in the strip* $0 \leq \text{Im}\,\lambda < \beta$ *and that* $|R_S(\lambda; A_0)|_X = 0(1)$ *as* $|\lambda| \to \infty$ *in the strip. Assume further that* $|u|_X \in L^2(0, \infty)$ *and* $|f(t)|_Y = 0(e^{-\beta t})$. *Then, for* $t \geq 1$ *and for any* $\varepsilon > 0$ *sufficiently small,*

$$u(t) = u_0(t) + u_1(t),$$

where u_0 *is a solution of* $L_0 u_0 = 0$ *and* u_1 *is a solution of* $L_0 u_1 = f$ *satisfying*

$$\int_1^\infty |e^{(\beta-\varepsilon)t}|u_1(t)|_X|^2 \, dt < \infty.$$

Proof. Set $v = \zeta_0 \cdot u$ and extend $f(t)$ by zero to $t < 0$. Then $L_0 v = f + g$ for all real t, where the support of g lies in the interval $[0, 1]$. Taking Fourier transforms, we get

$$\hat{v}(\lambda) = R_S(\lambda; A_0)[\hat{f}(\lambda) + \hat{g}(\lambda)]. \tag{6.23}$$

$\hat{g}(\lambda)$ is an entire function of exponential type and $\hat{f}(\lambda)$ is analytic for Im $\lambda < \beta$. Taking the inverse Fourier transform in (6.23) and using the residue theorem, we get $u(t) = u_0(t) + u_1(t)$, where

$$u_0(t) = i\sqrt{2\pi} \sum_{\text{Im } \lambda_j < \beta - \varepsilon} \operatorname*{Res}_{\lambda = \lambda_j} R_S(\lambda; A_0)e^{i\lambda t}[\hat{f}(\lambda) + \hat{g}(\lambda)],$$

$$u_1(t) = \frac{1}{\sqrt{2\pi}} \int_{\text{Im } \lambda = \beta - \varepsilon} e^{i\lambda t} R_S(\lambda; A_0)[\hat{f}(\lambda) + \hat{g}(\lambda)] \, d\lambda.$$

Note that we have assumed that $R_S(\lambda; A_0)$ has no poles on Im $\lambda = \beta - \varepsilon$; this is justified for otherwise we can slightly decrease ε. We leave it to the reader to verify that u_0, u_1 have the asserted properties.

LEMMA 6.5. *Let* u *be a solution of* $L_0 u = f$ *with* $|u|_X \in L^2(0, \infty)$. *Assume that* $R_S(\lambda; A_0)$ *is meromorphic in the strip* $0 \le$ Im $\lambda < \alpha$ *and* $|R_S(\lambda; A_0)|_X = 0(1)$ *as* $|\lambda| \to \infty$ *in the strip. Assume also that*

$$f(t) = \sum_{k=1}^n e^{-\beta_k t} \sum_{p=0}^{P_k} t^p f_{pk} + f_1(t), \tag{6.24}$$

where β_k *are positive numbers,*

$$\alpha > \max_k \beta_k, \qquad f_{pk} \in S, \qquad f_1(t) \in S,$$

and $|f_1(t)|_Y = 0(e^{-\alpha t})$. *Then, for any* $\varepsilon > 0$ *sufficiently small,*

$$u(t) = u_0(t) + \sum_{k=1}^n e^{-\beta_k t} p_k(t) + u_1(t), \tag{6.25}$$

where $u_0(t)$ *is a solution of* $L_0 u = 0$, $e^{-\beta_k t} p_k(t)$ *is a solution of*

$$L_0 u = e^{-\beta_k t} \sum_{p=0}^{p_k} t^p f_{pk}, \qquad (6.26)$$

$p_k(t)$ *is a polynomial in* t *with coefficients in* $D(A_0)$, *and* u_1 *satisfies* $L_0 u_1 = f_1$ *and*

$$\int_0^\infty |e^{-(\alpha-\varepsilon)} |u_1(t)|_X|^2 \, dt < \infty.$$

Proof.   Let u_{kp} be the solution of $L_0 u = e^{-\beta_k t} t^p f_{pk}$ given by Lemma 6.3. Set $u_k = \sum_{p=0}^{p_k} u_{kp}$, $v = \sum_{k=1}^m u_k$. Apply Lemma 6.4 to $w = u - v$.

We now turn to the case where $A(t)$ depends on t.

LEMMA 6.6.   *Let the conditions* (A_1)–(A_3) *hold and assume that* L_0 *satisfies the* ζ_0 *hypothesis. Assume further that the conditions* (i), (ii) *of Theorem 6.1 hold. Then, for any* $0 < \delta < \alpha$, *there is a* δ-*asymptotically approximate solution* u_δ *of*

$$Lu = e^{-\beta t} t^q f,$$

where $f \in S$, β *is a positive number and* q *a positive integer.* $u_\delta(t)$ *has the form*

$$u_\delta(t) = \sum_j e^{-\lambda_j t} p_j(t),$$

where $\lambda_j = \beta + \gamma$, $\gamma \in \Gamma(\alpha; \alpha_1, \ldots, \alpha_m)$, $\mathrm{Im}\,\lambda_j < \delta$ *and* $p_j(t)$ *is a polynomial in* t *with coefficients in* $D(A_0)$.

Proof.   We may assume that $\delta < \beta$, for otherwise we take $u_\delta \equiv 0$. Let v_0 be the solution of $L_0 v_0 = e^{-\beta t} t^q f$ given by Lemma 6.3. Suppose that v_k has been defined for $k \leq n$ and that it has the form

$$v_n = \sum_{k=1}^{M_n} e^{-\lambda_{nk} t} p_{nk}(t),$$

where the λ_{nk} satisfy: $\lambda_{nk} < \delta$ and $p_{nk}(t)$ are polynomials in t with coefficients in $D(A_0)$. We shall define v_{n+1} as follows: Set

$$f_{n+1}(t) = \sum_k{}' e^{-\tilde\lambda_{nk} t} \tilde p_{nk}(t), \qquad (6.27)$$

where

$$\sum_k e^{-\tilde\lambda_{nk} t} \tilde p_{nk}(t) = \sum_{j=1}^m e^{-(\lambda_{nk}+\alpha_j)t} A_j \, p_{nk}(t) \equiv \sum_{j=1}^m e^{-\alpha_j t} A_j v_n$$

and the prime in the sum of (6.27) indicates summation only over the terms with $\tilde{\lambda}_{nk} < \delta$. v_{n+1} is the solution of

$$L_0 v_{n+1} = f_{n+1}(t)$$

given by Lemma 6.3.

Having defined the v_n inductively for all n, we notice that $f_n = 0$ if $n > [(\delta - \beta)/\alpha_1]$. Denote by v_N the last nonzero v_n. Then

$$u_\delta = v_0 + v_1 + \cdots + v_N$$

is the desired approximate solution, as is easily verified using Lemma 6.4.

We can now state the final result:

THEOREM 6.3.   *Let* u *be a solution of* Lu $=$ f *with* $|u|_X \in L^2(0, \infty)$. *Let the assumptions* (A_1)–(A_3) *hold, and assume that* L_0 *satisfies the* ζ_0 *hypothesis and that the conditions* (i), (ii) *of Theorem 6.1 hold. Assume finally that* f *has the form* (6.24) *with* β_k *positive numbers,* $f_{pk} \in S$, $f_1(t) \in S$, *and that* $|f_1(t)|_Y = 0(e^{-(\alpha - \varepsilon)t})$ *for every* $\varepsilon > 0$. *Then, for any* $\varepsilon > 0$,

$$u(t) = u_0(t) + u_1(t) + w(t),$$

where $u_0(t)$ *is a finite sum of* $(\alpha - \varepsilon)$-*asymptotically approximate solutions of* Lu $= 0$, $u_1(t)$ *is an* $(\alpha - \varepsilon)$-*asymptotically approximate solution of* Lu $=$ f, *and* $w(t)$ *satisfies*

$$\int_0^\infty |e^{-(\alpha - \varepsilon)t}|w(t)|_X|^2 \, dt < \infty.$$

$u_0(t)$, $u_1(t)$ *both have the form*

$$\sum_j e^{i\lambda_j t} p_j(t) \qquad (finite \; sum), \tag{6.28}$$

where $p_j(t)$ *are polynomials in* t *with coefficients in* $D(A_0)$. *For* u_0, $\lambda_j = \tilde{\lambda}_j + i\gamma$, *where* $\gamma \in \Gamma(\alpha; \alpha_1, \ldots, \alpha_m)$ *and* $\tilde{\lambda}_j$ *is a pole of* $R_S(\lambda; A_0)$ *in* $0 \le \operatorname{Im} \lambda < \alpha$, *and, for* u_1, $\lambda_j = i(\beta_k + \gamma)$, *where* $\gamma \in \Gamma(\alpha; \alpha_1, \ldots, \alpha_m)$.

Proof.   Let u_{pk} be the $(\alpha - \varepsilon)$-asymptotically approximate solution given in Lemma 6.6 for the equation

$$Lu = e^{-\beta_k t} t^p f_{pk},$$

and set $u_1 = \sum_{k=1}^{n} \sum_{p=0}^{p_k} u_{pk}$. Then $v(t) = u(t) - u_1(t)$ satisfies $Lv = h$, where $|h(t)|_Y = 0(e^{-(\alpha-\varepsilon)})$. By Theorem 6.1 and the corollary to Theorem 6.2, $v(t) = u_0(t) + w(t)$, where $w(t)$ satisfies (6.28) and u_0 is a finite sum of $(\alpha - \varepsilon)$-asymptotically approximate solutions of $Lu = 0$. u_1 and u_0 both have the form asserted in the theorem. This completes the proof.

7 | ASYMPTOTIC BEHAVIOR OF SOLUTIONS OF ELLIPTIC EQUATIONS

Let Ω be a compact n_0-dimensional C^∞ manifold without boundary. It will actually be sufficient for our applications to consider the special case where Ω is the unit hypersphere in R^{n_0+1}. Consider a differential operator in $Q_a^b = \Omega \times \{a < t < b\}$,

$$A_0(\omega, D_\omega, D_t) = \sum_{k=0}^{2m} B_k(\omega, D_\omega)D_t^{2m-k}, \tag{7.1}$$

where ω is a variable point in Ω, $B_0(\omega, D_\omega) \equiv 1$ and $B_k(\omega, D_\omega)$ is a differential operator of order $\leq k$.

The proofs of Theorems 5.1, 5.2 can be extended to the present case. Thus, the inequality (5.7) holds for all $u \in C^{2m}(\Omega)$ and the operator $A_0(\omega, D_\omega, \lambda)$ maps $H^{2m}(\Omega)$ onto $L^2(\Omega)$ if λ is real and its absolute value is sufficiently large. Note that the norm of $H^j(\Omega)$ is defined by taking a finite number of coordinate neighborhoods $\Omega_k (1 \leq k \leq v)$ of Ω, and then setting

$$(\|u\|_j^\Omega)^2 = \sum_{k=1}^{v} (\|u\|_j^{\Omega_k})^2.$$

Set

$$X = H^{2m}(\Omega) \times H^{2m-1}(\Omega) \times \cdots \times H^1(\Omega),$$

$$Y = H^{2m-1}(\Omega) \times H^{2m-2}(\Omega) + \cdots \times H^0(\Omega).$$

One can reduce the equation

$$A_0(\omega, D_\omega, D_t)u(\omega, t) = f(\omega, t) \qquad (0 \leq t < \infty)$$

into a first-order system

$$D_t U - A_0 U = F$$

as in Section 5 (cf. (5.12)–(5.14)). The analog of Theorem 5.3 is then valid. Thus $R_S(\lambda; A_0)$ is meromorphic in any strip $|\text{Im } \lambda| \leq \gamma$ and $|R_S(\lambda; A_0)|_X = 0(1)$ as $|\lambda| \to \infty$ in the strip.

Consider now a more general situation where

$$Au \equiv A_0 u + \sum_{j=1}^{p} f_j(t)A_j u + g(t)B(t)u = f(t) \tag{7.2}$$

and

$$A_j = \sum_{k=0}^{2m} B_k^{(j)}(\omega, D_\omega)D_t^{2m-k} \qquad (0 \leq j \leq p),$$

$$B(t) = \sum_{k=1}^{2m} \tilde{B}_k(t, \omega, D_\omega)D_t^{2m-k},$$

where $B_0^{(j)} \equiv \delta_{0j}$, $B_k^{(j)}(\omega, D_\omega)$ is a differential operator of order $\leq k$ and all the coefficients are continuous functions on Ω, and $\tilde{B}_k(t, \omega, D_\omega)$ is a differential operator of order $\leq k$ with continuous and bounded coefficients in Q_0^∞.

Let

$$u_0 = u, \qquad D_t u_j = u_{j+1} \qquad (0 \leq j \leq 2m - 2),$$

$$U = (u_0, \ldots, u_{2m-1}), \qquad F = (0, \ldots, 0, f),$$

and define

$$A_0 U = \left(u_1, u_2, \ldots, u_{2m-1}, -\sum_{k=0}^{2m-1} B_{2m-k}^{(0)} u_k\right),$$

$$A_j U = \left(0, 0, \ldots, 0, -\sum_{k=0}^{2m-1} B_{2m-k}^{(j)} u_k\right) \qquad (1 \leq j \leq p),$$

$$B(t)U = \left(0, 0, \ldots, 0, -\sum_{k=0}^{2m-1} \tilde{B}_{2m-k} u_k\right).$$

Then the equation (7.2) is equivalent to the equation

$$D_t U - A_0 U - \sum_{j=1}^{p} f_j(t)A_j U - g(t)B(t)U = F. \tag{7.3}$$

The ranges of A_j $(1 \leq j \leq p)$ and B are in S and their domains contain $D(A_0)$.

PROBLEM. (1) Let $|f_i(t)| \to 0$, $|g(t)| \to 0$ as $t \to \infty$, and let A_0 be elliptic. If $|f|_Y$ and $|u(\cdot, t)|_0$ are in $L^2(0, \infty)$, then $|U|_X$ is in $L^2(0, \infty)$.

Consider now an elliptic equation

$$Au \equiv \sum_{|\alpha| \leq 2m} a_\alpha(x)D^\alpha u = f \tag{7.4}$$

in $R^n - B^n$, where B^n is the unit ball. We assume that in polar coordinates $(x = (r, \omega))$,

$$a_\alpha(x) = a_\alpha(0) + \sum_{j=1}^{l} a_j^\alpha(\omega)r^{-j} + r^{-(l+1)}b_\alpha(r, \omega), \tag{7.5}$$

$$f(x) = r^{-2m} \sum_{j=1}^{l} f_j(\omega)r^{-j} + r^{-2m-l-1}\tilde{f}(r, \omega), \tag{7.6}$$

and that $a_j^\alpha(\omega)$, $f_j(\omega)$ are continuous on $\Omega_n = \{x \in R^n; |x| = 1\}$ and $b_\alpha(r, \omega)$, $\tilde{f}(r, \omega)$ are continuous and bounded in $R^n - B^n$.

If we introduce the variable t by $r = e^t$, then we can write (7.4) in the form (7.3) with $p = l$, $f_j(t) = e^{-jt}$, $g(t) = e^{-(l+1)t}$ and

$$F = (0, \ldots, 0, h(\omega)e^{2mt}\tilde{f}),$$

where

$$h(\omega) = \sum_{|\alpha|=2m} a_\alpha \zeta^\alpha \neq 0 \qquad \left(\zeta^\alpha = \frac{x^\alpha}{r^{|\alpha|}}\right).$$

The function $u(t)$ defined by $u(r, \omega)$ is in $L^2((0, \infty); X)$ if $r^{-n/2}u(r, \omega)$ is in $L^2(R^n - B^n)$.

We can now apply Theorem 6.3 and conclude:

THEOREM 7.1. *Let* A *be an elliptic operator in* $R^n - B^n$ *and let the asymptotic expansions (7.5), (7.6) hold. If* u *is a solution of (7.4) and* $r^{-n/2}u \in L^2(R^n - B^n)$, *then for any* $0 < \varepsilon < 1$ *there is a finite number of exponents* λ_j *of the form* $\lambda_j = \tilde{\lambda}_j + ik_j$ $(1 \leq j \leq N)$, *where* k_j *is an integer and* $\tilde{\lambda}_j$ *is a pole of* $R_S(\lambda; A_0)$ *in the strip* $0 \leq \operatorname{Im} \lambda < \ell + 1$ *or* $\tilde{\lambda}_j = 0$, *such that*

$$u(r, \omega) = \sum_{j=1}^{N} r^{i\lambda_j} \sum_{k=0}^{M_j} a_{kj}(\omega)(\log r)^k + v(r, \omega), \tag{7.7}$$

$$r^{-n/2-\varepsilon+\ell+1}v \in L^2(R^n - B^n) \tag{7.8}$$

and $a_{kj} \in H^{2m}(\Omega_n)$.

We shall supplement Theorem 7.1 with the following result.

THEOREM 7.2. *Let* A_0 *be an elliptic operator with constant coefficients of order* 2m. *Then the poles of* $R_S(\lambda; A_0)$ *are of the form* $\lambda = k/i$, *where* k *is an integer and* $i = \sqrt{-1}$. *If* $n > 2m$, *then the numbers* $\lambda = k/i$ *with* $2m - n < k < 0$ *cannot be poles of* $R_S(\lambda; A_0)$.

Proof. Suppose λ is a pole of $R_S(\lambda; A_0)$. Then there is a nonzero function $u_0(\omega)$ such that $A_0(\omega, D_\omega, \lambda)u_0 = 0$. Hence the function $v_0 \equiv e^{i\lambda t}u_0(\omega)$ satisfies $A_0(\omega, D_\omega, D_t)v_0 = 0$ for all real t. Setting $\lambda = k/i$, $e^t = r$, it follows that the function $w(x) = r^k u_0(\omega)$ satisfies $A_0(D_x)w(x) = 0$ for all $x \neq 0$. Note that

$$D_x^j w(x) = r^{k-j} w_j(x), \qquad w_j(x) \text{ homogeneous of degree } 0. \tag{7.9}$$

DEFINITION. A solution $v(x)$ of $A_0(D_x)v = 0$ in $V - \{0\}$, where V is some neighborhood of 0 in R^n, is said to have a *pole of order* s at $x = 0$ (s positive integer) if

$$D_x^j v(x) = \begin{cases} o(|x|^{(2m-1-j)-s-1}) & \text{if } 2m - (s+1) \leq j \leq 2m - 1, \\ 0(|\log x|) & \text{if } 0 \leq j < 2m - (s+1), \end{cases} \tag{7.10}$$

where D_x^j is any jth derivative.

Consider such a function $v(x)$ and denote by S the set of all functions $u(x)$ analytic in a neighborhood of 0 such that

$$H(u) \equiv \lim_{\varepsilon \to 0} \int_{|x|=\varepsilon} B[u, v] \, dS_x$$

exists. Here $B[u, v]$ is a bilinear form associated with the operator A_0 (cf. Section I.3). S is a linear space and H is a linear functional on S.

PROBLEMS. (2) Prove: If u is a solution of $A_0 u = 0$ in $U - \{0\}$, where U is a neighborhood of 0, then $u \in S$ and

$$\int_{|x|=\varepsilon} B[u, v] \, dS_x$$

is independent of ε.

(3) Let u be an analytic function in a neighborhood of 0 satisfying $u(x) = 0(|x|^{s-n+2})$. Prove that $u \in S$ and $H(u) = 0$.

Now let u be any function analytic in a neighborhood of 0. Write $u = u_1 + u_2$, where u_1 is a polynomial of degree at most $s - n + 1$ and u_2 is as in Problem 3. It follows that if $u \in S$, then $u_1 \in S$ and

$$H(u) = H(u_1). \tag{7.11}$$

Consider the linear space P of all polynomials in x of degree $\leq s - n + 1$. We can extend H into a linear functional $\tilde{H}$ on P. But then $\tilde{H}$ must be a linear function of the coefficients of the polynomials w of P. For u_1 as above, these coefficients are essentially some derivatives of $u(x)$ at $x = 0$, of order $\leq s - n + 1$. Hence, after using (7.11) we get

$$H(u) = \sum_{|\alpha| \leq s-n+1} c_\alpha D_x^\alpha u(0) \tag{7.12}$$

for any function u in S, which is analytic in a neighborhood of 0.

Let $z \in R^n$, $|z| < \delta$. Apply (I.3.5) with $L = A_0^*$, $R = \{x; \varepsilon < |x| < \delta\}$, where K is a fundamental solution of A_0^*. Taking $\varepsilon \to 0$ and using the result of Problem 2, one finds that

$$v(z) = \int_{|x| = \delta} B[K(x - z), v(x)] \, dS_x + H(K(x - z)).$$

Using Theorem I.4.1, it follows that the integral on the right is an analytic function $v_0(z)$ for $|z| < \delta$.

By (7.12) we have

$$H(K(x - z)) = \sum_{|\alpha| \leq s-n+1} c_\alpha D_z^\alpha K(-z).$$

Combining these remarks and using Theorem I.4.1, we get

$$v(z) = v_0(z) + |z|^{2m-n-(s-n+1)}[A(z) \log |z| + B(z)], \tag{7.13}$$

where $A(z)$, $B(z)$ are homogeneous in z of degree 0.

Note that if (7.10) holds with $s < 0$ or with positive s, $s < n - 1$, then the previous considerations lead to (7.13) with $A \equiv B \equiv 0$.

Suppose now that k is not an integer. Then (7.9) shows that $w(x)$ has a pole of order $s = 2m - [k] - 1$. Comparing the representation (7.13) with the representation $w(z) = |z|^k u_0(\omega)$ and recalling that $u_0(\omega) \not\equiv 0$, we get a contradiction.

Suppose next that $n > 2m$, that $\lambda = k/i$, and that k is an integer with $2m - n < k < 0$. Then $w(x)$ has a pole of order $s = 2m - k - 1$, so that $s < n - 1$. As remarked above, the representation (7.13) then holds with $A \equiv B \equiv 0$. Thus $w(x)$ is analytic at $x = 0$. Since $w(x) = |x|^k u_0(\omega)$, $u_0 \not\equiv 0$, and $k < 0$, we get a contradiction. This completes the proof of Theorem 7.2.

8 | INTEGRAL EQUATIONS IN BANACH SPACE

Consider the integral equation

$$u(t) = k(t) - \int_0^t h(t - \tau)A(\tau)u(\tau)\,d\tau \tag{8.1}$$

for $0 \le t \le t_0$, where u, k are functions with values in a (complex) Banach space X, h is a complex-valued function, and $A(t)$ is an operator (unbounded in general) in X. We shall assume that $A(t)$ satisfies the conditions (B_1)–(B_3) of Section II.3.

If $h(t) \equiv 1$, then (8.1) is equivalent to

$$\frac{du}{dt} + A(t)u = f(t),$$

$$u(0) = k(0), \tag{8.2}$$

where $f(t) = dk(t)/dt$. In that case the theory developed in Part II applies. It is our purpose to show that most of the theory developed in Part II applies also to (8.1) if h satisfies some very general conditions. We shall assume:

(H) $h(0) > 0$, $h \in C^1[0, t_0)$, dh/dt is absolutely continuous in $[0, t_0)$ and $d^2h/dt^2 \in L^p(0, t_0)$ for some $p > 1$.

Note that the condition $h(0) > 0$ is essential. In fact, if $h(t) \equiv -1$, then (8.1) reduces to (8.2) with $A(t)$ replaced by $-A(t)$, for which the Cauchy problem cannot be solved in general.

We shall also assume:

(K) $k(0) \in D(A^\mu(0))$ for some $0 < \mu \le 1$, and dk/dt is uniformly Hölder continuous (exponent β) in $[0, t_0)$.

DEFINITION. A function $u(t)$ from $[0, t_0)$ into D_A is called a *solution* of (8.1) if $u(t)$ is continuous in $[0, t_0)$, if $A(t)u(t)$ is continuous in $(0, t_0)$, $\|A(t)u(t)\|$ is integrable in $(0, t_0 - \varepsilon)$ for any $\varepsilon > 0$, and, finally, if (8.1) is satisfied for $0 \le t < t_0$. If also $t^\delta\|A(t)u(t)\|$ remains bounded for some $0 \le \delta < 1$ as $t \to 0$, then we say that u is a *strong solution*.

THEOREM 8.1. *Assume that (B_1)–(B_3) (of Section II.3) and (H), (K) hold. Then there exists a unique strong solution of (8.1). $t^{1-\mu}\|A(t)u(t)\|$ remains bounded as $t \to 0$, $u(t)$ is uniformly Hölder continuous with exponent μ in $[0, t_0)$ and, moreover, if $k(0) \in D_A$, then $A(t)u(t)$ and $du(t)/dt$ are continuous in $[0, t_0)$.*

THEOREM 8.2. Let the assumptions of Theorem 8.1 hold and assume further that d^2h/dt^2 *is uniformly Hölder continuous in some interval* $[0, \delta)$ *with* $\delta > 0$. *Then there cannot exist more than one solution of* (8.1).

We shall give here only the main ideas of the proof of Theorem 8.1. We shall reduce (8.1) to an "integro-differential" equation whose form is a perturbation of the equation $du/dt + h(0)A(t)u = dk/dt$. If we differentiate (8.1) formally, we get

$$\frac{du}{dt} + h(0)A(t)u(t) = \frac{dk}{dt} - \int_0^t \frac{dh(t-\tau)}{dt} A(\tau)u(\tau) \, d\tau. \tag{8.3}$$

Writing the last term in the form

$$\int_0^t \frac{dh(t-\tau)}{dt} [A(t) - A(\tau)]u(\tau) \, d\tau - A(t) \int_0^t \frac{dh(t-\tau)}{dt} u(\tau) \, d\tau$$

and introducing the function

$$\tilde{u}(t) = u(t) + \frac{1}{h(0)} \int_0^t \frac{dh(t-\tau)}{dt} u(\tau) \, d\tau, \tag{8.4}$$

(8.3) takes the form

$$\frac{d\tilde{u}}{dt} + h(0)A(t)\tilde{u} = \frac{dk}{dt} + f(\tilde{u}; t), \tag{8.5}$$

where

$$f(\tilde{u}; t) = \int_0^t \frac{dh(t-\tau)}{dt} \{[A(t) - A(\tau)]A^{-1}(\tau)\}A(\tau)u(\tau) \, d\tau$$

$$+ \frac{dh(0)/dt}{h(0)} u(t) + \frac{1}{h(0)} \int_0^t \frac{d^2h(t-\tau)}{dt^2} u(\tau) \, d\tau. \tag{8.6}$$

Thus, the equation (8.1) reduces to the equation (8.5) plus the initial condition

$$\tilde{u}(0) = k(0). \tag{8.7}$$

Note that (8.4) is a Volterra integral equation of the usual type. It can therefore be solved by successive approximations. We thus find that

$$u(t) = \tilde{u}(t) + \int_0^t l(t-\tau)\tilde{u}(\tau) \, d\tau, \tag{8.8}$$

where $l(t)$ is a function depending only on $h(t)$ and is "as smooth as" $h(t)$.

It can now be shown that the system (8.5), (8.7) is equivalent to (8.1). To solve (8.5), (8.7), we employ the fundamental solution $U(t, s)$ of the equation $dv/dt + h(0)A(t)v = 0$. Then, formally, (8.5), (8.7) reduces to

$$\tilde{u}(t) = U(t, 0)k(0) + \int_0^t U(t, s)\frac{dk(s)}{ds}\, ds + \int_0^t U(t, s)f(\tilde{u}; s)\, ds. \qquad (8.9)$$

One now proves that there exists a unique solution $\tilde{u}$ of (8.9). In proving this, use is made of Theorem II.16.4 and of some estimates on $U(t, s)$ derived in Part II. Finally, the function u, given by (8.8), is shown to be the unique strong solution of (8.5), (8.7).

It can also be proved that there exists a fundamental solution $W(t, \tau)$. This is an operator-valued function that satisfies the equation

$$W(t, s) = I - \int_s^t h(t - \tau)A(\tau)W(\tau, s)\, d\tau \qquad (0 \le s \le t < t_0) \qquad (8.10)$$

in some weak sense. Note that the integral in (8.10) does not make sense in general, since $\|A(\tau)W(\tau, s)\|$ may be bounded from below by $c|\tau - s|^{-1}$ $(c > 0)$, as is the case when $h(t) \equiv 1$.

The solution of (8.1) can be represented in the form

$$u(t) = W(t, 0)k(0) + \int_0^t W(t, s)\frac{dk(s)}{ds}\, ds, \qquad (8.11)$$

provided dk/dt is in the domain of $A^\mu(t)$ and $A^\mu(t)(dk/dt)$ is continuous in $[0, t_0)$.

If $A(t)$ is independent of t, then $W(t, \tau) \equiv W(t - \tau)$ is a bounded operator. This is true also under the more general assumption that the domain of $A^\rho(t)$ is independent of t and

$$\|[A(t) - A(\tau)]A^{-\rho}(\tau)\| \le C|t - \tau|^\alpha;$$

here ρ is any positive number < 1.

If $A(t)$ is independent of t, one can also give necessary and sufficient conditions for $W(t)$ to belong to $L^p((0, \infty); B(X))$. Suppose, in particular, that $t\, dh/dt$ belongs to $L^1(0, \infty)$ and that $h(t)$ is monotone decreasing and nonnegative in $(0, \infty)$. Then $W(t)$ belongs to $L^1((0, \infty); B(X))$ if and only if $h(\infty) > 0$ and

$$\left\{-\frac{1}{\hat{h}(s)}\,;\, \operatorname{Re} s \ge 0\right\} \subset \rho(A),$$

where $\hat{h}$ is the Laplace transform of h.

For details concerning the topics of this section, the reader is referred to the papers of Friedman and Shinbrot [1], and Friedman [12].

9 | OPTIMAL CONTROL IN BANACH SPACE

We shall need the concept of Bochner integral for functions $x(t)$ from a real interval (a, b) into a separable Banach space X. A function $x(t)$ is called *measurable* if, for any bounded linear functional f, $f(x(t))$ is a measurable function (in the Lebesgue sense). A function $x(t)$ is said to be *countably-valued* if it assumes at most a countable set of values in X, and each value is assumed on a measurable subset of (a, b). It is known that $x(t)$ *is measurable if and only if there exists a sequence of countable-valued functions converging almost everywhere to* $x(t)$.

DEFINITION.   A countably-valued function $x(t)$ is said to be integrable (in the Bochner sense) in (a, b) if and only if $\|x(t)\|$ is Lebesgue integrable. Its integral over any measurable subset E of (a, b) is the number

$$\int_E x(t)\, dt \equiv \sum_{k=1}^{\infty} x_k\, \mu(E_k \cap E),$$

where $x(t) = x_k$ on E_k and $\mu(F)$ is the Lebesgue measure of F.

Note that the last series is absolutely convergent.

DEFINITION.   A function $x(t)$ is integrable (in the Bochner sense) in (a, b) if and only if there exists a sequence of countably-valued integrable functions $\{x_n(t)\}$ converging almost everywhere to $x(t)$ such that

$$\lim_{n \to \infty} \int_a^b \|x(t) - x_n(t)\|\, dt = 0. \tag{9.1}$$

We then define the integral of $x(t)$ over any measurable subset E of (a, b) by

$$\int_E x(t)\, dt = \lim_{n \to \infty} \int_E x_n(t)\, dt. \tag{9.2}$$

It can be shown that this definition is meaningful. Indeed, since $x(t)$ is measurable, $\|x(t)\|$ is measurable and thus the integral in (9.1) makes sense. The existence of the limit in (9.2) is easily verified, and so is the fact that the definition of the integral of $x(t)$ over E is independent of the particular sequence $\{x_n\}$.

It is known that $x(t)$ is integrable (Bochner) if and only if $x(t)$ is measurable and $\int_a^b \|x(t)\|\, dt < \infty$.

The standard rules for integrals are valid. In particular, if A is a closed linear operator and if $x(t)$ and $Ax(t)$ are integrable (Bochner), then

$$A \int_E x(t) \, dt = \int_E Ax(t) \, dt$$

for any measurable subset E.

PROBLEM.   (1) Prove that if $A(t)$ is a bounded operator in X for any t in a bounded interval $[a, b]$, strongly continuous in $t \in [a, b]$, and if $x(t)$ is integrable (Bochner) in (a, b), then also $A(t)x(t)$ is integrable (Bochner).

From now on all integrals of functions with values in X are taken in the Bochner sense. Note that in the definition of the space $L^p((a, b); X)$ (used in previous sections), the elements are functions $x(t)$ that are measurable and for which $\int_a^b \|x(t)\|^p \, dt < \infty$. Thus, for $p = 1$, these are precisely the functions integrable (Bochner) on (a, b).

Consider an evolution equation

$$\frac{du}{dt} + A(t)u = f(t) \tag{9.3}$$

in a separable real Banach space X, with an initial condition

$$u(0) = u_0 . \tag{9.4}$$

Let U be a fixed subset of X. Any function defined and integrable in some interval $(0, t_0)$ is called a *control*. If also $f(t) \in U$ for almost all t, then we call f an *admissible control*. We shall call U the *control set*.

We assume that either $A(t)$ is independent of t and generates a strongly continuous semigroup, or else (if it depends on t), it satisfies the conditions (B_1)–(B_3) of Section II.3 for all finite t_0. Thus in both cases there exists a fundamental solution $S(t, \tau)$ that is strongly continuous in (t, τ) for $0 \le \tau \le t < \infty$.

DEFINITION.   For any control $f(t)$ in $(0, t_0)$, the *solution* of (9.3), (9.4) is defined to be the function

$$u(t) \equiv S(t, 0)u_0 + \int_0^t S(t, \tau)f(\tau) \, d\tau \qquad (0 \le t \le t_0). \tag{9.5}$$

By the result of Problem 1, the last integral exists. By the results of Sections II.3–II.7, if $f(t)$ is uniformly Hölder continuous and if (B_1)–(B_3) hold, then $u(t)$ satisfies (9.3), (9.4) in the sense of Section II.3.

Let W be a fixed subset of X. We shall call W the *target set*.

Our basic assumption is the following:

(P) There exists an admissible control $f(t)$, $0 \le t \le t_0$, for which the corresponding solution $u(t)$ of (9.3), (9.4) reaches the target set at time t_0—that is, $u(t_0) \in W$.

We then ask: Does there exist an admissible control for which the corresponding solution reaches the target set at a minimal time? If so, we call this control an *optimal control*, and the pair (f, u) an *optimal solution*. The minimal time is called the *optimal time*.

The next question is: Is the optimal solution unique?

We first prove existence.

THEOREM 9.1. *Let* X *be a reflexive Banach space, let* U *be a bounded, closed convex set, and let* W *be a closed convex set. Assume further that the condition* (P) *holds. Then there exists an optimal solution.*

Proof. Let T be the infimum of all the times t_0 as in (P) and let (f_n, u_n) be a minimizing sequence—that is,

$$u_n(T_n) = S(T_n, 0)u_0 + \int_0^{T_n} S(T_n, \tau)f_n(\tau) \, d\tau \text{ is in } W, \qquad T_n \searrow T. \qquad (9.6)$$

Consider $\{f_n\}$ as a bounded sequence in $Y = L^2((0, T_1); X)$ (f_n is extended by 0 for $t > T_n$). Since X is reflexive, Y is also reflexive (by Theorem 8.20.5 of Edwards [1]). Hence there exists a subsequence (which we again denote by $\{f_n\}$) that is weakly convergent in Y to some element f. But then the right-hand side of (9.6) is easily seen to be weakly convergent in X to

$$y \equiv S(T, 0)u_0 + \int_0^T S(T, \tau)f(\tau) \, d\tau.$$

Hence also $\{u_n(T_n)\}$ is weakly convergent to y. If we show that $f(t) \in U$ for almost all t and that $y \in W$, then $f(t)$ is an optimal control and the proof is complete.

Now, as is well known, there exists a family of bounded linear functionals $\{f_\alpha\}$ on X and a set of real numbers $\{c_\alpha\}$ such that $z \in U$ if and only if $f_\alpha(z) \le c_\alpha$ for all α. Consider

$$w_{\varepsilon, m} \equiv \frac{1}{\varepsilon} \int_s^{s+\varepsilon} f_m(t) \, dt.$$

Since $f_m(t) \in U$ almost everywhere, $f_\alpha(w_{\varepsilon,m}) \le c_\alpha$. But $f_\alpha(w_{\varepsilon,m}) \to f_\alpha(w_\varepsilon(s))$ as $m \to \infty$, where

$$w_\varepsilon(s) = \frac{1}{\varepsilon} \int_s^{s+\varepsilon} f(t) \, dt.$$

Hence $f_\alpha(w_\varepsilon(s)) \leq c_\alpha$—that is, $w_\varepsilon(s) \in U$. As is well known, for almost all s, $w_\varepsilon(s) \to f(s)$ as $\varepsilon \to 0$. Since U is closed, $f(s) \in U$.

Next let $\{g_\alpha\}$, $\{d_\alpha\}$ be sets of bounded linear functionals and of real numbers, respectively, such that $z \in W$ if and only if $g_\alpha(z) \leq d_\alpha$ for all α. Then $g_\alpha(u_n(T_n)) \leq d_\alpha$ for all α. Since $\{u_n(T_n)\}$ converges to y weakly, $g_\alpha(y) \leq d_\alpha$ for all α. Hence $y \in W$.

To prove uniqueness we shall impose the following condition on the target set:

(**Q**) W is a closed convex set with nonempty interior.

We recall (see, for instance, Dunford and Schwartz [1; p. 417]):

LEMMA 9.1. Let V *be a closed convex set in a real Banach space* X *and let* W *be a subset of* X *satisfying* (**Q**)*. If* $V \cap (\text{int } W) = \phi$ *and* $V \cap W = \{z\}$*, then there exists a continuous linear functional* $g \neq 0$ *such that*

$$g(v) \leq g(z) \leq g(w) \qquad \text{for all } v \in V, w \in W.$$

DEFINITION. A convex set U is called *strictly convex* if whenever x, y, and $(x + y)/2$ belong to ∂U then $x = y$.

PROBLEM. (2) Prove that the unit ball in $L^p(\Omega)$ $(1 < p < \infty)$ is strictly convex.

DEFINITION. We say that $A(t)$ has the *backward uniqueness* property if for any bounded subinterval (a, b) of $(0, \infty)$ and for any solution $u(t)$ of $du/dt + A(t)u = 0$ in this interval, the relation $u(b) = 0$ implies $u(t) = 0$ for $a \leq t \leq b$. This condition can also be stated in terms of the equation $du/dt - A(t)u = 0$—that is, $x(a) = 0$ implies $x(t) = 0$ for $a \leq t \leq b$. We say that $A(t)$ has the *weak backward uniqueness* property if in any interval (a, b) and for any solution $u(t)$ of $du/dt + A(t)u = 0$ in this interval, if $u(t)$ vanishes on a set of positive measure then $u(a) = 0$.

For examples, see Section II.18.

We shall need the following assumption:

$$\|A^{-1}(s)[A(t) - A(\tau)]x\| \leq C |t - \tau|^\alpha \|x\| \qquad (x \in D_A) \qquad (9.7)$$

for some constants C, α with $0 < \alpha \leq 1$.

THEOREM 9.2. Let X *be a real Banach space. Let* U *be a convex set and let* W *satisfy the condition* (Q)*. Assume that* (9.7) *holds and that* A* *satisfies the weak backward uniqueness property. If* $(\hat{f}, \hat{u})$ *is an optimal solution, with optimal time* T*, then* $\hat{f}(t) \in \partial U$ *for almost all* $t \in (0, T)$*.*

Proof.   Consider the set

$$\Omega_T = \left\{ y \in X; \, y = S(T, 0)u_0 + \int_0^T S(T, \tau)f(\tau) \, d\tau, f \text{ any admissible control} \right\}.$$

Since U is a convex set, Ω_T is also a convex set. The target set W is also convex and $W \cap \Omega_T$ contains the point $z = \hat{u}(T)$. This point must lie on the boundary of W. Indeed, suppose $z \in \text{int } W$. Since $\|\hat{f}(t)\|$ is a bounded function, $\hat{u}(t)$ is easily seen to be continuous. Hence $\hat{u}(T - \varepsilon) \in W$ for some $\varepsilon > 0$ sufficiently small. But this contradicts the minimality of T.

The previous argument also shows that $\Omega_T \cap (\text{int } W) = \phi$.

For any $x \in X$ and any continuous linear functional g on X, we write $g(x)$ also in the form (x, g). (If X is a Hilbert space we take $(\quad, \quad)$ to be the scalar product.)

Applying Lemma 9.1 to the sets Ω_T, W, we get

$$(y, g) \le (\hat{u}(T), g) \qquad \text{for all } y \in \Omega_T. \tag{9.8}$$

It follows that

$$\left(\int_0^T S(T, \tau)f(\tau) \, d\tau, g \right) \le \left(\int_0^T S(T, \tau)\hat{f}(\tau) \, d\tau, g \right)$$

for any admissible control f. Hence

$$\int_0^T (f(\tau), S^*(T, \tau)g) \, d\tau \le \int_0^T (\hat{f}(\tau), S^*(T, \tau)g) \, d\tau. \tag{9.9}$$

Suppose now that the assertion of the theorem is false. Then there exists a subset Δ of $(0, T)$ having positive measure such that

$$\text{dist. } (\hat{f}(t), \partial U) \ge \delta > 0 \qquad \text{for all } t \in \Delta.$$

For every bounded measurable function $h(t)$ with support in Δ, the function $\hat{f}(t) + \varepsilon h(t)$ is then an admissible control if ε is any real number sufficiently small in absolute value. Substituting $f = \hat{f} + \varepsilon h$ into (9.9), we find that

$$\int_\Delta (h(\tau), S^*(T, \tau)g) \, d\tau = 0.$$

Since h is arbitrary, it follows that the function $w(t) \equiv S^*(T, t)g$ vanishes on Δ.

LEMMA 9.2.   *For any* $y \in X^*$, $S^*(T, t)y$ *is a solution of*

$$\frac{dw}{dt} - A^*(t)w = 0. \tag{9.10}$$

Assume that the lemma is true. Then $w(t)$ is a solution of $dw/dt - A^*w = 0$ in $(0, T)$. By the weak backward uniqueness property it follows that $w(T) = 0$—that is, $g = 0$, which is impossible.

Proof of Lemma 9.2.   By (II.6.16), if $x \in D_A$, then

$$\frac{S(t, \tau + \Delta\tau)x - S(t, \tau)x}{\Delta\tau} = S(t, \tau + \Delta\tau) \frac{I - S(\tau + \Delta\tau, \tau)}{\Delta\tau} x$$

$$= S(t, \tau + \Delta\tau) \frac{1}{\Delta\tau} \int_\tau^{\tau + \Delta\tau} A(\zeta)S(\zeta, \tau)x \, d\zeta.$$

Since the function $A(\zeta)S(\zeta, \tau)x$ is continuous in ζ for $\tau \le \zeta \le t_0$ (by Lemma II.6.2), it follows that $\partial S(t, \tau)x/\partial\tau$ exists and

$$\frac{\partial S(t, \tau)x}{\partial\tau} = S(t, \tau)A(\tau)x. \qquad (9.11)$$

As follows from problems (3)–(9) at the end of this section, $\partial S(t, \tau)/\partial\tau$ and $S(t, \tau)A(\tau)$ can be defined over all of X as bounded operators. Hence, if we apply $y \in X^*$ to both sides of (9.11), we find that

$$\frac{\partial S^*(t, \tau)y}{\partial\tau} - A^*(\tau)S^*(t, \tau)y = 0.$$

This completes the proof of the lemma.

From Theorem 9.2 we obtain:

COROLLARY 1.   If U is strictly convex, then there is at most one optimal solution.

Indeed, suppose (f_1, u_1) and (f_2, u_2) are two optimal solutions. It is easily seen that $((f_1 + f_2)/2, (u_1 + u_2)/2)$ is then also an optimal solution. Theorem 9.2 implies that $f_1(t), f_2(t)$ and $(f_1(t) + f_2(t))/2$ belong to ∂U for almost all t. Hence $f_1(t) = f_2(t)$ for almost all t, and $u_1(t) \equiv u_2(t)$.

The inequality (9.8) is a generalization of the Pontryagin maximum principle in the calculus of variations. From this inequality and the argument following it, we actually get:

COROLLARY 2.   Let X be a real Hilbert space. If $(\hat{f}, \hat{u})$ is an optimal control, then

$$\hat{f}(t) = \lambda(t)S^*(T, t)g, \qquad (9.12)$$

where $\lambda(t)$ is real-valued and $\lambda(t)S^*(T, t)g \in \partial U$ almost everywhere.

If U is symmetric with respect to the origin (that is, $x \in U$ implies $-x \in U$) then $\lambda(t)$ must be positive-valued. In particular:

COROLLARY 3. *Let* X *be a real Hilbert space. If* U *is the unit ball, then*

$$\hat{f}(t) = \frac{S^*(T, t)g}{\|S^*(T, t)g\|}.$$

(9.13)

Note that $\hat{f}(t)$ given by (9.13) is a continuous function.

Theorems 9.1, 9.2 can be applied to parabolic equations

$$\frac{\partial u}{\partial t} + A(x, t, D)u = f(x, t)$$

(9.14)

with coefficients in a cylinder $Q_\infty = \Omega \times (0, \infty)$, where Ω is a bounded domain in R^n and A is strongly elliptic of order $2m$. Together with (9.14) we impose the conditions

$$\frac{\partial^j u}{\partial \nu^j} = 0 \qquad \text{for } x \in \partial\Omega \qquad (0 \leq j \leq m - 1),$$

(9.15)

$$u(x, 0) = 0 \qquad \text{on } \Omega.$$

(9.16)

We consider all controls $f(x, t)$ satisfying

$$\int_\Omega \rho(x)|f(x, t)|^q \, dx \leq 1 \qquad (1 < q < \infty),$$

(9.17)

where $\rho(x)$ is a positive continuous function on $\bar{\Omega}$. From Theorems 9.1, 9.2 it follows that (under certain smoothness assumptions on A and $\partial\Omega$) if the condition (P) holds and if the weak backward uniqueness property is valid, then there exists a unique optimal solution. The weak backward uniqueness property is satisfied if $A(x, t, D)$ is independent of t. The backward uniqueness property is valid if $A(x, t, D) = A_0(x, t, D) + A_1(x, t, D)$, where $A_0(x, t, D)$ is formally selfadjoint, $A_1(x, t, D)$ is of order $\leq m$, and the coefficients are smooth; see Lions and Malgrange [1], and Lees and Protter [1].

Actually the proof of Theorem 9.2 extends to the case where the target set is taken in some space $L^s(\Omega)$ with $1/s > 1/q - 2m/n$.

Problems where the control function is not in the differential equation but in the boundary conditions can also be treated by the same method (see Friedman [8]). Related problems are considered in Friedman [10], [11].

The proof of Theorem 9.2 fails in the interesting case where W consists of just one point. If $A(t)$ is independent of t, then the assertion of Theorem 9.2 is still valid. If $A(t)$ depends on t, then the assertion holds under additional assumptions. For details, see Fattorini [1], [2] and Friedman [9].

In the following problems, $A(t)$ satisfies the conditions (B_1)–(B_3) of Section II.3 and (9.7).

PROBLEM.   (3) Prove that

$$\|A^{-\beta}(t)e^{-(t-\tau)A(\tau)}A^{\gamma}(\tau)\| \le C|t-\tau|^{\beta-\gamma} \qquad (0 \le \beta \le \gamma < 1+\alpha). \quad (9.18)$$

[*Hint:* Use (II.14.11) for $A^*(t)$.]

(4) Consider the integral equation

$$W(t,\tau) = \overline{A^{-\beta}(t)e^{-(t-\tau)A(\tau)}A(\tau)} + \int_{\tau}^{t} W(t,s)\overline{A^{-1}(s)[A(\tau)-A(s)]A(\tau)}e^{-(s-\tau)A(\tau)}\,ds,$$

where $0 < \beta \le 1$ and where $\overline{V}$ means the closure of V. Prove that the equation has a unique solution and

$$\|W(t,\tau)\| \le C|t-\tau|^{1-\beta}.$$

(5) Show that $W(t,\tau) = \overline{A^{-\beta}(t)U(t,\tau)A(\tau)}$.
(6) Prove:

$$\|\overline{A^{-\beta}(t)U(t,\tau)A^{\gamma}(\tau)}\| \le C|t-\tau|^{\beta-\gamma} \qquad (0 < \beta \le \gamma < 1+\alpha). \quad (9.19)$$

[*Hint:* Use (II.4.2) and (9.18).]

(7) Prove (9.19) for $\beta = 0, 0 < \gamma < 1$.
[*Hint:* Use (II.4.23) and (9.18).]

(8) Use (II.6.16) and Problems (6), (7) to prove (9.19) for $\beta = 0, 0 < \gamma < 1+\alpha$.

From the last result it follows that $\overline{U(t,\tau)A(\tau)}$ is a bounded operator for $t > \tau$. With the notation $S(t,\tau) = U(t,\tau)$ of the present section we have that $\overline{S(t,\tau)A(\tau)}$ is a bounded operator for $t > \tau$. From (9.11) and an argument involving the mean value theorem we see also that $S(t,\tau)$ is Lipschitz continuous in τ. Hence $\partial S(t,\tau)x/\partial\tau$ exists not only for $x \in D_A$ but for all x, and is a bounded operator.

Bibliographical Remarks

Part 1.

The concepts of weak and strong derivatives and of mollifiers were introduced by Sobolev [1], [2] and Friedrichs [1]. Theorem 6.3 is due to Meyers and Serrin [1]. Theorem 8.1 was proved by Nirenberg [1]. Theorems 10.2 and 9.1 include all the Sobolev inequalities; most of these inequalities were first derived in Sobolev [1]. Theorem 9.3 was proved by Nirenberg [2] and, independently, by Gagliardo [1], [2] (except for $a = j/m$). Theorem 11.2 is due to Sobolev [1], [2]. Theorem 12.1 is due to Gårding [1]. Theorem 14.1 was given and first used by Lax and Milgram [1]. Regularity theorems for weak solutions were proved by several authors. In the interior, this was done by Friedrichs [2], John [2], Browder [1], Lax [1], Nirenberg [1], and others. For regularity of solutions of equations more general than elliptic see Hörmander [1] and the references given there. Regularity on the boundary for the Dirichlet problem was first proved by Nirenberg [1] and Browder [2].

The outline of Sections 14–17 follows several previous expositions—in particular, Nirenberg [1], Bers–John–Schecter [1], Friedman [5], and Agmon [3].

The proof of Theorem 16.1 is due to Nirenberg [1]. The proof of interior regularity for elliptic equations given in the second part of Section 16 was given by Nirenberg [2] and it is a variant of the proof given by Lax [1]. Lemma 17.2 is due to Agmon [1], [3]. Other related lemmas (but not as convenient) were given by other authors, such as Nirenberg [2]. The L^2 estimates of Theorem 17.2 were proved by various authors. The analogous L^p estimates for the Dirichlet problem were derived by Koselev [1], Slobodeckii [1], and Browder [5]. The L^2 estimates for general elliptic boundary problems were derived by Schechter [1], Peetre [1], and others. Finally, the L^p estimates for general elliptic boundary problems were derived by Agmon, Douglis, and Nirenberg [1] (and for systems of equations in [2]); see also Browder [4]. Theorem 18.2 and (its consequence) Theorem 18.3 were proved in Agmon–Douglis–Nirenberg [1] and, more simply, in Agmon [2] and Agmon–Nirenberg [1].

Part 2.

For more details on semigroups, see Hille–Phillips [1]. The results of Section 2 are due to Yosida [1] and to Hille–Phillips [1]. (See also Solomiak [1].) The results of Sections 3–8 were derived independently by Sobolevski [1] and Tanabe [1], [2], [3]. Theorem 11.1 is due to Agmon and Nirenberg [1] and Theorem 11.2 is due to Friedman [7]. The results of Section 12 were proved by Friedman [4]. For the case of parabolic equations, Tanabe [5], [6] has extended Theorems 11.2, 12.1 to time-dependent boundary conditions. Kato and Tanabe [1] proved existence of solutions for evolution equations as in Sections 3–8 but with the domain of $A(t)$ depending on t; in [7] Tanabe proved differentiability theorems for the solution. Theorem 13.1 was proved by Tanabe [4]. For parabolic equations (also in noncylindrical domains) the asymptotic behavior as $t \to \infty$ was studied by Friedman [1], [2].

Fractional powers of operators were studied by several authors—in particular, Yosida [2], Kato [1]. The results of Sections 14, 15 and Theorems 16.1, 16.2 were derived by Sobolevski [1]. Theorems 16.6, 16.7 were derived in Friedman [6]. Theorem 18.1 is due to Agmon and Nirenberg [1]. Results of the same type as the corollary to Theorem 18.1, with weaker assumptions, were derived for parabolic equations by Lees and Protter [1] and by Lions and Malgrange [1]. Theorem 19.1 is due to Krein and Prozorovskaja [1]. (See also Agmon–Nirenberg [1].)

Part 3.

The results of Chapter 1 for more general elliptic systems were proved by Morrey and Nirenberg [1]. Special cases were derived by various authors. Theorem 2.1 is due to Komatzu [1]. (Another proof, by Sobolevski [1], has a gap (in deriving (2.50).) Theorem 2.2 is due to Friedman [7]. Theorem 3.1 is a special case of an analyticity theorem for more general equations, proved by Tanabe [5] and (more generally) by Cavallucci [1]. The proof of Theorem 4.1 is due to Ogawa [1]. Similar results were previously proved by Agmon and Nirenberg [1], who also proved Theorems 4.2, 4.3. The case where $\phi \in L^p$, $1 \le p \le 2$, was first proved by Cohen and Lees [1].

The results of Section 5 are due to Agmon and Nirenberg [1]. In that paper they have also derived results on the asymptotic behavior of solutions of evolution equations in case $A(t)$ is independent of t and $f(t) \equiv 0$. The results of Section 6, which extend some of their work to $A(t)$ depending on t and to nonhomogeneous equations, are due to Pazy [1]. The application given in Section 7 is also due to Pazy [1]. The proof of Theorem 7.2 is based on John [1]. In the special case where A_0 is the Laplacian, the result of Theorem 7.1 (supplemented by Theorem 7.2) was previously proved, by a different method, by Meyers [1].

The results stated in Section 8 are due to Friedman and Shinbrot [1]. The results of Section 9 are due to Friedman [8].

Bibliography

S. AGMON

[1] "The L_p approach to the Dirichlet problem, I: Regularity theorems," *Ann. Scula Norm. Sup. Pisa*, **13** (1959), 405–448.

[2] "On the eigenfunctions and on the eigenvalues of general elliptic boundary value problems," *Comm. Pure Appl. Math.*, **15** (1962), 119–147.

[3] *Lectures on Elliptic Boundary Value Problems*. Princeton, N.J.: D.Van Nostrand Co., Inc., 1965.

[4] "General elliptic boundary value problems," *to appear*.

S. AGMON, A. DOUGLIS, AND N. NIRENBERG

[1] "Estimates near the boundary for solutions of elliptic partial differential equations satisfying general boundary conditions, I," *Comm. Pure Appl. Math.*, **12** (1959), 623–727.

[2] "Estimates near the boundary for solutions of elliptic partial differential equations satisfying general boundary conditions, II," *Comm. Pure Appl. Math.*, **7** (1964), 35–92.

S. AGMON AND L. NIRENBERG

[1] "Properties of solutions of ordinary differential equations in Banach space," *Comm. Pure Appl. Math.*, **16** (1963), 121–239.

L. BERS, F. JOHN, AND M. SCHECTER

[1] *Partial Differential Equations*. New York: Interscience Publishers, 1964.

E. E. BROWDER

[1] "Strongly elliptic systems of differential equations," pp. 15–51 in *Contributions to the Theory of Partial Differential Equations*, Annals of Mathematics Studies, no. 33. Princeton, N.J.: Princeton University Press, 1954.

[2] "On the regularity properties of solutions of elliptic differential equations," *Comm. Pure Appl. Math.*, **9** (1956), 351–361.

[3] "Estimates and existence theorems for elliptic boundary value problems," *Proc. Nat. Acad. Sci. U.S.A.*, **45** (1959), 365–372.

[4] "A priori estimates for solutions of elliptic boundary value problems, I, II, III," *Neder. Akad. Wetensch. Indag. Math.*, **22** (1960), 149–159, 160–169; **23** (1961), 404–410.

[5] "On the spectral theory of elliptic differential operators, I," *Math. Ann.*, **142** (1961), 20–130.

A. P. CALDERON AND A. ZYGMUND

[1] "On singular integrals," *Amer. J. Math.*, **79** (1956), 289–309.

A. CAVALLUCCI

[1] "Sulla proprietè differenziali delle solizioni delle equazioni quasi-ellittiche," *Ann. Mat. Pura Appl.* (Ser. IV), **67** (1965), 143–168.

P. J. COHEN AND M. LEES

[1] "Asymptotic decay of solutions of differential inequalities," *Pacific J. Math.*, **11** (1961), 1235–1249.

N. DUNFORD AND J. T. SCHWARTZ

[1] *Linear Operators, Part I: General Theory.* New York: Interscience Publishers, 1964.

R. E. EDWARDS

[1] *Functional Analysis.* New York: Holt, Rinehart and Winston, Inc., 1965.

H. O. FATTORINI

[1] "Time-optimal control of solutions of differential equations," *J. SIAM Control*, Ser. A, **2** (1964), 54–59.

[2] Some observations on a paper by A. Friedman, *J. Math. Analys. Appl.*, to appear.

A. FRIEDMAN

[1] "Convergence of solutions of parabolic equations to a steady state," *J. Math. and Mech.*, **8** (1959), 57–76.

[2] "Asymptotic behavior of solutions of parabolic equations of any order," *Acta Math.*, **106** (1961), 1–43.

[3] *Generalized Functions and Partial Differential Equations.* Englewood Cliffs, N.J.: Prentice-Hall, Inc., 1963.

[4] "Uniqueness of solutions of ordinary differential inequalities in Hilbert space," *Archive Rat. Mech. Anal.*, **17** (1964), 353–357.

[5] *Partial Differential Equations of Parabolic Type.* Englewood Cliffs, N.J.: Prentice-Hall, Inc., 1964.

[6] "Remarks on nonlinear parabolic equations," *Proc. Symposia Appl. Math.*, **17** (Amer. Math Soc., 1965), 3–23.

[7] "Differentiability of solutions of ordinary differential equations in Hilbert space," *Pacific J. Math.*, **16** (1966), 267–271.

[8] "Optimal control for parabolic equations," *J. Math. Analys. Appl.*, **18** (1967), 479–491.

[9] "Optimal control in Banach spaces," *J. Math. Analys. Appl.*, **19** (1967), 35–55.

[10] "Optimal control in Banach space with fixed end-points," *J. Math. Analys. Appl.*, **24** (1968), 161–181.

[11] "Differential games of pursuit in Banach games," *J. Math. Analys. Appl.*,**25** (1969), 93–113.

[12] "Monotonicity solution of Volterra integral equations in Banach space," *Trans. Amer. Math. Soc.*, (1969) *to appear.*

A. FRIEDMAN AND M. SHINBROT

[1] "Volterra integral equations in Banach space," *Trans. Amer. Math. Soc.*, **126** (1967), 131–179.

K. O. FRIEDRICHS

[1] "The identity of weak and strong extensions of differential operators," *Trans. Amer. Math. Soc.*, **55** (1944), 132–155.

[2] "On the differentiability of solutions of linear elliptic differential equations," *Comm. Pure Appl. Math.*, **6** (1953), 299–326.

E. GAGLIARDO

[1] "Proprietà di alcune classi di funzioni in più variabili," *Ricerche di Mat.*, **7** (1958), 102–137.

[2] Ulteriori proprietà di alcune classi di funzioni in più variabili, *Ricerche di Mat.*, **8** (1959), 24–51.

L. GÅRDING

[1] "Dirichlet's problem for linear elliptic partial differential equations," *Math. Scand.*, **1** (1953), 55–72.

E. HILLE AND R. S. PHILLIPS

[1] *Functional Analysis and Semi-Groups*, Amer. Math. Soc. Colloq. Publ., vol. 31 Providence, R.I.: Amer. Math. Soc., 1957.

L. HÖRMANDER

[1] *Linear Partial Differential Equations*. Berlin: Springer-Verlag, 1963.

F. JOHN

[1] "General properties of solutions of linear elliptic partial differential equations," *Proceedings of the Symposium on Spectral Theory and Differential Problems*, Stillwater, Okla. 1951, 113–175.

[2] "Derivatives of continuous weak solutions of linear elliptic equations," *Comm. Pure Appl. Math.*, **6** (1953), 327–335.

[3] *Plane Waves and Spherical Means Applied to Partial Differential Equations*. New York: Interscience Publishers, 1955.

T. KATO

[1] "Fractional powers of dissipative operators," *J. Math. Soc. Japan*, **13** (1961), 246–274.

T. KATO AND H. TANABE

[1] "On the abstract evolution equation," *Osaka Math. J.*, **14** (1962), 107–133.

H. KOMATZU

[1] "Abstract analyticity in time and unique continuation property of solutions of parabolic equations," *J. Fac. Sci. Univ. Tokyo*, Sec. I, **9** (1961), 1–11.

A. I. KOSELEV

[1] "A priori L_p estimates and generalized solutions of elliptic equations and systems," *Uspehi Math. Nauk*, **13** (1958), 29–89.

S. G. KREIN AND O. I. PROZOROVSKAJA

[1] "Analytic semigroups and incorrect problems for evolutionary equations," *Dokl. Akad. Nauk SSSR*, **13** (1960), 277–280.

P. D. LAX

[1] "On Cauchy's problem for hyperbolic equations and the differentiability of solutions of elliptic equations," *Comm. Pure Appl. Math.*, **8** (1955), 615–633.

P. D. LAX AND A. MILGRAM

[1] "Parabolic equations," pp. 167–190 in *Contributions to the Theory of Partial Differential Equations, Ann. Math. Studies*, Annals of Mathematics Studies, no. 33. Princeton, N.J.: Princeton University Press, 1954.

M. LEES AND M. H. PROTTER

[1] "Unique continuation for parabolic differential equations and inequalities," *Duke Math. J.*, **28** (1961), 369–382.

J. L. LIONS AND B. MALGRANGE

[1] "Sur l'unicité rétrograde dans les problèmes mixtes paraboliques," *Math. Scand.*, **8** (1960), 277–286.

N. MEYERS

[1] "An expansion about infinity of solutions of linear elliptic equations," *J. Math. and Mech.*, **12** (1963), 247–264.

N. G. MEYERS AND J. SERRIN

[1] "$H = W$," *Proc. Nat. Acad. Sci. U.S.A.*, **51** (1964), 1055–1056.

C. B. MORREY AND L. NIRENBERG

[1] "On the analyticity of the solutions of linear elliptic systems of partial differential equations," *Comm. Pure Appl. Math.*, **10** (1957), 271–290.

L. NIRENBERG

[1] "On elliptic partial differential equations," *Ann. Scuo. Norm. Sup. Pisa*, **13**(3) (1959), 1–48.

[2] "Remarks on strongly elliptic partial differential equations," *Comm. Pure Appl. Math.*, **8** (1955), 648–674.

H. OGAWA

[1] "Lower bounds for solutions of differential inequalities in Hilbert space," *Proc. Amer. Math. Soc.*, **16** (1965), 1241–1243.

A. PAZY

[1] "Asymptotic expansions of solutions of ordinary differential equations in Hilbert space," *Archive Rat. Mech. Anal.*, **24** (1967), 193–218.

J. PEETRE

[1] "Another approach to elliptic boundary value problems," *Comm. Pure Appl. Math.*, **14** (1961), 711–713.

J. SCHAUDER

[1] "Der Fixpunktsatz in Funtionalräumen," *Studia Math.*, **2** (1930), 171–180.

M. SCHECHTER

[1] "General boundary value problems for elliptic partial differential equations," *Comm. Pure Appl. Math.*, **12** (1959), 457–482.

L. N. SLOBODECKII

[1] "Estimates in L_p of solutions of elliptic systems," *Doklady Akad. Nauk SSSR*, **123** (1958), 616–619.

S. L. SOBOLEY

[1] "On a theorem of functional analysis," *Mat. Sbornik* (N.S.), **4** (1938), 471–497.

[2] *Application of Functional Analysis in Mathematical Physics*, Amer. Math Soc. Translations, vol. 7. Providence, R.I., 1963.

P. E. SOBOLEVSKI

[1] "On equations of parabolic type in a Banach space," *Trudy Moscov. Mat. Obšč.*, **10** (1961), 297–350.

M. Z. SOLOMIAK

[1] "The application of the semigroup theory to the study of differential equations in Banach spaces," *Doklady Akad. Nauk SSSR*, **122** (1958), 766–769.

H. TANABE

[1] "A class of the equations of evolution in a Banach space," *Osaka Math. J.*, **11** (1959), 121–145.

[2] "Remarks on the equations of evolution in a Banach space," *Osaka Math. J.*, **12** (1960), 145–166.

[3] "On the equations of evolution in a Banach space," *Osaka Math. J.*, **12** (1960), 363–376.

[4] "Convergence to a stationary state of the solutions of some kind of differential equations in a Banach space," *Proc. Japan Acad.*, **37** (1961), 127–130.

[5] "On differentiability and analyticity of solutions of weighted elliptic boundary value problems," *Osaka J. Math.*, **2** (1965), 163–190.

[6] "Note on uniqueness of solutions of differential inequalities of parabolic type," *Osaka J. Math.*, **2** (1965), 191–204.

[7] "On regularity of solutions of abstract differential equations of parabolic type in Banach space," *J. Math. Soc. Japan*, **19** (1967), 521–542.

K. YOSIDA

[1] "On the differentiability of semigroups of linear operators," *Proc. Japan Acad.*, **34** (1958), 337–340.

[2] "Fractional powers of infinitesimal generators and analyticity of the semigroup generated by them," *Proc. Japan Acad.*, **36** (1960), 86–89.

Index